Methods in Enzymology

Volume XL
HORMONE ACTION
Part E
Nuclear Structure and Function

METHODS IN ENZYMOLOGY

EDITORS-IN-CHIEF

Sidney P. Colowick   Nathan O. Kaplan

Methods in Enzymology

Volume XL

Hormone Action

Part E

Nuclear Structure and Function

EDITED BY

Bert W. O'Malley

DEPARTMENT OF CELL BIOLOGY
BAYLOR COLLEGE OF MEDICINE
TEXAS MEDICAL CENTER
HOUSTON, TEXAS

Joel G. Hardman

DEPARTMENT OF PHYSIOLOGY
VANDERBILT UNIVERSITY SCHOOL OF MEDICINE
NASHVILLE, TENNESSEE

1975

ACADEMIC PRESS   New York   San Francisco   London
A Subsidiary of Harcourt Brace Jovanovich, Publishers

COPYRIGHT © 1975, BY ACADEMIC PRESS, INC.
ALL RIGHTS RESERVED.
NO PART OF THIS PUBLICATION MAY BE REPRODUCED OR
TRANSMITTED IN ANY FORM OR BY ANY MEANS, ELECTRONIC
OR MECHANICAL, INCLUDING PHOTOCOPY, RECORDING, OR ANY
INFORMATION STORAGE AND RETRIEVAL SYSTEM, WITHOUT
PERMISSION IN WRITING FROM THE PUBLISHER.

ACADEMIC PRESS, INC.
111 Fifth Avenue, New York, New York 10003

United Kingdom Edition published by
ACADEMIC PRESS, INC. (LONDON) LTD.
24/28 Oval Road, London NW1

Library of Congress Cataloging in Publication Data

O'Malley, Bert W
    Hormone Action.

    (Methods in enzymology; v. 36-)
    Includes bibliographical references and indexes.
    CONTENTS: pt. A. Steroid hormones.–
pt. C. Cyclic nucleotides.–pt. D. Isolated cells,
tissues, and organ systems.–pt. E. Nuclear structure
and function.
        1.   Enzymes.   2.   Hormones.   3.   Cyclic nucleotides.
I.   Hardman, Joel G., joint author.   II.   Title.
III.   Series: Methods in enzymology; v. 36 [etc.]
[DNLM: 1.   Cell nucleus.   2.   Hormones.   3.   Nucleotides, Cyclic.   W1 ME9615K v. 40 / QH595 H812]
QP601.C733 vol. 36, etc. [QP601] 574.1'925'08s
ISBN 0–12–181940–X (v. 40)   [574.1'92]   74-10710

PRINTED IN THE UNITED STATES OF AMERICA

Table of Contents

Contributors to Volume XL. vii

Preface ix

Volumes in Series. xi

Section I. The Cell Nucleus and Cell Division

1. Techniques of Localization of Proteins and Nucleoproteins in the Cell Nucleus by High Resolution Autoradiography and Cytochemistry	M. Bouteille, A. M. Dupuy-Coin, and G. Moyne	3
2. A Nuclear System for DNA Replication from HeLa Cells	Daniel L. Friedman	41
3. Analysis of the Action of Chemical Agents upon the Cell Division Cycle	James B. Kurz and Daniel L. Friedman	44
4. Methods for Analysis of Cell Cycles *in Vivo*	S. H. Socher	58
5. Electron Microscopy and Autoradiography of Chromosomes	Elton Stubblefield	63
6. Parallel Isolation Procedures for Metaphase Chromosomes, Mitotic Apparatus, and Nuclei	Wayne Wray	75

Section II. The Cell Nucleus and Chromatin Proteins

7. Fractionation of Chromatin	DeLill S. Nasser and Brian J. McCarthy	93
8. Isolation of Template Active and Inactive Regions of Chromatin	James Bonner, Joel Gottesfeld, William Garrard, Ronald Billing, and Lynda Uphouse	97
9. Methods for Analysis of Histones	Lubomir S. Hnilica	102
10. Methods for the Assessment of Selective Histone Phosphorylation	Rod Balhorn and Roger Chalkley	138
11. Methods for Isolation and Characterization of Nonhistone Chromosomal Proteins	Sarah C. R. Elgin	144
12. Methods for Isolation and Characterization of Chromosomal Nonhistone Proteins	A. J. MacGillivray, D. Rickwood, A. Cameron, D. Carroll, C. J. Ingles, R. J. Krauze, and J. Paul	160

✓ 13. Methods for Isolation and Characterization of Acidic Chromatin Proteins	ELIZABETH M. WILSON AND THOMAS C. SPELSBERG	171
✓ 14. Methods for Analysis of Phosphorylated Acidic Chromatin Protein Interactions with DNA	LEWIS J. KLEINSMITH AND VALERIE M. KISH	177
15. Immunochemical Characteristics of Chromosomal Proteins	F. CHYTIL	191
16. Chromatin Protein Kinases	VALERIE M. KISH AND LEWIS J. KLEINSMITH	198
17. Circular Dichroism Analysis of Nucleoprotein Complexes	THOMAS E. WAGNER, VAUGHN VANDEGRIFT, AND DEXTER S. MOORE	209

Section III. General Methods for Evaluating Hormone Effects

18. Methods for Analysis of Enzyme Synthesis and Degradation in Animal Tissues	ROBERT T. SCHIMKE	241
19. Enrichment of Polysomes Synthesizing a Specific Protein by Use of Affinity Chromatography	E. BRAD THOMPSON AND J. V. MILLER, JR.	266
20. The Use of Inhibitors in the Study of Hormone Mechanisms in Cell Culture	WALTER A. SCOTT AND GORDON M. TOMKINS	273
21. The Design of Double Label Radioisotope Experiments	EDWIN D. BRANSOME, JR.	293
22. Use of Antibodies to Nucleosides and Nucleotides in Studies of Nucleic Acids in Cells	B. F. ERLANGER, W. J. KLEIN, JR., V. G. DEV, R. R. SCHRECK, AND O. J. MILLER	302
23. Techniques for the Study of Hormone Effect on Collagens	DOROTHY H. HENNEMAN AND GEORGE NICHOLS, JR.	307
AUTHOR INDEX		379
SUBJECT INDEX		394

Contributors to Volume XL

Article numbers are in parentheses following the names of contributors.
Affiliations listed are current.

ROD BALHORN (10), *Department of Biochemistry, University of Iowa, Iowa City, Illinois*

RONALD BILLING (8), *Department of Surgery, University of California School of Medicine, Los Angeles, California*

JAMES BONNER (8), *Division of Biology, California Institute of Technology, Pasadena, California*

M. BOUTEILLE (1), *Laboratoire de Pathologie Cellulaire, Biomédicale des Cordeliers, Ecole de Médecine, Paris, France*

EDWIN D. BRANSOME, JR. (21), *Division of Metabolic and Endocrine Disease, Department of Medicine, Medical College of Georgia, Augusta, Georgia*

A. CAMERON (12), *The Beatson Institute for Cancer Research, Royal Beatson Memorial Hospital, Glasgow, Scotland*

D. CARROLL (12), *Department of Microbiology, University of Utah Medical School, Salt Lake City, Utah*

ROGER CHALKLEY (10), *Department of Biochemistry, University of Iowa, Iowa City, Illinois*

F. CHYTIL (15), *Department of Biochemistry, School of Medicine, Vanderbilt University, Nashville, Tennessee*

V. G. DEV (22), *Department of Human Genetics and Development, Columbia University, New York, New York*

A. M. DUPUY-COIN (1), *Laboratoire de Pathologie Cellulaire, Biomédicale des Cordeliers, Ecole de Médecine, Paris, France*

SARAH C. R. ELGIN (11), *Department of Biochemistry and Molecular Biology, Harvard University, Cambridge, Massachusetts*

B. F. ERLANGER (22), *Department of Microbiology, Columbia University, New York, New York*

DANIEL L. FRIEDMAN (2, 3), *Department of Molecular Biology, Vanderbilt University, Nashville, Tennessee*

WILLIAM GARRARD (8), *Biochemistry Department, Southwestern Medical School, Dallas, Texas*

JOEL GOTTESFELD (8), *Division of Biology, California Institute of Technology, Pasadena, California*

DOROTHY H. HENNEMAN (23), *Division of Health Sciences, University of Delaware, Newark, Delaware*

LUBOMIR S. HNILICA (9), *Department of Biochemistry, M. D. Anderson Hospital, and Tumor Institute, Houston, Texas*

C. J. INGLES (12), *C. H. Best Institute, Banting and Best Department of Medical Research, University of Toronto, Toronto, Ontario, Canada*

VALERIE M. KISH (14, 16), *Worcester Foundation for Experimental Biology, Shrewsbury, Massachusetts*

W. J. KLEIN, JR. (22), *Department of Medicine, University of Rochester School of Medicine and Dentistry, and Rochester General Hospital, Rochester, New York*

LEWIS J. KLEINSMITH (14, 16), *Department of Zoology, The University of Michigan, Ann Arbor, Michigan*

R. J. KRAUZE (12), *Serum and Vaccine Institute, Warsaw, Poland*

JAMES B. KURZ (3), *Department of Molecular Biology, Vanderbilt University, Nashville, Tennessee*

BRIAN J. MCCARTHY (7), *Department of Biochemistry and Biophysics, University of California, San Francisco, California*

A. J. MACGILLIVRAY (12), *The Beatson Institute for Cancer Research, Royal Beatson Memorial Hospital, Glasgow, Scotland*

J. V. MILLER, JR. (19), *Department of Medicine, University of New Mexico Medical School, Albuquerque, New Mexico*

O. J. MILLER (22), *Departments of Human Genetics and Development and Obstetrics and Gynecology, Columbia University, New York, New York*

DEXTER S. MOORE (17), *Department of Chemistry, and Chemical Biodynamics Laboratory, University of California Berkeley, California*

G. MOYNE (1), *Institut de Recherches Scientifiques sur le Cancer Villejuif, France*

DELILL S. NASSER (7), *Department of Biochemistry and Biophysics, University of California, San Francisco, California*

GEORGE NICHOLS, JR. (23), *Department of Medicine, Harvard Medical School, Boston, Massachusetts*

J. PAUL (12), *The Beatson Institute for Cancer Research, Royal Beatson Memmorial Hospital, Glasgow, Scotland*

D. RICKWOOD (12), *The Beatson Institute for Cancer Research, Royal Beatson Memorial Hospital, Glasgow, United Kingdom*

ROBERT T. SCHIMKE (18), *Department of Biological Sciences, Stanford University, Stanford, California*

R. R. SCHRECK (22), *Department of Human Genetics and Development, Columbia University, New York, New York*

WALTER A. SCOTT (20), *Department of Microbiology, The Johns Hopkins University School of Medicine, Baltimore, Maryland*

S. H. SOCHER (4), *Department of Cell Biology, Baylor College of Medicine, Houston, Texas*

THOMAS C. SPELSBERG (13), *Department of Molecular Medicine, Mayo Clinic, Rochester, Minnesota*

ELTON STUBBLEFIELD (5), *Section of Cell Biology, Department of Biology, The University of Texas System Cancer Center, M. D. Anderson Hospital and Tumor Institute, Houston, Texas*

E. BRAD THOMPSON (19), *Laboratory of Biochemistry, National Cancer Institute, Bethesda, Maryland*

GORDON M. TOMKINS (20), *Department of Biochemistry and Biophysics, University of California San Francisco, California*

LYNDA UPHOUSE (8), *Department of Psychology, Yale University, New Haven, Connecticut*

VAUGHN VANDEGRIFT (17), *Department of Chemistry, Illinois State University, Normal, Illinois*

THOMAS E. WAGNER (17), *Department of Chemistry, Ohio University, Athens, Ohio*

ELIZABETH M. WILSON (13), *Department of Biochemistry, Vanderbilt Medical School, Nashville, Tennessee*

WAYNE WRAY (6), *Department of Cell Biology, Baylor College of Medicine, Houston, Texas*

Preface

During the past five-year period, work in the field of "steroid hormone action" has repeatedly pointed to the cell nucleus as the primary site of action. This is not illogical since steroid hormones are thought to exert a direct influence, perhaps through their receptor proteins, on genomic function in eukaryotic cells. Although we cannot make a similar strong case for the mechanism of action of protein hormones, it seems that at least indirectly these hormones must also effect synthesis and processing of nuclear constituents.

In this volume, we have attempted to collect a series of methodologies which would allow investigators to assess hormone effects on nuclear structure, DNA replication, mitosis, and gene transcription. Techniques for assessing nucleic acid metabolism per se are conspicuously absent from this volume but are thoroughly covered in other volumes of "Methods in Enzymology." In considering gene function and specific gene restriction in eukaryotes, it seems almost certain that these processes will ultimately be related to the composition and location of chromosomal proteins. For this reason, we have emphasized techniques which can be utilized to study chromatin composition, structure, and function. This volume also contains a collection of general techniques and analytical approaches important to those involved in the study of hormone action but which have not been included in the earlier volumes of this series because they did not fall clearly within the previously defined categories.

Omissions have inevitably occurred—some because potential authors were overcommitted, some because of editorial oversight, some because of the timing of new developments relative to the publication deadline. Some apparent omissions have been covered in previous volumes of "Methods in Enzymology."

We thank Drs. S. P. Colowick and N. O. Kaplan who originated the idea for and encouraged the compilation of this volume. We thank the staff of Academic Press for their help and advice. We especially thank the contributing authors for their patience and full cooperation and for carrying out the research that made this volume possible.

Bert W. O'Malley
Joel G. Hardman

METHODS IN ENZYMOLOGY

EDITED BY

Sidney P. Colowick and Nathan O. Kaplan

VANDERBILT UNIVERSITY
SCHOOL OF MEDICINE
NASHVILLE, TENNESSEE

DEPARTMENT OF CHEMISTRY
UNIVERSITY OF CALIFORNIA
AT SAN DIEGO
LA JOLLA, CALIFORNIA

I. Preparation and Assay of Enzymes
II. Preparation and Assay of Enzymes
III. Preparation and Assay of Substrates
IV. Special Techniques for the Enzymologist
V. Preparation and Assay of Enzymes
VI. Preparation and Assay of Enzymes (*Continued*)
Preparation and Assay of Substrates
Special Techniques
VII. Cumulative Subject Index

METHODS IN ENZYMOLOGY

EDITORS-IN-CHIEF

Sidney P. Colowick   Nathan O. Kaplan

VOLUME VIII. Complex Carbohydrates
Edited by ELIZABETH F. NEUFELD AND VICTOR GINSBURG

VOLUME IX. Carbohydrate Metabolism
Edited by WILLIS A. WOOD

VOLUME X. Oxidation and Phosphorylation
Edited by RONALD W. ESTABROOK AND MAYNARD E. PULLMAN

VOLUME XI. Enzyme Structure
Edited by C. H. W. HIRS

VOLUME XII. Nucleic Acids (Part A and B)
Edited by LAWRENCE GROSSMAN AND KIVIE MOLDAVE

VOLUME XIII. Citric Acid Cycle
Edited by J. M. LOWENSTEIN

VOLUME XIV. Lipids
Edited by J. M. LOWENSTEIN

VOLUME XV. Steroids and Terpenoids
Edited by RAYMOND B. CLAYTON

VOLUME XVI. Fast Reactions
Edited by KENNETH KUSTIN

VOLUME XVII. Metabolism of Amino Acids and Amines (Parts A and B)
Edited by HERBERT TABOR AND CELIA WHITE TABOR

VOLUME XVIII. Vitamins and Coenzymes (Parts A, B, and C)
Edited by DONALD B. MCCORMICK AND LEMUEL D. WRIGHT

VOLUME XIX. Proteolytic Enzymes
Edited by GERTRUDE E. PERLMANN AND LASZLO LORAND

VOLUME XX. Nucleic Acids and Protein Synthesis (Part C)
Edited by KIVIE MOLDAVE AND LAWRENCE GROSSMAN

VOLUME XXI. Nucleic Acids (Part D)
Edited by LAWRENCE GROSSMAN AND KIVIE MOLDAVE

VOLUME XXII. Enzyme Purification and Related Techniques
Edited by WILLIAM B. JAKOBY

VOLUME XXIII. Photosynthesis (Part A)
Edited by ANTHONY SAN PIETRO

VOLUME XXIV. Photosynthesis and Nitrogen Fixation (Part B)
Edited by ANTHONY SAN PIETRO

VOLUME XXV. Enzyme Structure (Part B)
Edited by C. H. W. HIRS AND SERGE N. TIMASHEFF

VOLUME XXVI. Enzyme Structure (Part C)
Edited by C. H. W. HIRS AND SERGE N. TIMASHEFF

VOLUME XXVII. Enzyme Structure (Part D)
Edited by C. H. W. HIRS AND SERGE N. TIMASHEFF

VOLUME XXVIII. Complex Carbohydrates (Part B)
Edited by VICTOR GINSBURG

VOLUME XXIX. Nucleic Acids and Protein Synthesis (Part E)
Edited by LAWRENCE GROSSMAN AND KIVIE MOLDAVE

VOLUME XXX. Nucleic Acids and Protein Synthesis (Part F)
Edited by KIVIE MOLDAVE AND LAWRENCE GROSSMAN

VOLUME XXXI. Biomembranes (Part A)
Edited by SIDNEY FLEISCHER AND LESTER PACKER

VOLUME XXXII. Biomembranes (Part B)
Edited by SIDNEY FLEISCHER AND LESTER PACKER

VOLUME XXXIII. Cumulative Subject Index Volumes I–XXX
Edited by MARTHA G. DENNIS AND EDWARD A. DENNIS

VOLUME XXXIV. Affinity Techniques (Enzyme Purification: Part B)
Edited by WILLIAM B. JAKOBY AND MEIR WILCHEK

VOLUME XXXV. Lipids (Part B)
Edited by JOHN M. LOWENSTEIN

VOLUME XXXVI. Hormone Action (Part A: Steroid Hormones)
Edited by BERT W. O'MALLEY AND JOEL G. HARDMAN

VOLUME XXXVII. Hormone Action (Part B: Peptide Hormones)
Edited by BERT W. O'MALLEY AND JOEL G. HARDMAN

VOLUME XXXVIII. Hormone Action (Part C: Cyclic Nucleotides)
Edited by JOEL G. HARDMAN AND BERT W. O'MALLEY

VOLUME XXXIX. Hormone Action (Part D: Isolated Cells, Tissues, and Organ Systems)
Edited by JOEL G. HARDMAN AND BERT W. O'MALLEY

VOLUME XL. Hormone Action (Part E: Nuclear Structure and Function)
Edited by BERT W. O'MALLEY AND JOEL G. HARDMAN

VOLUME 41. Carbohydrate Metabolism (Part B)
Edited by W. A. WOOD

VOLUME 42. Carbohydrate Metabolism (Part C)
Edited by W. A. WOOD

VOLUME 43. Antibiotics
Edited by JOHN H. HASH

Methods in Enzymology

Volume XL
HORMONE ACTION
Part E
Nuclear Structure and Function

Section I
The Cell Nucleus and Cell Division

[1] Techniques of Localization of Proteins and Nucleoproteins in the Cell Nucleus by High Resolution Autoradiography and Cytochemistry

By M. BOUTEILLE, A. M. DUPUY-COIN, and G. MOYNE

Although a few components of the cell nucleus, e.g., the nucleolus, can be identified with routine electron microscopy (EM) most of the knowledge on the *chemical nature* of nuclear structures has been gained through rather complex techniques. If one is to obtain such information on a given nuclear structure, whether normal or pathologic, one has to face a number of technical difficulties. The treatment of the fresh tissue is different according to the desired goal. The fixatives for EM autoradiography (ARG) and EM cytochemistry have to be prepared very carefully and always used fresh. They vary according to the technique employed. Generally speaking, it is necessary to have a clear idea of the technique to be used *before* removing the sample. To facilitate this choice, we have drawn in Fig. 1 the pathway to follow from the sampling time down to the biological interpretation.

Basic Procedure

Glass Vessel Washing. It is of considerable importance never to mix the vessels used for EM ARG with the ones employed for EM cytochemistry. All reactions in which heavy metals are involved must be carried out in the same vessels for all experiments. For instance, vessels for the EM ARG developer must not be used for the fixative and vice versa. Never let any vessel dry; after the experiment, immerse it immediately in water. Wash each vessel separately (not all the glassware in a large sink) with a detergent. Rinse 20 times or so in running water, 2 times in distilled water, and finally in absolute ethanol. Dry in an oven at 60°.

Sampling. EM ARG can be performed on sections from routinely processed material, provided that the cells have been exposed to the labeled precursor, and if care has been taken to employ the proper fixative, as discussed below. On the contrary most EM cytochemical procedures require that the reaction be carried out or initiated before embedding. Among the reactions which can be made on ultrathin sections, including enzymatic digestions, a number require special fixation, therefore routine material cannot be used. The decision of how to fix and embed the material must be made after careful consideration of the methods to be subsequently employed.

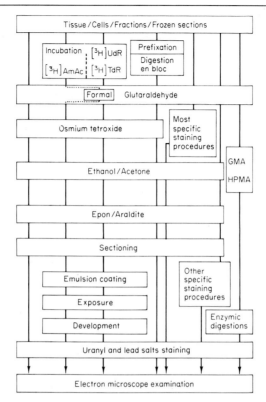

Fig. 1. Various pathways to be followed from the sampling to the electron microscope examination. Note that many techniques need special fixation conditions and that some of them involve technical steps even before the fixation. [^{3}H]AmAc, [^{3}H]UdR, and ^{3}[H]TdR: ^{3}H-labeled amino acid, uridine, and thymidine, respectively; GMA, glycol methacrylate; HPMA, hydroxypropyl glycol methacrylate.

Fixation in Glutaraldehyde. Use for all morphological studies, prior to postfixation with osmium tetroxide. Avoid for EM ARG with amino acids, since free labeled amino acids are artificially bound to the structures. Better than formaldehyde for EM ARG with labeled uridine or thymidine.[1] Both fixatives are good for enzymatic digestions. We use a freshly made 1.66% glutaraldehyde solution in phosphate buffer pH 7.2–7.4. For morphological purposes fix at 4° for 60 minutes, 15–30 minutes for enzymatic digestion. Rinse 24 hours with 4 changes.

Fixation in Formaldehyde. This is good for all purposes and is required for EM ARG with amino acids. Never use formalin or stored formaldehyde. Heat paraformaldehyde powder in distilled water 40% (w/v)

[1] A. Monneron and Y. Moulé, *Exp. Cell Res.* **51**, 531 (1968).

until the solution reaches 60°. Then add NaOH until the temperature drops. Heat again up to 60° and continue until adding NaOH does not drop the temperature. Cool and filter. Dilute 10:90 (v/v) in the buffer.

A simpler preparation is to heat 4% (v/v) paraformaldehyde in the buffer until the solution is clear. Cool and filter. Fix for 4 hours for enzymatic digestions and 24 hours for best morphological preservation. Rinse 24 hours.

Osmium Tetroxide Fixation. This step is necessary for morphological studies, and for EM ARG, since the chemotactic effect is negligible. In this case, however, carboning the section is advisable. We use a 2% (w/v) solution in the buffer. Fix for 1 hour after the usual rinse following aldehyde fixation. Dehydration can be carried out immediately.

This method is to be avoided for most cytochemical reactions, especially for enzymatic digestions, which are prevented by the deposits of osmium. Enzymatic digestions, however, may be carried out on osmicated material if the sections are treated with periodic acid or hydrogen peroxide (see below). The results are usually as good, sometimes better, but the interpretation is more difficult.

Electron Microscope Autoradiography

At first glance, EM ARG is the best tool for investigating the protein and nucleic acid distribution in the nucleus, since the method offers high specificity, excellent contrast, and fair resolution. EM ARG provides *quantitative data*, contrary to most cytochemical procedures. Time-course experiments can be performed, resulting in *kinetic data*. Finally, EM ARG can be employed in combination with EM cytochemistry, which gives considerable information. However, it must be noted that after short pulses EM ARG reveals only newly formed material of a given molecular species, not the complete distribution of this molecule. When longer pulses or continuous labeling are performed, one should pay attention to the possible shift of the activity from one region of the nucleus to another. Therefore one must interpret EM ARG data with prudence when information on the distribution of nuclear molecular species is sought.

We will first describe the EM ARG procedures that are the most widely used for all cytological purposes, so that the reader can choose the one that best fits his project. Detailed information on less often employed techniques can be found elsewhere.[2] We will describe the method that we have developed in our laboratory, by combining the procedures that are adapted more specifically to nuclear studies. Finally, we will sum-

[2] J. Jacob, *Int. Rev. Cytol.* **30**, 91 (1971).

marize the labeling conditions which are thought to be best indicated to localize newly formed RNA, DNA, or proteins, since to date there is no general agreement on the duration of the pulse and the time of labeling during the cell cycle. These problems have been extensively dealt with in another study.[3]

Sectioning. Medium gold sections, about 1200 Å thick, can be used for the combination of good contrast with fair resolution in the EM ARG procedure.[4] The sections must be made by the same operator with the same ultramicrotome, preferably automatic. The thickness of the sections can be controlled by interferometry, if one is especially interested in resolution. Place the sections on the bright side of Formvar-coated grids for single grid coating techniques. Layering with carbon is unnecessary if one uses Ilford L4 emulsion, unless the sections are stained with uranyl and lead salts prior to coating.

Coating. Use Ilford L4 emulsion for best reproducibility, simplicity of use, and consistency of resolution. This emulsion offers rather large crystals, but of fairly constant diameters (0.18 μm). Use Gevaert NUC 307 emulsion for studies that need very high resolution[5] since the crystals are only 0.07 μm in diameter. However, this emulsion has been reported to be difficult to handle, and reproducible techniques for coating, exposure, and development are less easy to work out. The same can be said of the Kodak NTE emulsion.[6]

When a simple and rapid, routine procedure is preferred, the "loop method" with pregelled film can be chosen.[7] Attach 3–4 grids edge to edge with a double-faced sticky tape (Scotch tape 400) placed on a glass slide. Melt 10 g of Ilford L4 emulsion in 20 ml of distilled water in a water bath at 45° for 15 minutes. Stir thoroughly but slowly; cool the emulsion in an ice bath for a few minutes. Equilibrate the emulsion at room temperature for 30 minutes. Dip a platinum or nickel circular loop, 4 cm in diameter in the emulsion. Carefully withdraw the loop, in which the film of emulsion gels immediately. Place the film on the grid. Consistent results can be obtained with practice, particularly if one often checks the interference color of the film, which should be purple.[6,8] However, this method is hard to standardize, and only practice will allow one to obtain a high yield of crystal monolayers.

[3] M. Bouteille, M. Laval, and A. M. Dupuy-Coin, *in* "The Cell Nucleus" (H. Busch, ed.), Vol. I, p. 5. Academic Press, New York, 1974.
[4] M. M. Salpeter, L. Bachmann, and E. E. Salpeter, *J. Cell Biol.* **41**, 1 (1969).
[5] P. Granboulan, *J. Roy. Microsc. Soc.* **81**, 165 (1963).
[6] M. M. Salpeter and L. Bachmann, *J. Cell Biol.* **22**, 469 (1964).
[7] L. G. Caro and R. P. Van Tubergen, *J. Cell Biol.* **15**, 173 (1962).
[8] L. Bachmann and M. M. Salpeter, *Lab. Invest.* **14**, 1041 (1965).

The "dipping technique"[9,10] is easier to learn and is more reliable in the hands of the beginner. Wash glass slides in detergent, tap water, distilled water, and 95% ethanol, several hours in each. Mark the slides with two crosses at 2.5 cm from the end of the slide, 1 cm apart from each other. Dip the slides in a filtered 2% (w/v) solution of Parlodion in isoamyl acetate; let them dry at a 30° angle, the crossed side down. Redipping the bottom of the slide after drying is advised to make the final removal of the sections easier. With a platinum loop, place 3–5 ultrathin sections on the upper side of the slides over each cross. If staining is wanted, stain as usual with fresh uranyl and lead salts solutions by dipping the slides and layer with carbon prior to coating. Dilute the Ilford L4 emulsion in the darkroom 1 volume to 4 volumes of water and place in a water bath at 40° for 1 hour with gentle stirring every 15 minutes. Dip the slides one by one, and withdraw slowly. Dipping and withdrawing must be done always at the same speed by the same operator, since this is an important factor causing variations in the thickness of the coated film. More reproducible results can be obtained by using a semiautomatic device.[11] Let the slides dry in a vertical position. After exposure and developing, the sections must be removed from the slides before they start to dry. Scrape both sides of the slides with a razor blade and remove the bottom edge of the Parlodion film (which was dipped twice in the Parlodion solution). Strip the film off the slide on the surface of the water and place grids over the two groups of sections on the floating film. Pick up the preparation with a filter paper, cut around the Parlodion and let it dry. This last operation is time consuming and requires skill and practice. The methods of washing slides cause considerable variations in the easiness of section removal. There is usually some loss of preparations. The fact that this operation takes place at the end of the whole ARG procedure, including exposure, is a severe limitation. In addition, the sections must be stained before the coating, and therefore carbon layered, since after removal of the preparations the sections are placed in sandwich between the Parlodion layer and the grid. This can be avoided, however, by the following procedure of recovering the sections. Prior to the complete stripping of the Parlodion film off the slide, immerse two grids and place them on the upper side of the slide over the crosses. Stick the Parlodion film again over the glass slide, so that the sections come over the grids. Remove the slide and let it dry. Staining should be made when the preparations are not completely dry. Another difficulty

[9] P. Granboulan, *Symp. Int. Soc. Cell Biol.* **4**, 43–63 (1965).
[10] N. Granboulan, *in* "Methods in Virology" (K. Maramorosch and H. Koprowski, eds.), Vol. 3, pp. 617–637. Academic Press, New York, 1967.
[11] B. M. Kopriwa, *J. Histochem. Cytochem.* **15**, 501 (1967).

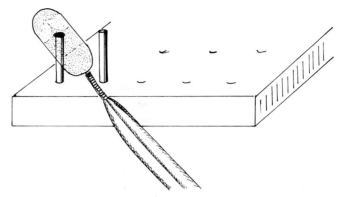

FIG. 2. Coating single grids with the golden zone of the emulsion film in a platinum loop.

of the "dipping technique" is that large variations in thickness of the emulsion occur from the bottom to the top of the immersed slide. Conditions must be standardized so that the purple zone coincides with the place of the sections. This is difficult to achieve, especially when a large number of slides are coated in the same session. As in the "loop technique," the interference color cannot be judged in the dark room during the coating, but only on check slides which are coated and examined in the light.

All these drawbacks can be avoided by viewing the interference color of the emulsion film directly in the safelight before coating each grid.[12] In our hands the best results have been obtained with the following procedure.[13] Prepare the emulsion the day before the coating session. Mark a clean jar at 3 and 4 cm from the bottom with black ink. In the safelight, fill with distilled water up to the lower mark and pour Ilford L4 emulsion into the water until the level reaches the upper mark. This will yield a 1:4 dilution (v/v) which allows the subsequent coating operations to take place at room temperature. Melt the emulsion in a water bath at 40° for 1 hour and stir thoroughly but slowly every 15 minutes. Store the emulsion in a light-safe box overnight, this stabilizes the emulsion. The next day, melt again the emulsion for 30–60 minutes at 40°. Let the emulsion cool down to room temperature. Better results will be obtained if the room is at about 23–25° with a high degree of humidity. Place the grids on top of glass or metal poles, 3 mm in diameter, 20 mm in height, placed 25 mm apart from each other on a plate in 4 rows of 5 poles (Fig. 2). Dip a platinum loop 10 × 300 mm into the emulsion,

[12] G. Haase and G. Jung, *Naturwissenschaften* **51**, 404 (1964).
[13] M. Bouteille, to be published.

and let it gel vertically, in front of the safelight. A gray zone develops at the top of the loop. When observed in reflecting light, this zone displays a metallic sheen, and in daylight it corresponds roughly to the golden interference zone of the gelled film. When placed onto a grid and examined in the electron microscope, such films present densely packed monolayers of crystals with a reproducibility close to 100%. As this appears as an absolute criterion for the monolayers, it is only necessary to modify the properties of the emulsion in order to obtain this zone before coating each grid. If the golden zone develops too slowly, the film gels before the zone forms and the emulsion should be warmed up, for a few minutes each time, in a 40° water bath until the zone forms properly. If the emulsion is too warm, the golden zone will appear rapidly, but the film will also break. The emulsion should then be cooled in an ice bath. This does not occur usually with the 1:4 dilution at 25°. After coating, breathe gently on the grid to attach firmly the emulsion to the specimen, pick up the grid with forceps, and place it edge to edge on a double-faced sticky tape. As the whole procedure ensures that a monolayer is formed before coating any grid, it results in highly reproducible coating for all grids. This in turn allows strictly standardized methods of development and fixation. However, this method requires skill and practice before a good yield of monolayers is obtained. At the beginning it is advisable to use five serial, equivalent grids for each time point of the experiments, since it may happen that some of the five have been missed. This method is more time-consuming than the techniques described above. However, monolayers are easier to obtain than with the usual loop technique, and the problem of stripping the films in the dipping technique is eliminated. Although it is not advisable for occasional EM ARG studies, it is probably the best method of coating for quantitative work in which high reproducibility of the sensitivity, resolution, and grain size is sought.

Exposure. Place the slides in boxes with a desiccant and seal the boxes with black "electric tape." Store at 4°. No significant fading is observed with Ilford L4 emulsions for periods of exposure of 1 year or more, and storage in inert gas atmosphere is not required. No carbon layering is needed either if the sections are not stained before coating.

Photographic Processing. For years chemical developers have been the most widely used, which resulted in silver grains appearing as coiled filaments. When Ilford L4 emulsion is developed with D-19, large filaments, up to 0.4 μm in "diameter," are produced. Smaller grains are obtained with Microdol-X, but in both cases, the grains are much larger than the halide crystals, which partially impairs the resolution, and the complexity and the surface of the grains make the identification of the cell background difficult, or even impossible when small organelles or granules are studied; this is especially inappropriate for the cell nucleus, where the chemical nature of a number of granules and bodies are under intense EM ARG and cytochemical study.[3]

The physical developer p-phenylenediamine produces much smaller, comma-shaped grains[7] but is a less reliable and less sensitive developer than the preceding ones.[11] A technique combining the use of the physical developer Elon–ascorbic acid with gold latensification was introduced for the Kodak NTE emulsion[6] and adapted to the Ilford L4 emulsion.[14] This technique produces very small grains, with an absolute sensitivity 5 times as high as that obtained with Microdol-X developer, and a very easy identification of the cell background. To obtain the gold solution, add 1 ml of a 4° stored, 2% (w/v) $HAuCl_4$ solution in 99 ml of distilled water. Adjust pH at 7.0 with NaOH, add 250 mg of KCNS and 300 mg of KBr. Make up to 500 ml with boiled distilled water. To make the developer, add to some boiled distilled water 225 mg of Elon (Metol), 1.5 g of ascorbic acid, 2.5 g of borax, 0.5 g of KBr and make up to 500 ml with boiled distilled water. Use the gold solution and developer always at the same time after dissolution, preferably 1 hour. Immerse the grids 5 minutes in the gold solution and 7.5 minutes in the developer. Fix for 2 minutes in the usual fixer and wash. The only disturbing point with this method is that many silver grains look like fragments of larger ones which seem not completely developed. This results in a large number of grains over a small organelle or area, and may mislead by suggesting a high activity in terms of number of grains per surface area, whereas with another technique the number of grains would have been very small. This requires quantitative studies that are difficult to perform because of the large variations in shape and size of the grains.

Combination of gold latensification with the recently introduced Phenidon developer[15] was found to be possible[16] (Fig. 3). When this developing technique is used after single grids coating as described above, highly reproducible results were obtained in our hands in terms of size and shape of the silver grains. Make the gold solution as described above. To obtain the Phenidon developer, mix with a magnetic stirrer without heating 600 ml of doubly distilled water, and add each reagent after complete dissolution of the preceding one, that is, 30 minutes in our routine procedure. Add 12 g of ascorbic acid, 2 g of Phenidon (1-phenyl-3-pyrazolidone), 4.8 g of potassium bromide, 10.4 g of sodium carbonate, 80 g of anhydrous

[14] E. Wisse and A. D. Tates, *Electron Microsc. Proc. 4th Eur. Reg. Conf., Rome, 1968* p. 465 (1968).
[15] H. Lettré and N. Paweletz, *Naturwissenschaften* **53**, 269 (1966).
[16] M. Bouteille, *Exp. Cell Res.* **69**, 135 (1971).

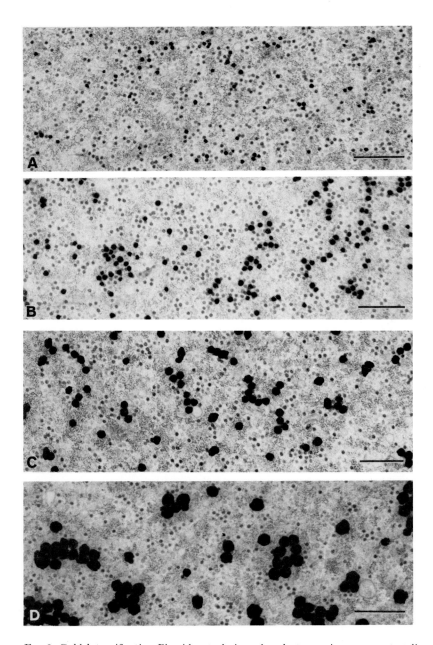

FIG. 3. Gold latensification-Phenidon technique for electron microscope autoradiograph development. The size and the number of the silver grains depend upon various parameters. This figure illustrates size variations according to the temperature of development. Compare with the phages labeled with [^{3}H]thymidine, which are 540 Å in diameter. The silver grains are hardly distinguished from the phages in (A), temperature of development 10°. Their size increases when the temperature is raised to 15° (B), 22° (C), and 30° (D).

sodium sulfite, 48 g of potassium thiocyanate. Make up to 800 ml with bidistilled water. Filter and use the same day. Do not use the solution if turbid, which occurs when complete dissolution of some reagent was not obtained. Make the fixer with 210 g of sodium thiosulfate in 700 ml of distilled water. The most reproducible results are observed with 5 minutes gold latensification, 30 seconds washing in distilled water, 2 minutes development, and 5 minutes fixation. Wash 3 times for 5 minutes. All these operations must be done at 18°. If not, significant variations in size, shape, and number of grains will occur. This is achieved by cooling running water to about 15°C with a cooling unit, and heating the water in the water bath with an ultrathermostat ± 0.1°. When the temperature is stabilized, there is no variation for hours and a large number of grids can be developed in the same session. It is also advisable to use the fresh solution at the same given time after dissolution of the last chemical. Under these conditions, relatively small and remarkably "spherical" grains are obtained. The use of gold latensification produces about 2–3 times as many grains as with the Phenidon developer alone. The shape of the grains facilitates the identification of cell structures and the distinction from precipitates or from the products of cytochemical reactions.[16] Finally, the grain counting is made much easier and faster.

Labeling Conditions for DNA. Recently three methods have been proposed which will allow direct DNA localization without previous incorporation of [^{3}H]thymidine. This would be of great interest, since then the shift of activity from one part of the nucleus to another would be avoided. Unfortunately, at this point these methods are under investigation in several laboratories, so that no definite technique can be described here. They will only be mentioned as promising methods which are presently being developed.

    1. [^{3}H]Actinomycin D binding has been recently introduced at the EM level.[17] Cultured cells are incubated with this tritiated antibiotic, which has a remarkable affinity for native, double-stranded DNA.

    2. EM ARG analysis of hybridization *in situ* has also been proposed with tritiated 28 S RNA in *Xenopus* ovaries[18] and with Shope virus complementary [^{3}H]RNA in Shope papilloma.[19] High specific activities, particularly pure nucleic acids, and a large number of DNA copies are probably required for this method, which has an immense range of applications.

[17] R. Simard, *J. Cell Biol.* **35**, 716 (1967).
[18] J. Jacob, K. Todd, M. L. Birnstiel, and A. Bird, *Biochim. Biophys. Acta* **228**, 761 (1971).
[19] O. Croissant, C. Dauguet, P. Jeanteur, and G. Orth, *C. R. Acad. Sci. Ser. D* **274**, 614 (1972).

3. Incubation of ultrathin sections with [³H]dATP and purified terminal deoxynucleotidyltransferase resulted in DNA localization by apparently adding dAMP to the 3'-OH ends of DNA at the surface of the sections.[20]

However, *incorporation of* [³H]thymidine is the most widely used technique. As a complete review on the experiments with [³H]thymidine for localization of newly formed DNA is available elsewhere,[3] consideration will be given here only to the way the incorporation must be conducted. If long pulses (more than 15 minutes) are given to nonsynchronized cells, the resulting labeling will be found in most cases over the whole nucleus, with a higher density in the nucleolus and the condensed chromatin. This is probably the best technique for investigation of the *DNA content of an unknown organelle*, if it is not considered as a possible site of replication (Fig. 4). When short pulses were compared with either long pulses or chases, the results were contradictory, and it is not clear whether the peripheral chromatin is labeled first and the diffuse chromatin second, or vice versa. The use of synchronizing drugs may be harmful to the cells and the sites of replication may be altered. The general impression is that the diffuse chromatin is labeled first at the early S period, and the distribution becomes more homogeneous as the S period proceeds. Cells synchronized preferably by other means than drugs (e.g., metaphase selection) should therefore be exposed to short pulses of [³H]thymidine in order to determine whether a given organelle is a *site of replication*.

Labeling Conditions for RNA. A number of investigations have dealt with this question.[3] It is now possible to state that in order to investigate the RNA content of a given organelle and the possibility of this organelle being a site of transcription, one must vary the time and conditions of labeling. If the organelle considered is related to the nucleolus, one must take into consideration that the initial transcription probably takes place at the junction between the nucleolar chromatin and the nucleolar fibrils. This can be visualized by short pulses (2 minutes or less). Nucleolar fibrils will be labeled in about 15 minutes and the nucleolar granules in half an hour. The use of chases is recommended to facilitate quantitative analysis of such small structures. In respect to this, actinomycin D or another drug inducing nucleolar segregation might help the differentiation of the various nucleolar zones.[21]

If the structure under investigation is extranucleolar, one can determine whether this organelle has a *RNA component*, which will be best studied after long pulses (1 hour or more) when an equilibrium has been

[20] S. Fakan and S. P. Modak, *Exp. Cell Res.* **77**, 95 (1973).
[21] M. Geuskens and W. Bernhard, *Exp. Cell Res.* **44**, 579 (1966).

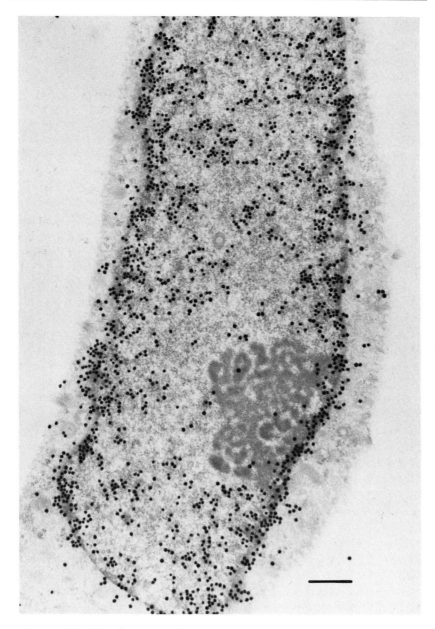

Fig. 4. Fibroblast in culture, labeled with [^{3}H]thymidine for 24 hours. Most of the labeling is found in the condensed chromatin, and less activity is visible in the dispersed chromatin or the nucleolus. Autoradiogram, gold latensification-Phenidon. ×12,000.

reached, or whether it is a *site of transcription*. In the latter case, it should be labeled after short pulses, 2 minutes or less, since the perichromatin region, which is supposed to be the main site of extranucleolar transcription, is also labeled within this time.

Using isolated nuclei, one is also able to obtain information on the type of RNA that the organelle considered contains. Incubation with tritiated triphosphates in the presence of Mg^{2+} at low ionic strength results in activating the RNA polymerase which controls ribosome-type RNA synthesis, while with Mn^{2+} and $(NH_4)_2SO_4$, the heterogeneous RNA synthesis is activated. In the first case, the activity is found over the nucleolus (Fig. 5), and in the second one, throughout the nucleus. An organelle carrying RNA from either origin may be expected to be labeled only when the corresponding conditions are met.

Labeling Conditions for Proteins. The problem is more complex than with nucleic acids because the question of the nuclear or cytoplasmic origin of nuclear proteins is still controversial.[3] The bulk of the nuclear proteins, however, is now believed to originate from cytoplasmic protein synthesis. To investigate the *protein content* of a nuclear organelle by tritiated amino acids incorporation in whole cells, it is advised to use a rather long pulse (30 minutes or more) for two reasons: first the migration of labeled proteins from the cytoplasm into the nucleus is rather slow and reaches an equilibrium 2 hours after the beginning of the pulse[22]; second, the nuclear activity is low compared to the cytoplasmic labeling (Fig. 6).

A new hypothesis has been proposed that some proteins may be elaborated by the nucleus itself,[23] on the basis of experiments with leucine incorporation into purified nuclear fractions. Recent investigations have shown that the EM ARG distribution of the labeled material was not random in the nucleus.[24,25] It is therefore reasonable to imagine a situation in which an organelle is labeled after incubation in isolated nuclei, but not in whole cells. This could then be interpreted as indicating the nuclear origin of its protein content.

Quantitative Analysis of the Results. In a few cases the activity in the structure under investigation is so high that the interpretation is obvious. In most instances, however, quantitative analysis is required to avoid being misled by occasional high activities in certain parts of the cell. This is of particular importance when the labeling is low, in which case a statistical study of the grain distribution is necessary to establish

[22] M. Bouteille, *Exp. Cell Res.* **74**, 343 (1972).
[23] K. M. Anderson, *Int. J. Biochem.* **3**, 449 (1972).
[24] M. Laval and M. Bouteille, *Exp. Cell Res.* **76**, 337 (1973).
[25] M. Laval and M. Bouteille, *Exp. Cell Res.* **79**, 391 (1973).

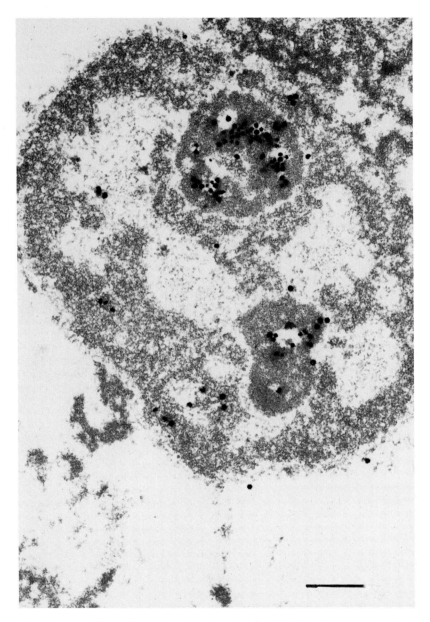

FIG. 5. Isolated rat liver nucleus, incubated for RNA polymerase activity, at low ionic strength, with [³H]UTP for 5 minutes. Almost all the grains are found over the nucleoli. Autoradiogram, gold latensification-Phenidon. ×15,000.

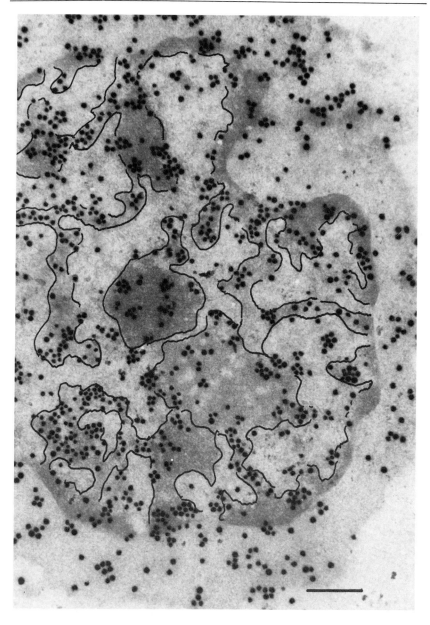

FIG. 6. Blast cell from the spleen of an hyperimmunized rabbit. Isolated spleen cells have been incubated with [^{3}H]leucine. The bulk of the activity is observed along the borderline between the condensed and the dispersed chromatin, where most of the replication and extranucleolar transcription processes are thought to take place. Autoradiogram, gold latensification–Phenidon. ×15,000.

the kinetics of the labeling.[26] Counting the number of grains per organelle is not representative of the results, since a large organelle of low specific activity may show a total number of grains much higher than a smaller one with a higher specific activity. In the EM ARG method the results usually cannot be expressed in terms of specific activity, i.e., disintegrations per minute per milligram (dpm/mg) of the labeled product, but as a number of grains. However, as the number of grains is related to the number of dpm, and assuming that the labeled product is uniformly distributed in the organelle, the figure closest to specific activity is the *number of grains per unit volume;* that is, if the sections are of uniform thickness (see above) per unit area. This quantity, which is the *grain density,* must be calculated on a large number of organelles, photographed randomly (usually 100) from at least 3 different blocks of tissue. It should be expressed per 1 μm^2 or 100 μm^2 with the standard error of the mean, and the number of organelles counted should be mentioned.

When the structure is large enough, counting the number of grains per unit area may be sufficient. However, we are mostly concerned in the present review with small organelles or granules, a few hundreds of nanometers in diameter. Even if they are strongly labeled, this will not be visualized on the autoradiograms and the *density distribution* must be studied.[4] First, determine the HD value (distance from the hot source in which half of the grains fall) either experimentally or by comparison with published data.[4] Photograph a large number of cells containing the organelle considered. The cells must be photographed randomly; for instance, the first 100 cells encountered, whatever the appearance of the cell. If the organelle is a disc in ultrathin section, count the grains which fall in an annulus, 1 HD in radius around the outer limit of the organelle. [If the organelle is linear (membrane, etc.), count the grains falling in a band, 1 HD in width.] Count all the grains falling in this circle in all the 100 organelles photographed. The corresponding figure will be +1 HD. Operate in the same manner for an annulus, 1 HD in diameter, around the first one. This will be called +2 HD. Repeat the operation for annuli +3, 4, . . . , not more than 10 HD, and then for annuli inside the organelles, that is −1, 2, . . . , HD. Now use a clear plastic sheet on which points have been regularly drawn with India ink in rows at a constant distance from each other (preferably at a distance in millimeters corresponding to 1 μm in the cells, in order to express easily the density as the number of grains per square micrometer). Place the plastic sheet randomly over each organelle, and count the number of points falling in each annulus inside and outside the organelles. The quotient of the number of grains by the number of points falling in each annulus for the 100 organelles will be the grain density in grains per square mi-

[26] M. A. Williams, *Advan. Opt. Electron Microsc.* 3, 219 (1969).

crometer. Express the densities in an histogram with the origin 0 between the +1 HD and −1 HD cases, and compare with the published curves.[4] If the organelle is more active than the surroundings, the curve of density will be higher inside (−HD) than outside (+HD) the organelle. If there is no difference, the curve will be horizontal. If the organelle is less radioactive, the curve will be higher in the positive annuli. The significance of the results will be expressed as the statistical accuracy of all the sample, which is

$$\text{grains/points} = x \pm [(1/\text{grains}) + (1/\text{points}) \times 100\%]^{1/2}$$

This type of investigation has recently permitted us to demonstrate the protein content of nuclear organelles, 0.3 μm in diameter, called "nuclear bodies."[27] Even very small granules, such as the perichromatin granules, could also be investigated by this technique, since their size is constant and their number is high enough to permit statistical studies.

Electron Microscope Cytochemistry

EM Cytochemistry of Nuclear Proteins

EM cytochemical detection of the specific proteins, enzymes, is outside the scope of this review. Therefore, we will describe the methods for nuclear proteins and nucleic acids in general. The methods will be treated according to their claimed specificity, whether assumed or experimentally established. The degree of specificity of each technique is the fundamental problem of cytochemistry, and in respect to this, the reader is referred to each original article.

Histones can be stained after acrolein fixation (Fig. 7) which leaves free the aldehyde group of acrolein when bound to histidine. The aldehyde is subsequently revealed with silver impregnation.[28] Fix in 5–10% acrolein in phosphate buffer for 15–25 minutes. Embed as usual in glycolmethacrylate.[29] Prepare the silver nitrate solution by adjusting the pH of 10 ml of $AgNO_3$, 0.3%, to pH = 7.5–9 with 5% borax and complete to 30 ml with distilled water. Float free sections for 30–60 minutes at 60°. The reaction is rather selective for histones in the nucleus, but other proteins (pancreatic granules, collagen) may also stain. The silver reactions in general offer good contrast, but the resolution is impaired by the size of the silver grains (30–200 Å in diameter). Arginine-rich histones have been claimed to stain with another silver reaction.[30] Fix in neutral for-

---

[27] A. M. Dupuy-Coin and M. Bouteille, *Exp. Cell Res.* in press.
[28] V. Marinozzi, *J. Roy. Microsc. Soc.* **81**, 141 (1963).
[29] E. H. Leduc and W. Bernhard, *J. Ultrastruct. Res.* **19**, 196 (1967).
[30] E. K. MacRay and G. D. Meetz, *J. Cell Biol.* **45**, 235 (1970).

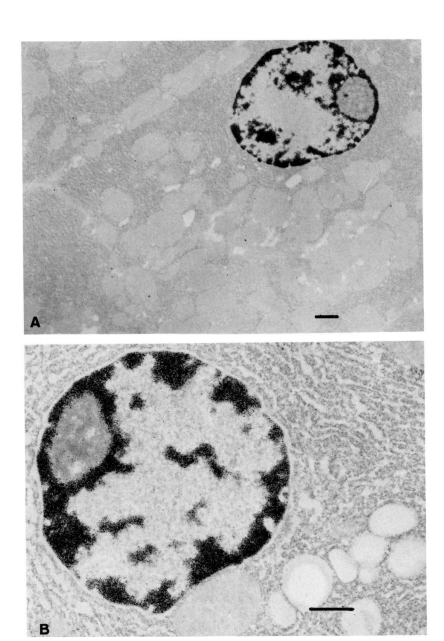

Fig. 7. Pancreatic cells, fixed with formalin-acrolein, embedded in glycol methacrylate, stained for DNP with the silver-methenamine technique for 30 minutes at pH 8.0–8.5. (A) Low magnification, showing the very high contrast of the nucleus as compared with the cytoplasm. (B) At higher magnification the size of the grains, which limits the resolution, is apparent. (A) ×6000, (B) ×12,000. Courtesy of Dr. V. Marinozzi, Rome, Italy.

malin for 3 hours. Prepare the ammoniacal silver nitrate solution by progressive addition of 10% $AgNO_3$ to ammonium hydroxide until the solution becomes almost clear. Stain the blocks of tissue for 5 minutes, wash in distilled water, treat for 5 minutes in 3% formalin until the blocks turn brownish, and embed in Epon. This technique results in deposits of reduced metallic silver. Its specificity, mechanism, and reproducibility are not yet elucidated.

Nucleic acid-associated proteins have been reported to stain with the argentaffin reaction (without further reduction).[31] Fix with 2% glutaraldehyde in 0.1 M phosphate buffer pH 7.3 containing 0.22 M sucrose. Postfix or not with OsO_4, and embed in Araldite. Mount the section on titanium grids. Immerse the grids in fresh 5% $AgNO_3$ in doubly distilled water for 30 minutes, pick up the grids with plastic forceps, and rinse. The specificity of the reaction has been studied by means of enzymatic digestions.

To obviate the poor resolution of silver reactions, phosphotungstic acid (PTA) has also been employed. Neutral PTA after glutaraldehyde fixation has been applied to various nuclear organelles.[32] Fix in 3% cold glutaraldehyde in a 0.1 M phosphate buffer with $CaCl_2$ and $MgCl_2$, rinse 3 times for 20 minutes in the buffer. Dehydrate as usual in ethanol, treat the blocks in 1% PTA in ice cold absolute ethanol overnight. Rinse twice in ethanol, embed in Epon. This selective, but not specific, technique has been applied to many materials.[33,34]

Colloidal iron has also been used to reveal basic proteins, taking advantage of their affinity for an electronegative solution of colloidal iron at high pH.[35] Fix in 10% formalin in phosphate buffer, pH 7.4, for 2–3 hours at 4°. Incubate at 37°, 20 μm frozen sections with 0.01% DNase at pH 6.8 in 1 mM $MgCl_2$ for 2 hours. Treat the blocks in the solution of colloidal iron prepared by adding potassium ferrocyanide to an electropositive solution of ferric hydroxide (see the original paper for details) for 15 hours at 4°. Wash before and after the reaction with $M/35$ phosphate adjusted to pH 4.2 with $M/15$ citric acid, and then to 10.5 with 0.1 N NaOH. Embed in Epon. The specificity has been carefully tested by several inhibitors and reactions *in vitro*. The contrast is high, and the resolution of the final deposits is rather good. This reaction, although complex, seems the most reliable at this point (Fig. 8).

Contrasting with the fair number of reactions which are intended to

[31] J. W. Smith and R. J. Stuart, *J. Cell Sci.* **9**, 253 (1971).
[32] W. F. Sheridan and R. J. Barrnett, *J. Ultrastruct. Res.* **27**, 216 (1969).
[33] P. Esponda and G. Gimenez-Martin, *Chromosoma* **38**, 405 (1972).
[34] L. F. La Cour and B. Wells, *Cytologia* **36**, 111 (1971).
[35] E. Puvion and P. Blanquet, *J. Microsc. (Paris)* **12**, 171 (1971).

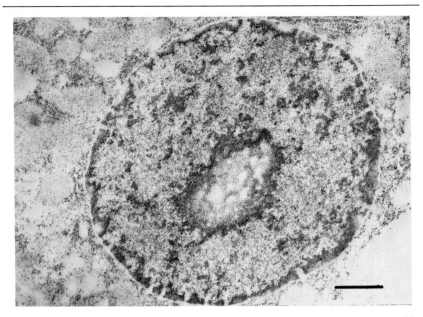

Fig. 8. Rat liver cell, stained with colloidal iron at pH 10.5 for histones. The staining is observed in both condensed and dispersed chromatin. ×12,000. Courtesy of Dr. E. Puvion, Lille, France.

stain specifically basic proteins, there is only one study in which acidic proteins are claimed to be stained.[36,37] Fix tissue cultures under precise conditions in 2% glutaraldehyde in 0.1 M cacodylate buffer for 30 minutes. After centrifugation, wash the pellets in Millonig buffer and postfix in OsO_4, 1% in Millonig buffer. Wash and embed in Epon. Stain with lead citrate as usual. The potential specificity of the reaction is under study by means of enzymatic digestions.

EM Cytochemistry of Nucleic Acids

When one starts to investigate a structure that may contain nucleic acids, one has to choose one of two pathways. Either use methods which stain both nucleic acids and then methods selective for RNP and DNP, or start with the selective methods, in order to omit one step. The state of EM cytochemistry has not reached the point where it can offer investi-

[36] L. Recher, J. Whitescarver, and L. Briggs, *J. Ultrastruct. Res.* **29**, 1 (1969).
[37] L. Recher, N. Parry, J. Whitescarver, and L. Briggs, *J. Ultrastruct. Res.* **38**, 398 (1972).

gators a large number of specific tests, as in light microscopy. It must be emphasized that more than one reaction has to be employed before the chemical nature of the structure considered can be characterized without uncertainty.

1. *Methods Staining Both Nucleic Acids.* The basophilic properties of nucleic acids due to the electronegative phosphate groups have long been utilized in attempts to localize them at the EM level.[38,39] The highest specificity combined with satisfactory contrast has been achieved with *indium* after blocking most groups other than phosphates reacting with trivalent *indium*.[40] Fix small blocks in 10% acrolein for 30 minutes at 4°. Dehydrate at 4° in graded acetone solutions, add increasing amounts of pyridine to the highest concentration. Treat for 2 hours in pyridine saturated with $LiBH_4$ at 4° made a few hours earlier. Wash 3 times in pyridine at room temperature, treat overnight with pyridine containing 40% acetic anhydride saturated with anhydrous potassium or sodium acetate for acetylation. Wash $3\times$ 10 minutes in pyridine, 5 minutes in pyridine–acetone (1:1), $3\times$ 10 minutes in acetone at 4°. Treat for 2 hours at 4° in acetone containing 25 mg/ml of anhydrous $InCl_3$. Wash in acetone $2\times$ 15 minutes, and embed in methacrylate or Vestopal (not in epoxy resins). The reaction is highly specific for all nuclear and cytoplasmic components of known nucleic acid nature. The contrast and the resolution are also good, and in several papers this reaction has been reported to be useful. Another heavy metal, *bismuth*, which binds only to phosphate groups has also been proposed for staining nucleic acids.[41]

Methods that do not use the basophilic properties of nucleic acids have been reported: the usual uranyl and lead salts under special conditions of fixation and processing,[42,43] sodium tungstate,[44] and silver and other reactions.[45]

Recently, two well established methods have been described in detail for the demonstration of the nucleic acids at the EM level. Although they require some practice, they are surely worth trying either before the methods directed to DNP or RNP or simultaneously for confirmation.

[38] H. Swift and E. Rasch, *J. Histochem. Cytochem.* **6**, 391 (1958).
[39] B. Mundkur, *Exp. Cell Res.* **25**, 1 (1961).
[40] M. L. Watson and W. G. Aldridge, *J. Biophys. Biochem. Cytol.* **11**, 257 (1961).
[41] P. Albersheim and U. Killias, *J. Cell Biol.* **17**, 93 (1963).
[42] H. E. Huxley and G. Zubay, *J. Biophys. Biochem. Cytol.* **11**, 273 (1961).
[43] V. Marinozzi and A. Gautier, *J. Ultrastruct. Res.* **7**, 436 (1962).
[44] H. Swift and B. J. Adams, *J. Histochem. Cytochem.* **14**, 744 (1966).
[45] J. P. Thiéry, *Electron Microsc. Proc. 6th Int. Congr. Kyoto 1966*, Vol. II, p. 73. Maruzen, Tokyo.

The first one is based on the specificity of *acriflavine* for nucleic acids. This has been utilized at the EM level by coupling with the electron dense phosphotungstate.[46,47] This complex is commercially available. To carry out the reaction, fix in phosphate-buffered 10% formalin or 3% glutaraldehyde in *s*-collidine buffer, pH 7.3. Wash in the same buffer with 5% sucrose. After partial dehydration in ethanol, immerse the blocks in the acriflavine-PTA solution in 50% ethanol for 2 hours at 4°, at a given concentration (OD: 0.86 at 4610 nm). Dehydrate in acetone and embed in Durcupan. After sectioning, pick up the sections without delay. The reaction has the same selectivity as acriflavine (that is nucleic acids and myelin), is rather simple to perform, and offers satisfactory contrast.

In the second method, *thallium* ethylate has been introduced as a good contrasting agent.[48,49] Fix as usual with glutaraldehyde and/or osmium, embed in Epon. Float the sections on hydrogen peroxide to remove the osmium, if necessary. To prepare the staining solution, immerse 2 g of clean metallic thallium as small fragments (pay attention to the high toxicity of thallium) in 25 ml of absolute ethanol in a 25-ml vial, and seal. Use after 1–2 days and until the solution becomes cloudy (10–20 days). The thallium can be used again. To stain, immerse copper grids with sections in 1 ml of absolute ethanol. Add 1 drop of staining solution and 1 drop of water. Treat for 10 minutes, wash very briefly in ethanol. If necessary, stabilize the stain by heating the grids for 3 minutes at 200°. This reaction is specific for hydroxyl groups, so that RNA, DNA, and polysaccharides are stained.

2. Proposed Methods Staining the RNP. When one of the preceding methods has shown that the structure under investigation contains nucleic acids, one must turn to methods for RNP and DNP. Unfortunately, there is no specific method for RNA at this point, and EM cytochemical distinction between RNA and RNP is not yet possible. Only the combination with EM ARG will help in this respect.

A simple but rather unspecific staining method consists of staining aldehyde-fixed material with *lead salts*.[43] As described originally, fix with 10% formalin in phosphate or Michaelis buffer, pH 6.9–7.3 for 60 minutes at 4°, dehydrate in acetone, embed in Vestopal, and stain the sections with lead hydroxide according to Watson.[50] This reaction is not specific, but the contrast of RNP components in the nucleus is enhanced to a

---

[46] V. Chan-Curtis, W. D. Belt, and C. T. Ladoulis, *J. Histochem. Cytochem.* **18**, 609 (1970).

[47] V. Chan-Curtis, M. Beer, and T. Koller, *J. Histochem. Cytochem.* **18**, 628 (1970).

[48] P. Mentré, *J. Microsc. (Paris)* **11**, 79 (1971).

[49] P. Mentré, *J. Microsc. (Paris)* **14**, 251 (1972).

[50] M. L. Watson, *J. Biophys. Biochem. Cytol.* **4**, 727 (1958).

large extent. The use of lead citrate[51,52] after any aldehyde fixation will produce the same effect with far better reproducibility.

The EDTA staining method has been recently introduced[53,54] as one of the simplest techniques available in nuclear EM cytochemistry (Fig. 9). It is sometimes recommended to start with this technique when a structure to be studied is supposed to contain nucleic acids, since it offers the advantage of a negative image for chromatin and a positive one for RNP. Fix with an aldehyde as usual (no osmium) and embed in Epon or GMA. Prepare the EDTA solution as follows: add 50 ml of distilled water to 7.44 g of EDTA (sodium salt). Dissolve the EDTA by slow addition of 1 N NaOH to pH 7.0 and complete with distilled water. This yields a 0.2 M Na-EDTA solution, pH 7.0. Float grid mounted sections on the solution for various periods of time. The duration of treatment depends upon the tissue, the organelle under investigation, the thickness of the section and varies from one sample to another. The criterion is the bleaching of the stain of the chromatin, as judged after poststaining with lead citrate (Reynolds for 1 minute). This is obtained after 5–40 minutes of treatment, usually around 15 minutes for tissue fixed in glutaraldehyde for 60 minutes, and around 30 minutes for cell cultures after the same fixation. Therefore it is recommended to treat a first series of sections for 5, 10, 25, . . . , 40 minutes. Then, treat another series around the time when the chromatin has been bleached out whereas the rest of the nucleus, including the nucleolus, keeps its contrast. At this point, a variation of ±1 minute of treatment may be critical. When the exact time of treatment is known, there will be no variation for the sample considered unless (1) there are variations in the thickness of the sections, (2) the thickness throughout the sections is not even, and (3) the ultrastructure of the nuclei, mainly the condensation of the chromatin is variable from one part of the section to another (which occurs in tissue containing cells at various degrees of differentiation). It is not difficult to obtain reproducible results if one follows the above instructions. Then the problem of the interpretation arises. The specificity of the reaction in the nucleus is restricted to the bleaching of the chromatin. The structures which keep their contrast are known or believed to the ribonucleoproteins in most cells examined.[55] However, nuclear proteinaceous components or carbohydrates, if any, are expected to keep also their contrast, so that the interpretation of the results should be cautious. In the cyto-

[51] J. H. Venable and R. Coggeshall, *J. Cell Biol.* **25**, 407 (1965).
[52] E. S. Reynolds, *J. Cell Biol.* **17**, 208 (1963).
[53] W. Bernhard, *C. R. Acad. Sci.* **267**, 2170 (1968).
[54] W. Bernhard, *J. Ultrastruct. Res.* **27**, 250 (1969).
[55] A. Monneron and W. Bernhard, *J. Ultrastruct. Res.* **27**, 266 (1969).

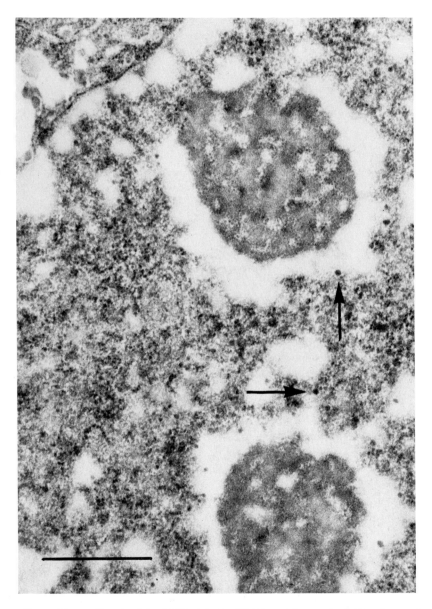

Fig. 9. Liver cell, stained for RNP with the EDTA staining method. The chromatin is entirely bleached out, while the nucleolus and the interchromatin substances retain the contrast. Several perichromatin granules (arrows) are visible. ×30,000.

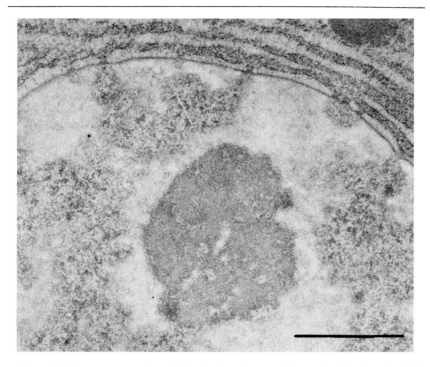

FIG. 10. Mouse pancreatic cell, hydrolyzed on section and stained with PTA for 2 minutes (HAPTA technique). The chromatin is not stained, whereas the nucleolus, the interchromatin substance, and most cytoplasmic organelles are visible. ×30.000. Courtesy of Dr. A. Gautier, Lausanne, Switzerland.

plasm, several components, such as zymogen granules, glycogen, desmosomes, also retain their contrast. The reaction must therefore be regarded as a preferential method for nuclear RNP, and, in any case, the only one available at the present time. Its usefulness has been demonstrated in a variety of cells and has considerably increased our knowledge of the nuclear RNP.[3] It is especially useful to decide whether an organelle which has the staining affinities of nucleic acids contains DNP or RNP.

A negative image of chromatin has been also obtained by a different mechanism with phosphotungstic acid used as follows (HAPTA technique).[56] Fix with an aldehyde, embed in Epon or Vestopal, float sections on 5 N HCl for 45 minutes at 20°. Rinse in water. Immerse grids in 1% PTA in 80% alcohol, pH 3.5 for 2 minutes at 20°. Wash rapidly in water. The mechanism is unknown and must be related to differential hydrolysis (Fig. 10).

[56] A. Gautier, Electron Microsc. Proc. 4th Eur. Reg. Conf. Rome, 1968 Vol. II, p. 81 (1968).

3. *Various Methods Available to Reveal DNA.* The complexity, contrast, specificity, and resolution of these methods vary to a large extent. The first step is to take advantage of the high affinity of uranyl salts for DNA and DNP, as a counterpart of the special affinity of lead salts for RNP (see above). The best results have been obtained after formalin or acrolein fixation (no osmium), and staining with 0.5% uranyl acetate in Michaelis buffer pH 5.[42,43] This simple technique may be useful in certain circumstances, when nonosmicated tissue is available, by comparison with lead stainable material. Although simple to apply, this technique cannot be considered as specific enough, and the recent development of the Feulgen-like techniques at the EM level has made it somewhat obsolete. The same can probably be said also of another nonspecific but selective method that enhances considerably the contrast of chromatin. This is based on the strong affinity of the basic protein RNase for DNA.[57,58] The contrast is due to the combination of RNA digestion by the enzyme and binding of the enzyme to the DNA. The contrast obtained may be emphasized by uranyl acetate. Another way of increasing the contrast of chromatin is to employ *phosphotungstic acid (PTA)*.[59] Fix in aldehyde. Treat the blocks in 1 N HCl for 10–15 minutes at 60°. Rinse in distilled water for 30 minutes. Dehydrate in ethanol, embed in Epon or Araldite. Stain sections for 15–30 minutes with 1% PTA adjusted to pH 7.0 with 1 N NaOH. The specificity of the reaction has been tested with nucleic acid extraction, but is not understood yet nor completely established. The contrast is good and the resolution excellent. This reaction is advisable, considering its simplicity.

By increasing the complexity of the technique utilized, important gains in specificity can often be obtained. Tissue treated by the classical Feulgen method can be observed in the electron microscope, but the contrast is too low. Modifications of the Feulgen method have been recently proposed in order to adapt it for ultrastructural studies. This reaction is based on the hydrolysis with 1 N HCl at 60° or 5 N HCl at room temperature, which breaks preferentially the bond between purine bases and the C-1 of deoxyribofuranose. The latter subsequently presents aldehydelike groups that react with the Schiff reagent. This yields a complex which is red in light microscopy but rather electron lucid. To enhance the contrast without increasing complexity of the reaction, one should use *silver* reagents in place of the Schiff reagent. Silver combines with aldehydes and displays granular deposits of high contrast, but of poor defini-

[57] Y. Yotsuyanagi, *C. R. Acad. Sci.* **250**, 1522 (1960).
[58] Y. Yotsuyanagi and C. Guerrier, *C. R. Acad. Sci.* **260**, 2344 (1965).
[59] V. Marinozzi and M. Derenzini, *Electron Microsc. Proc. 5th Eur. Congr. Manchester 1972*, p. 288.

tion and low reproducibility. One of the most recent and successful methods is the following.[60-62] Fix in glutaraldehyde as usual, and embed in Epon. Hydrolyze sections with 5 N HCl for 60 minutes at 20°. Float the sections on silver-methenamine[63] for 60 minutes at 60°. If needed, counterstain with uranyl acetate. Obtain a nonhydrolyzed control. The interpretation is made difficult by the fact that some cell components are spontaneously argentophilic without previous hydrolysis. By comparison with nonhydrolyzed sections, material which stains only after hydrolysis may be regarded as DNA. A recent modification of this type of reaction has introduced thiocarbohydrazide (TCH) as a Schiff-like reagent.[64] Hydrolyze sections with HCl and treat with TCH in acetic solution. Reveal the TCH, which is electron lucid, by silver proteinate (Fig. 11).

The recent introduction of thallium ethylate[48,49] as a contrasting agent in Feulgen-like reactions,[65,66] increases the complexity, but also the specificity and resolution, of the reaction. As this reagent binds to hydroxyl groups, they have to be blocked by acetylation prior to the application of the Schiff reagent. Fix in 1.6% glutaraldehyde in phosphate buffer, pH 7.3 for 1 hour at 4°. Rinse in the same buffer 4× for 10 minutes each time to wash out soluble aldehydes. To acetylate the OH groups,[40] dehydrate in graded acetone solution, then add increasing amounts of pyridine. Rinse in pyridine 3× for 10 minutes each. Treat overnight in pyridine–acetic anhydride 60:40 at 45°. Wash in pyridine 3× for 10 minutes; in pyridine–acetone 1:1 for 5 minutes; in acetone 3× for 10 minutes. Embed in Epon. To hydrolyze, float sections on gold grids on N 5 HCl for 10–30 minutes at room temperature. Wash thoroughly in distilled water. Dry for 20 minutes or more. Stain with thallium ethylate as shown above[49] or as described more recently.[66] Details of 30 Å have been observed; this makes the resolution rather high. The specificity of the reaction has been tested by enzymatic digestion and by treatment of a variety of DNA containing organelles and organisms, and has been shown to reveal all types of DNA except for mitochondrial DNA. The contrast is satisfactory. The reproducibility depends on practice. Despite the apparent complexity, this method probably is a good compromise at

---

[60] D. Peters, *Electron Microsc. Proc. 6th Int. Congr. Kyoto, 1966* Vol. II, p. 195. Maruzen, Tokyo.

[61] D. Peters and H. Giese, *Acta Histochem.*, Suppl. 10, 119 (1969).

[62] D. Peters and H. Giese, *7ème Congr. Int. Microsc. Electron. Grenoble, 1970*, Vol. I, p. 557.

[63] R. Korson, *J. Histochem. Cytochem.* **12,** 875 (1964).

[64] J. P. Thiéry, *J. Microsc. (Paris)* **14,** 95 (Abstract) (1972).

[65] G. Moyne, *C. R. Acad. Sci. Ser. D* **274,** 247 (1972).

[66] G. Moyne, *J. Ultrastruct. Res.* **45,** 102 (1973).

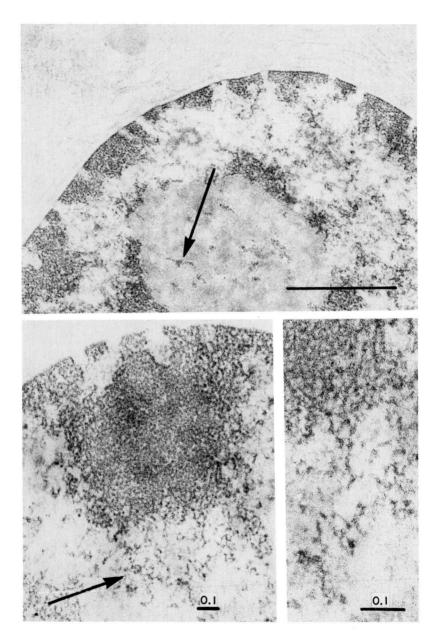

FIG. 11. Rat pancreas stained for DNA. Hydrolysis on sections, treatment with thiocarbohydrazide and silver proteinate. (A) At low magnification, only the peripheral chromatin and the peri- and intra- (arrow) nucleolar chromatin are stained. (B) At higher magnification, the DNP fibers are observed as condensed peripheral chromatin and as loosened chromatin (arrow) in the interchromatin region. ×67,000. (C) At very high magnification the chromatin fibers are well individualized. (A) ×30,800, (B) ×67,000, (C) ×126,000. Courtesy of Dr. J. P. Thiéry, Ivry-sur-Seine, France.

the present time for resolution, specificity, contrast, and reproducibility (Fig. 12).

In order to reach maximum specificity, i.e., that of the Feulgen reaction itself, even at the expense of contrast, a variety of Schiff-like agents

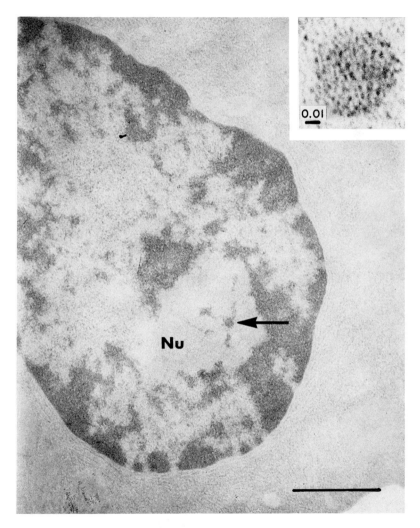

FIG. 12. Rat pancreatic cell, stained for DNA by acetylation, Feulgen reaction and thallium ethylate. The peripheral, perinucleolar, and intranucleolar (arrow) chromatin are stained. The cytoplasmic and nucleolar (Nu) RNP are not visible. ×24,000. Inset: Simian adenovirus 7, stained by the same method. The proteinaceous capside is not visible. The DNA fibers are observed individually. ×400,000.

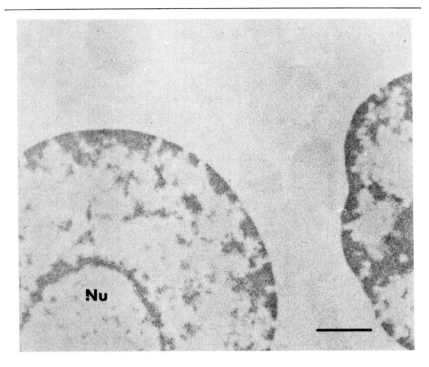

FIG. 13. Rat pancreas cell, stained for DNA with osmium ammine. Only the peripheral, perinucleolar, and intranucleolar chromatin are visible. Nu, nucleolus. ×15,000. Courtesy of Drs. A. Gautier and R. Cogliati, Lausanne, Switzerland.

have been proposed in the last three years.[67-70] *Acriflavine* previously treated with SO_2 can be used for 60 minutes on sections hydrolyzed with 5 N HCl at 20° for 15 minutes, after glutaraldehyde fixation and Epon embedding. *Ruthenium red* also treated with SO_2 is used for 60 minutes on sections from glutaraldehyde-fixed, Epon embedded, and HCl-hydrolyzed material. The *inorganic ammine of osmium* which has been synthesized,[70] may also be used in Feulgen-like reactions after bleaching in SO_2. Fix in 1.6% glutaraldehyde in phosphate buffer. Embed in Epon. Hydrolyze floating or gold grid-mounted sections with 5 N HCl at 20° for 25 minutes. Treat 0.1 g of osmium ammine in 10 ml distilled water by SO_2 for 30 minutes. Let sedimentation occur and use the lower part of the yellow supernatant as stain. Float sections or grids on the supernatant for

[67] A. Gautier and M. Schreyer, *7th Congr. Int. Microsc. Electron., Grenoble, 1970*, Vol. I, p. 559.
[68] A. Gautier, M. Schreyer, and R. Cogliati, *Experientia* **26**, 693 (1970).
[69] A. Gautier, M. Schreyer, R. Cogliati, and J. Fakanova, *Experientia* **27**, 735 (1971).
[70] R. Cogliati and A. Gautier, *C. R. Acad. Sci. Ser. D* **276**, 3041 (1973).

45 minutes at 20° (Fig. 13). All these reactions offer a specificity comparable to that of the Feulgen reaction itself. The last one is very promising. Staining of the same structure with the osmium amine and the Schiff–thallium ethylate technique may be taken as a good indication that it actually contains DNA.

One of the main points which has not been yet elucidated as far as EM cytochemistry is concerned is the distribution of the disperse euchromatin, as opposed to the conspicuous condensed chromatin. In this respect it is worth describing the recent EM studies on the sites of fixation of *acridine orange* onto the DNA. This method is thought to reveal the derepressed portion of the genome.[71,72] Fix cells with 5% glutaraldehyde in medium 199 (Gibco), pH 6.5 for 2 hours at 4°. Treat with 1 mM acridine orange, pH 7.2. Wash three times in medium 199, incubate at 37° for 30 minutes in Eagle's medium with 0.8 mM Mg^{2+} and 1 mg/ml of DNase. Postfix with 1% OsO_4, dehydrate in ethanol and embed in Epon. Stain sections with 5% uranyl acetate. The cells (lymphocytes) may also be treated without fixation. Incubate aliquots of 10^6 cells/ml in the dark for 30 minutes at 37° in autologous plasma with heparin, 4 U/ml; penicilin, 0.06 mg/ml; streptomycin, 0.06 mg/ml; and 1 mM acridine orange, fix in 5% glutaraldehyde in medium 199 for 30 minutes at 4°, and wash 3 times. Incubate in DNase and embed as above. The product of the reaction is visible as large, 1000 Å particles, deposited on the euchromatin only. This reaction might be extremely useful in the study of gene activation, maturation, and differentiation processes.

Enzymatic Digestions of Nucleoproteins and Proteins

General Methodology

One of the most important methods now available for revealing the chemical nature of cell components is based on differential hydrolysis with specific enzymes. The main advantage is that this method can be combined with both specific staining methods and EM ARG. However, this technique is successful only when the required conditions of fixation, embedding, and incubation have been carefully considered. It is useless to spend a great deal of time trying to digest sections of overfixed or not appropriately embedded material. The following points must be stressed. The frequency and occurrence of the organelle to be studied must be high enough. If not, the usually poor ultrastructure of the sur-

---

[71] J. H. Frenster, *Cancer Res.* **31**, 1128 (1971).
[72] J. H. Frenster, *Nature (London) New Biol.* **236**, 175 (1972).

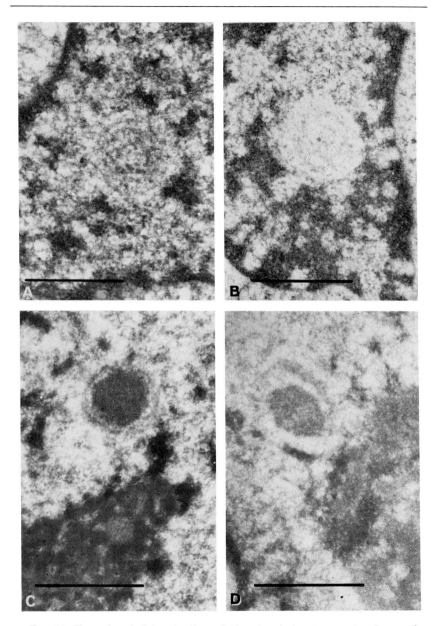

Fig. 14. Example of determination of the chemical components of a nuclear organelle by enzymatic digestions. "Simple nuclear bodies" in control sections of glycol methacrylate-embedded material are fibrillar in structure and offer low contrast (A). After digestion with pronase (B), the contrast disappears almost entirely, indicating their protein content. "Granular nuclear bodies" in control sections consist

rounding cell components will make identification impossible. The size must also be sufficiently large, and the shape should be characteristic. Otherwise, after complete digestion, the empty space left will be difficult to identify. Aldehyde fixation without postosmication is required for most digestions. For digestion on ultrathin sections, the tissue should be embedded in a water-soluble plastic, not in epoxy resins. In sections, DNase digestion will be impossible unless the tissue has been fixed in formalin. The interpretation of results is easier if digestion has been carried out with a single enzyme. However, as most nucleic acids are bound to proteins, previous digestion with a protease may unmask nucleic acid, which then will be sensitive to nucleases.

When the organelle is expected to consist of proteins or nucleoproteins (Fig. 14), start with *pronase* which is the most powerful enzyme. Then try *nucleases* alone, then Pronase followed by nucleases. If the organelle is sensitive to Pronase, which is a broad-spectrum protease, *pepsin*, which is more active on acid and aromatic amino acid bonds, and *trypsin*, which breaks preferentially the peptide bonds of lysine and arginine, should help to characterize roughly acid and basic proteins, respectively. For each of these enzymes, a full range of concentrations and incubation periods must be tried first—for instance 0.1 and 1%, each for 15, 30, 60 minutes and 2, 4, 8, and 16 hours. When the time of overdigestion, with heavy tissue destruction, is known, use shorter incubations and lower concentrations until the most effective digestion and the best preservation are simultaneously obtained. It is essential to carry out control experiments for the same periods of incubation in the same solution, that is same reagents without the enzyme, same pH, and preferably same osmolarity as obtained by addition of protein. Careful comparison between the digestions and the control are absolutely necessary to distinguish between specific hydrolysis and unspecific extraction by the solution. In brief, the right incubation time and concentration for a given organelle and a given enzyme are those where complete digestion of the organelle is obtained with the minimum alteration of the tissue and with complete absence of extraction in the control. Attention must be paid, after short fixation and GMA embedding, to the ultrastructure of the cells, which may be quite different from that in routine conditions of fixation and

---

of a peripheral capsule and a core of central granules (C). After RNase digestion (D), the peripheral capsule is not altered, whereas the central granules have their contrast significantly decreased. The contrast of the nucleolus in (D) is also diminished as compared to the control (C). This indicates the RNA content of the granules. (A) and (B) ×27,000; (C) and (D) ×30,000. From A. M. Dupuy-Coin and M. Bouteille, *J. Ultrastruct. Res.* **40**, 55 (1972) by permission.

embedding. Finally, the sensitivity of the unknown organelle must be compared in the section with the sensitivity of nuclear organelles as known from previous studies.[3] For instance, a suspected RNA-containing structure should behave with various enzymes as does the nucleolus. In the case of generally unfamiliar material, parallel experiments with such well known material as pancreas or liver are recommended as a criterion of enzyme sensitivity.

Procedure for Enzymatic Digestions

For incubation on small blocks, frozen sections of tissue, cell suspensions, or isolated cell fractions, fix the cells with 10% formalin (or better paraformaldehyde) in 0.2 M phosphate buffer, or 2% glutaraldehyde for 15–30 minutes. Incubate with the enzymatic solution and postfix with the aldehyde and/or OsO_4. Some authors treat the cell with 5% TCA or PCA for 10–15 minutes at 4° in order to reduce the period of incubation by solubilization of the hydrolyzed substances. Dehydrate and embed in Epon, Araldite, or Vestopal. The advantages of this technique are easy penetration of the enzyme because of the absence of embedding medium, more reproducible results, chiefly with nucleases (this method is recommended for DNase digestion), better preservation of the ultrastructure because the tissue is postfixed with osmium and embedded in usual epoxy resins.

Enzymatic hydrolysis on ultrathin sections from embedded material is now a widely used method,[73] which offers many advantages: fresh tissue is not required, and tissue embedded years before can be used. The hydrolysis, when obtained, can be observed throughout the whole section. The specificity of the digestion can be judged by serial control sections incubated for the same period of time in the same solution without the enzyme. It is strongly advised to embed in glycol methacrylate[29] or hydroxypropyl methacrylate,[74,75] which are water-soluble media. The penetration of solution is much easier than in other resins. Fix tissue or cells in 1.6% glutaraldehyde in phosphate buffer. For easy digestion but poor preservation fix for 15–30 minutes. In most cases a good digestion will be obtained after the usual fixation for 60 minutes. If formaldehyde is to be used, fix for 1–4 hours, but most digestions will be successful after 12–24 hours of fixation. Some enzymes have been shown to work on Epon-

[73] E. H. Leduc and W. Bernhard, *Symp. Int. Soc. Cell Biol.* **1**, 21 (1961).
[74] D. C. Pease, "Histological Techniques for Electron Microscopy." Academic Press, New York, 1960.
[75] E. H. Leduc and S. J. Holt, *J. Cell Biol.* **26**, 137 (1965).

FIG. 15. Transferring sections with plastic rings according to Marinozzi.

embedded tissue.[76] However, this is not recommended if fresh tissue can be obtained since then GMA embedding is possible. If only Epon-embedded tissue is available, pronase is the enzyme which is the most likely to be effective. Other proteases may occasionally be active, but one should not expect results with nucleases.

To carry out the incubations,[77] obtain pale gold sections of regular thickness with a diamond knife. Pick up the sections from the water-filled boat with rings of thin sheets of cellulose acetate, 3 mm of internal diameter and about 5 mm of external diameter, carried with forceps (Fig. 15). Float the rings upon the preheated enzymatic solution in porcelain depression plates or watch glasses. If the incubation time is to be long, place the plate in a petri dish with moistened filter paper, and change the solution if necessary. Wash 3 times with the same procedure in distilled water: first at incubation temperature, and let cool down to room temperature before the next wash in order to avoid spatial alteration in the sections. Long periods of washing enhance the digestion, but may also render the results less specific. Pick up the sections with Formvar-coated grids by placing the grid Formvar side down *over* the sections in the opening of the ring. Stain as usual.

Usual Concentrations and Periods of Incubation

The following figures are the most common in the literature.

1. *Pronase.* En bloc: 0.1–0.25% in distilled water adjusted to pH 7.3–7.4 with 0.01 N NaOH for 2–4 hours at 20 or 37°. On ultrathin sections of GMA embedded tissue: after aldehyde fixation and embedding in hydrosoluble medium, 0.01, 0.1, 0.5, and 1% in distilled water adjusted as above for 5 minutes to 4 hours. The best results have usually been obtained with rather low concentrations (0.01%) and long periods of in-

---

[76] A. Monneron and W. Bernhard, *J. Microsc. (Paris)* **5**, 697 (1966).
[77] V. Marinozzi, *J. Ultrastruct. Res.* **10**, 433 (1964).

cubation (20–60 minutes)[78] On ultrathin sections of Epon-embedded material[76] use 0.5% pronase in the same solution at 40° for 30 minutes to 24 hours. After glutaraldehyde fixation, the time of incubation is 10–24 hours. If the tissue has been osmicated, first oxidize the sections with 110 volumes of 10% H_2O_2 for 10–20 minutes or 10% periodic acid for 20–30 minutes, or saturated potassium periodate for 30–60 minutes. Then shorter times of incubation, 30 minutes to 2 hours, will usually be sufficient. Under these conditions, it is in most cases possible to hydrolyze proteinaceous components of the cell. This enzyme is probably the most effective of all for digestion in ultrathin sections.

2. Pepsin. En bloc: 0.1–1% in a solution of 0.1 N HCl, pH 1.5 for 10 minutes to 2 hours at room temperature or 37°. On ultrathin sections of GMA-embedded tissue, 0.1–0.5% (usually 0.5%) in the same solution, for 5 minutes to 4 hours at 37–38°.

3. Trypsin. En bloc: 0.1–0.3% in phosphate or 0.05 Tris buffer or in distilled water, pH 8.0 for 2–4 hours at 20 or 37°. On ultrathin sections: 0.1–0.5% (usually 0.3%) in distilled water adjusted to pH 8.0 with NaOH or 10 mM $CaCl_2$, pH 8.0 or 50 mM Tris buffer with 5 mM $CaCl_2$, pH 8.0 at 37° for 5 minutes to 24 hours.

4. Papain. En bloc: 0.25–1.0% in 10 mM phosphate buffer, pH 5.5 or distilled water, pH 6.5 with 5 mM cysteine at 37° for 2 hours. On ultrathin sections, 0.5% in distilled water, pH 6.5 with cysteine or in 0.2 mM EDTA and 0.2 mM KCN, pH 5.4, for 1–4 hours at 37°.

5. α-Chymotrypsin. On ultrathin sections, 0.5% in distilled water adjusted to pH 7.8 for 1–4 hours at 37°.

6. Ribonuclease. En bloc: 0.1% in distilled water adjusted to pH 6.8 or 10 mM Tris buffer, pH 6.8, or phosphate buffer pH 7.4 or acetate buffer pH 5.0, heated at 80° for 10 minutes to remove DNase activity, and adjusted to pH 6.5 with NaOH for 1–4 hours at 20, 37, 40, or 60°. On ultrathin sections, 0.1–1% (usually 0.1%) in the same solutions for 1–48 hours (usually 1–4 hours) at 37°, 38°, or 54° (usually 37°). Some authors have added 20 mM Mg^{2+} to the solution. Note that RNase has an optimum activity at about 60°.

7. Deoxyribonuclease. En bloc: 0.05–0.1% (usually 0.1%) in distilled water adjusted to pH 6.5, usually with 3 mM magnesium sulfate, pH 6.8, for 1–4 hours (usually 1 hour), at 20° or 37°. On ultrathin sections, the results are unsatisfactory and not reproducible. The best results have been observed after short formalin fixation, 10% for 15 minutes at 4° (no glutaraldehyde), embedding in GMA, and treatment in the same solutions for 10 minutes to 4 hours at 37°.

[78] A. Monneron, *J. Microsc. (Paris)* **5**, 583 (1966).

Interpretation of the Results

When a complete digestion with no alteration in the controls has been obtained, one must pay attention to the main causes of error. First, few enzymes can be considered as absolutely pure. Most of them have side effects—for instance, DNase activity in RNase preparations, or protease activities in nucleases. This may facilitate the extraction to such a point that the chemical nature of the organelle may be mistaken. Second, enzymatic extraction in ultrathin sections is the result of two actions: (1) enzymatic hydrolysis resulting in specific bond breaking, and (2) solubilization of the breakdown products, which in some instances may produce a hole in the preparation. An RNA core might be dissolved although only its proteinaceous coat was specifically hydrolyzed by a protease. A misinterpretation could then be made that the core is proteinaceous also. On the contrary, when the organelle under study is not hydrolyzed by RNase, for instance, this does not mean that there is no RNA present. In fact (1) the ratio of RNA to protein may be too small to allow dissolution of the organelle, (2) a RNA core may be protected against the enzyme by a protein coat, and (3) if the size of the organelle is much smaller than the thickness of the section, the penetration of the enzyme may be impaired unless the organelle is sectioned at least tangentially. Examples of these various possibilities have been shown with viruses.[79] Therefore, the results of enzymatic digestions must always be combined with other techniques.

Final Remarks

It is of considerable importance to carry out several of the methods described above before drawing any conclusion regarding the chemical nature of a structure under investigation. Sensitivity to Pronase hydrolysis, for instance, must be confirmed by [^{3}H]leucine incorporation. In fact only the combination of (1) incorporation of the corresponding precursor, (2) specific staining affinity, and (3) sensitivity to the appropriate enzyme can be considered as a full line of evidence. As an ideal example, a "cartoon" has been drawn on the behavior of an RNA containing organelle when submitted to the various procedures described (Fig. 16). Although results of this sort can only be obtained after a long and hard work, they are certainly worthwhile, since in most cases the nuclear organelles under study cannot be isolated and analyzed biochemically. This is especially useful when the ultrastructure of the nucleus has been deeply

---

[79] W. Bernhard and P. Tournier, *Cold Spring Harbor Symp. Quant. Biol.* **22**, 67 (1962).

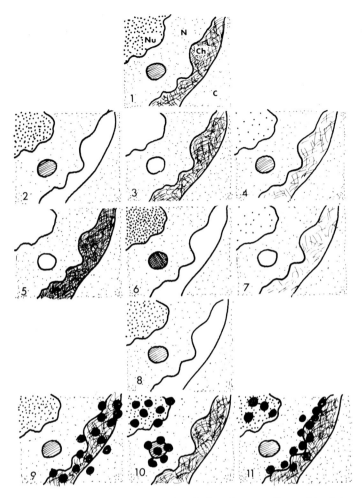

FIG. 16. Practical steps of the cytochemical and autoradiographic determination of the chemical components in an organelle supposed to consist of RNA. (1) Control section, where the organelle is figured in the nucleus (N), between the nucleolus (Nu), and the chromatin; C, cytoplasm. (2) the organelle is not affected by DNase, while the chromatin is hydrolyzed. (3) RNase digests both the organelle and the nucleolus. (4) Pronase attacks all the proteins of the nucleoproteins, but not the organelle tested. (5) Feulgen-like techniques reveal only the chromatin. (6) techniques staining RNP would increase the contrast of the nucleolus and the organelle. (7) Methods staining proteins do not stain the organelle. (8) After treatment by the EDTA staining method, the chromatin is bleached out, whereas both the nucleolus and the organelle retain their contrast. (9) Incubation with [^{3}H]thymidine for autoradiography; the grains are found mostly over the chromatin. (10) Incubation with [^{3}H]uridine. The activity is observed both in the nucleolus and the organelle. (11) Incubation with [^{3}H]leucine. The nucleolus and the borderline between the condensed and the dispersed chromatin are labeled, while the organelle is not. In practice, such clear-cut differences cannot be expected, so that only the combination of some of these techniques will bring sufficient evidence for the determination of the chemical nature of the organelle under investigation.

altered by an exogenous factor, whether viral, pathological or drug-induced. One only need consider the great deal of information which has been already gained on the cell nucleus[3] to realize the degree of efficiency that EM cytochemistry and autoradiography have reached at the present time.

Acknowledgments

A number of the methods described here have been developed in Dr. W. Bernhard's laboratory, and result from his continuous interest in increasing the range of action of electron microscopy. Many technical points which make a technique successful are due to the exceptional skill of Mrs. M. J. Burglen. The authors are also indebted to Dr. R. Bretton for helpful suggestions and Mrs. C. Taligault for preparation of the manuscript. This work was supported by C.N.R.S. and INSERM, France.

[2] A Nuclear System for DNA Replication from HeLa Cells

By DANIEL L. FRIEDMAN

Isolated nuclei, utilizing endogenous template, appear to continue the process of DNA replication that was proceeding *in vivo* before their isolation and may, under the proper conditions, be able to initiate new synthesis.[1,2] The apparent permeability of the nuclei to a variety of substances, such as nucleotides and proteins, makes this system useful for studies on the mechanism of DNA replication as well as the effects of chemical agents upon replication. The present discussion will be confined to the system from HeLa cells. Similar systems have been described from several other tissues.[3-6]

Procedure

Cell Cycle Synchronization. HeLa cells, grown in suspension culture to a concentration of $2-3 \times 10^5$ cells/ml, are synchronized in DNA syn-

---

[1] D. L. Friedman and G. C. Mueller, *Biochim. Biophys. Acta* **161**, 455 (1968).
[2] K. V. Kumar and D. L. Friedman, *Nature (London) New Biol.* **239**, 74 (1972).
[3] W. E. Lynch, R. F. Brown, T. Umeda, S. G. Langreth, and I. Lieberman, *J. Biol. Chem.* **245**, 3911 (1970).
[4] C. Teng, D. P. Bloch, and R. Roychoudhury, *Biochim. Biophys. Acta* **224**, 232 (1970).
[5] W. R. Kidwell, *Biochim. Biophys. Acta* **269**, 51 (1972).
[6] E. N. Brewer and H. P. Rusch, *Biochem. Biophys. Res. Commun.* **21**, 235 (1965).

thesis by addition of amethopterin (1 μM) and adenosine (50 μM).[7] Amethopterin induces a thymine-less state without affecting RNA or protein synthesis. Adenosine prevents the adverse effects of amethopterin on purine biosynthesis. Cells that were not in DNA synthesis at the time of addition of these agents progress through the cell cycle and accumulate at the beginning of the next DNA synthetic phase. After 16 hours, thymidine (10 $\mu g/10^6$ cells) is added, producing a synchronous wave of DNA synthesis. Cell homogenates are prepared 2.5–3.5 hours later, when the synthesis of DNA is maximal.

Preparation of Nuclear Homogenates. The cultures are cooled at 0° for 10 minutes and centrifuged for 10 minutes at 250 g. The pellet is suspended in buffer I (2 mM MgCl$_2$ in 10 mM potassium phosphate, pH 7.7) to a concentration of 2×10^6 cells/ml buffer. The cells are sedimented for 10 minutes at 600 g, and resuspended in the same buffer to a concentration of 7×10^7 cells/ml. After incubating for 10 minutes at 0° to allow swelling to occur, the cells are lysed in a loose-fitting Dounce homogenizer. (The number of strokes for complete lysis of cells and minimal breakage of nuclei must be predetermined, since it varies with the homogenizer.) The lysate is immediately diluted with an equal volume of buffer II (0.12 M KCl–0.12 M Tris·HCl, pH 8.0).

Purification of Nuclei. To obtain functional purified nuclei in high yield, the following procedure has been used. After homogenization the nuclei are sedimented at 1000 g for 5 minutes and rapidly suspended by vigorous pipetting in a solution of bovine serum albumin (18 mg/ml) in buffer I. The suspension is sedimented at 400 g for 5 minutes, resuspended in the same solution, and again sedimented at 400 g for 5 minutes. The nuclei are resuspended in the same solution, incubated for 4 minutes at 37°, cooled at 0° for 1 min, dispersed, and sedimented at 400 g for 5 minutes. The nuclei are finally suspended in a solution of bovine serum albumin (18 mg/ml) dissolved in a 1:1 mixture of buffer I and buffer II. Microscopic examination reveals a relatively clean preparation of nuclei with less than 2% whole cells. The nuclei are round and regular with prominent nucleoli. The overall yield of nuclei is about 50%.

Assay for DNA Synthesis. The assay mixture consists of 0.1 ml of diluted homogenate or purified nuclear suspension and 0.05 ml of a mixture containing 15 mM ATP, 25 5 mM MgCl$_2$, and 1.5 mM each of dATP, dGTP, dCTP, and [^{3}H]dTTP (6 μCi/μmole). The reaction mixture is incubated without shaking for 20 minutes at 37°. It is then vigorously pipetted to disperse clumps, and a 0.05-ml aliquot is distributed evenly on a 2.5-cm disk of Whatman 3 MM filter paper,[8] which had been pre-

[7] R. R. Rueckert and G. C. Mueller, *Cancer Res.* **20**, 1584 (1960).
[8] F. J. Bollum, *J. Biol. Chem.* **234**, 2733 (1959).

pared by soaking in 7% TCA in ether and drying in air. As soon as the aliquot has soaked into the disk (no more than 10 seconds), it is placed into cold 7% TCA (in water). All of the disks from an experiment are soaked together (10–20 ml per disk) for 30 minutes with occasional stirring. The TCA is decanted and replaced with fresh cold TCA (10 ml/disk). After at least 30 minutes more, this procedure is repeated. The disks are then washed with TCA 3 more times at 5-minute intervals. The TCA is replaced with cold 95% ethanol, and after 5 minutes the disks are removed and allowed to dry. The radioactivity of each disk is determined by liquid scintillation counting in 5 ml of scintillation fluid containing toluene as solvent. The reaction is approximately linear for 20 minutes, after which the rate gradually decays.

Properties of the System

Deoxyribonucleotide Requirement. All four deoxyribonucleotides are required for maximal activity. With crude nuclei, deletion of either dGTP or dCTP reduces the incorporation of [^{3}H]TTP to 20% that of the complete system; in purified nuclei, incorporation is decreased to less than 5%. Deletion of dATP reduces activity to 80% and 60% in crude and purified preparations, respectively. The half-maximal TTP concentration in purified nuclei is between 0.5 and 1 μM.

Activators and Inhibitors. In both crude and purified preparations there is a strong stimulation by ATP, which varies between 3- and 6-fold. This requirement is not replaced by GTP, CTP, UTP, or dATP. Mg^{2+} is absolutely required for synthesis, with an optimal concentration of 10 mM. Higher concentrations inhibit.

In crude preparations, synthesis is stimulated 2-fold by NaCl, KCl, or NH_4Cl at peak concentration of 40 mM. Concentrations of 80 mM are inhibitory. Similar effects are observed with purified nuclei, but the optimal concentration for activation is 60 mM. p-Hydroxymercuribenzoate (0.5 mM) and N-ethylmaleimide (2 mM) essentially completely inhibit the activity. This can be shown to be due to a sensitive step within the nucleus.

Requirement for Cytosol. While purified nuclei suspended in buffered albumin solution are quite active, the cytosol stimulates incorporation by 2- to 4-fold. The stimulatory component resides in the 100,000 g supernatant and can be purified by $(NH_4)_2SO_4$ precipitation. It separates into 4 peaks on gel filtration[9] and is sensitive to N-ethylmaleimide.

---

[9] G. C. Mueller, *in* "The Cell Cycle and Cancer" (R. Baserga, ed.), p. 296. Dekker, New York, 1971.

pH Optimum. The influence of pH has been studied in crude preparations. The optimum pH was 8.0, with a rapid fall off of activity above or below this pH.

Product of in Vitro Incorporation. At least 80% of the product of nuclear synthesis is nonextractable in phenol or chloroform, a result also observed with replicating DNA of the intact cell.[10] If whole cells are pretreated with bromodeoxyuridine prior to nuclear isolation, most of the labeled *in vitro* product sediments in CsCl with hybrid DNA,[11] indicating that *in vitro* replications is at or near the *in vivo* replication sites. Other data suggest that at least a portion of the synthesis *in vitro* is by way of short pieces of DNA, which are later joined together.[11] These and other studies indicate that synthesis in the nuclear system is similar, if not identical, to that observed in whole cells.

[10] D. L. Friedman and G. C. Mueller, *Biochim. Biophys. Acta* **174**, 253 (1969).
[11] W. R. Kidwell and G. C. Mueller, *Biochem. Biophys. Res. Commun.* **36**, 756 (1969).

[3] Analysis of the Action of Chemical Agents upon the Cell Division Cycle

By JAMES B. KURZ and DANIEL L. FRIEDMAN

The term cell cycle is defined here as the ordered sequence of events that occur between one cell division and the next. The cell division cycle (CDC) pertains to those particular events which are necessary for cell division. The CDC generally is depicted as in Fig. 1, with four periods[1]: mitosis (M); DNA synthesis (S); a gap following mitosis and preceding DNA synthesis (G_1); and a gap following DNA synthesis and preceding mitosis (G_2). It should be pointed out that this scheme represents only one aspect of the CDC, which may be termed the DNA replication-chromosome condensation cycle. The CDC can actually be thought of as many cycles consisting of events (or markers) whose occurrence is required for division in addition to those of the DNA replication–chromosome condensation cycle. Each cellular component has its own cycle during each generation. The interactions and control circuits that exist between the various cycles are complex and largely unknown. The events of different cycles may occur in series (fixed pattern) or in parallel with one another (variable pattern).[2] While the methods discussed here gen-

[1] A. Howard and S. R. Pelc, *Heredity* **6**, Suppl. "Symposium on Chromosome Breakage," p. 261 (1953).
[2] J. M. Mitchison, *in* "The Cell Cycle, Gene-Enzyme Interactions" (G. M. Padilla, G. L. Whitson, and I. L. Cameron, eds.), p. 361. Academic Press, New York, 1969.

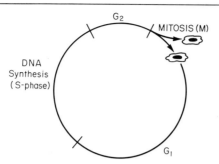

Fig. 1. Model of the cell division cycle.

erally will be based upon the model in Fig. 1, the same principles may be applied to markers other than DNA synthesis and mitosis.

The action of chemical agents upon the cell division cycle may be defined in terms of three parameters: the "execution point,"[3] the "termination point,"[3] and the "terminal state." The first two are points in time, measured in relation to an unperturbed cell cycle and will be expressed in minutes, counting backward from cytokinesis. The execution point (or "transition point"[4]), the time when the agent stimulates or impedes an event, is the time in the cell cycle at which the affected event is normally executed. For the present discussion the termination point is defined in relation to Fig. 1 as the point in time at which the action of the agent terminates with respect to the DNA synthesis–chromosome condensation cycle. In practice it may be difficult to pinpoint this time with more precision than to determine the stage (G_1, S, G_2, M) which is affected. The termination point may or may not be the same as the execution point, depending on how the affected event relates to DNA synthesis and mitosis. The terminal state is defined as the final state of the cells resulting from the action of an agent which completely blocks an event in the CDC. This will result in the slowing or stopping of many of the cycles within the CDC, while other cycles might continue normally. A full description of this final state is obviously impossible since it would be necessary to determine the final advancement of all the cycles within the CDC; however, partial descriptions can be useful for an understanding of the action of the agent. The terminal state will not be discussed further.

The following discussion is limited to analysis of populations of cul-

---

[3] L. H. Hartwell, J. Culotti, and B. Reid, *Proc. Nat. Acad. Sci. U.S.* **66**, 352 (1970).
[4] J. M. Mitchison, "The Biology of the Cell Cycle," p. 204. University Press, London, 1971.

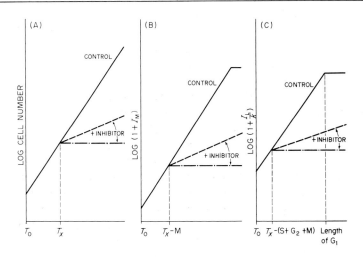

FIG. 2. Determination of execution point by (A) cell density curves, (B) I_M in presence of mitotic blocker, (C) I_L in presence of mitotic blocker. Curves are theoretical and assume no variation in cell cycle phase lengths. In each case, curves are presented for partial (- - -) and complete (-·-·) blocking agents. The control curves in (A) and (B) can be used to determine the population doubling time (T'). The plateau point of the control curve in (C) gives the length of G_1 assuming that the mitotic inhibitor immediately blocks cell division. The abbreviations M, S, G_2, and G_1 designate the length of these stages.

tured cells. Analysis of cell cycle parameters in single cells are described elsewhere.[5] Routine methods for handling cultured cells have been described in a previous volume.[6]

Determination of Execution Point

In Exponential Cell Cultures

1. By Cell Density Analysis. The time in the cell cycle at which an agent acts to impede a function required for cell division can be determined simply from cell density curves with exponentially growing populations.[7] Complete blockage of the function X will cause the log of cell number to reach a plateau after T_x minutes (Fig. 2A). A partial block will lead to a change in slope after T_x minutes. Prior to this time the

---

[5] C. Vendrely, *in* "The Cell Cycle and Cancer" (R. Baserga, ed.), p. 227. Dekker, New York, 1971.

[6] See Vol. 5 this series, [8].

[7] R. A. Tobey, D. F. Petersen, E. C. Anderson, and T. T. Puck, *Biophys. J.* **6**, 567 (1966).

log of the cell number will increase linearly at the same rate as the control, since the cells which have passed the event are unaffected by the agent and will divide. T_x will be equal to the time of execution, counted backward from cytokinesis to the event.

To perform this measurement, the chemical agent, suspended in saline or media, is added to cultures in their exponential growth phase; control cultures are treated in identical fashion, but without the agent. The cultures are sampled at frequent intervals for cell number determinations. A sampling system using a syringe to remove aliquots may be devised in order to decrease exposure of cells to air and to reduce other possible artifacts caused by multiple sampling. Since accurate determinations are critical, if possible cells should be counted with an electronic particle counter with appropriate corrections for doublets or larger cell aggregates.

This method will characterize only the last point of action of the agent prior to cytokinesis. If there are other effects of the agent they will not be identified. The method is simple and will give an accurate value providing there is little variation in the cell cycle times from cell to cell within the population. If the variation is great, the cell density curve will show a more gradual decay to the plateau level or the new slope. Extrapolation of the new slope to the control curve will approximate the average T_x.

2. By Mitotic Index (I_M) Analysis in the Presence of a Mitotic Inhibitor. The same principle, i.e., studying the kinetics of passage of cells through a certain marker, may be applied to mitosis in the presence of a mitotic inhibitor.[8] The chemical agent to be tested is added to cultures together with Colcemid (usually 0.025 µg/ml) or other mitotic inhibitors. At appropriate time intervals the cultures are sampled as above, and the mitotic index is determined. Mitotic cells may be identified and quantitated after gently squashing a concentrated preparation of cells with a coverslip followed by visualization with phase contrast microscopy. Alternatively, cells may be fixed, stained, and observed by light microscopy at a later time.[9] If one then plots log $[1 + I_M]$ versus time, a straight line will be obtained in the control cultures.[8] Completely or partially blocked cells will reach a plateau or a change in slope at a time which characterizes the execution point, counted backward from mitosis (Fig. 2B).

There are several advantages of this method over measurements of cell number. (1) The observed marker is slightly closer to the time of inhibition, thereby removing imprecision caused by variability in M.

---

[8] T. T. Puck and J. Steffen, *Biophys. J.* **3**, 379 (1963).
[9] D. M. Prescott, *Methods Cell Physiol.* **1**, 365 (1964).

(2) If the agent has two actions, one of which terminates in mitosis, this method allows one to observe the nonmitotic action. (3) If the agent blocks near to mitosis, the point of action will be more easily determined than with cell density curves since the percent change will be larger. Before this method is used, however, it must be established that the mitotic inhibitor blocks mitosis completely and that it does not affect other stages of the cell cycle. Other disadvantages of the method are that it requires tedious microscopic analysis and that it is sometimes difficult to recognize the early or late stages of mitosis.

If normal cytokinesis occurs in the presence of the agent, a similar analysis may be obtained without use of mitotic inhibitor. The mitotic index and cell number are measured at intervals, and a calculation is made of the fraction of cells from the original population which have entered mitosis (see Appendix I).

3. By Analysis of Fraction of Labeled Cells (Labeling Index or I_L) in Presence of a Mitotic Inhibitor. If the execution is in the G_1 phase, it is possible to apply the above methods using DNA synthesis as marker, thus allowing more accurate determinations of execution point. This method will give the execution point in relation to the S period and will therefore delineate T_x, if the length of the other stages are known. [^{3}H]-Thymidine (approximately 0.1 μCi/ml)[10-12] and Colcemid are added to cells, after which samples are removed periodically. Slides are prepared and processed for autoradiography[9] and the I_L is determined microscopically. The experimental I_L curve will deviate from the control curve at a time equal to the length of time preceding S phase that the agent acted to block traverse through the cycle (Fig. 2C). Assuming no variation in cell cycle stage lengths, straight-line control curves are obtained by plotting log $(1 + I_L/K)$ vs. time,[8] where $K = G_2/T$. The method is obviously feasible only in cases where the termination point also precedes S phase. The advantages and disadvantages of this method are similar to those described above in the mitotic index method.

Determination of Execution Point Using Synchronized Cell Cultures

Methods of synchronizing cells have been described in a previous volume of "Methods of Enzymology."[13] It is possible to determine the execution point by adding the agent at various times to a synchronized

[10] R. M. Drew and R. B. Painter, *Radiat. Res.* **11**, 535 (1959).
[11] G. F. Whitmore and S. Gulyas, *Science* **151**, 691 (1966).
[12] M. O. Krause, *J. Cell. Physiol.* **70**, 141 (1967).
[13] E. Robbins, this series, Vol. 32, [56].

population of cells.[14,15] As cells progress through a highly synchronized cell cycle, that point at which addition of the agent no longer has an effect on a subsequent marker (e.g., cell division, mitotic index in the presence of a mitotic inhibitor, or initiation of DNA synthesis) gives an indication of the execution point. Since perfect synchrony is seldom attainable, it is necessary to express the results as an arbitrary average time, e.g., the time at which 50% of the cells are affected by the agent.

If one knows the approximate execution point, it is also possible to utilize synchronized cells in a manner similar to that described for exponential cells (see Section 1 above), that is, by observing the time it takes treated cultures to deviate from controls. Timely addition of the agent will result in a more easily observed difference from controls than that observed in exponential cultures.[7]

Measurement of the Termination Point

In the Absence of Steady-State Cell Cycle Conditions

The most common methods for determining the termination point utilize synchronized populations of cells. Generally the agent is added at the beginning of a stage and the lengths of that and subsequent stages are determined. A change in duration of a stage would imply a termination point within that stage. The investigator should be attuned to the possibility that the execution point of an agent may reside in one stage and the termination point in another. The major problems in these determinations are those relating to the synchronization procedure, e.g., unbalanced growth, side effects of the synchronizing agent, and poor quality of the synchronization. It is therefore desirable to perform studies with synchronized populations which begin traversing the cell cycle at two different points. At the present time the most easily obtained points are mitosis and the beginning of S phase. The following discussion will consider methods for determining termination points for each of the four stages.

1. *Mitosis.* A variety of methods for analyzing changing mitotic indices have been described in order to determine the length of mitosis in synchronized cells.[7,14,16,17] The derivation of a simple and precise analy-

---

[14] P. N. Rao and J. Engelberg, *in* "Cell Synchrony, Studies in Biosynthetic Regulation" (I. L. Cameron and G. M. Padilla, eds.), p. 332. Academic Press, New York, 1966.

[15] Y. Doida and S. Okada, *Cell Tissue Kinet.* **5**, 15 (1972).

[16] H. Firket and P. Mahieu, *Exp. Cell Res.* **45**, 11 (1966).

[17] G. Galavazi and D. Bootsma, *Exp. Cell Res.* **41**, 438 (1966).

sis follows. The method requires the measurement of mitotic index and cell number at intervals following addition of the chemical agent to synchronized cells. The average length of mitosis is determined by integration of a curve of the corrected mitotic index versus time.

Assume that N_0 cells pass through mitosis with a mean length of mitosis, M_{ave}. If the experiment is of short duration, such that no cell progresses to mitosis a second time, all mitotic cells that are observed will be from the original stock of N_0 cells. Observed mitotic indices at time t, $I_{M_{ob}}(t)$, may be adjusted for increasing cell number to give corrected mitotic indices, $I_{M_c}(t)$. The corrected values will represent the fraction of original N_0 cells that are in mitosis. The correction is simply made by

$$I_{M_c}(t) = I_{M_{ob}}(t)N(t)/N_0 \qquad (1)$$

where $N(t)$ is the concentration of cells at time t. If the observation period is partitioned into a set of Δt's, the total number of cell-hours spent in mitosis, $N_0 M_{ave}$, is approximated by $\Sigma N_0 I_{M_c}(t) \Delta t$. Keeping t within Δt, the approximation improves as Δt decreases and in the limit of $\Delta t = dt \simeq 0$,

$$N_0 M_{ave} = \int_{t_b}^{t_a} N_0 I_{M_c}(t) dt \qquad (2)$$

where t_b = a time before any cells reach mitosis; t_a = a time after the last cell leaves mitosis. Dividing by N_0,

$$M_{ave} = \int_{t_b}^{t_a} I_{M_c}(t) \, dt \qquad (3)$$

Thus, integration of the curve of corrected mitotic index versus time will give M. The graph may be easily integrated by cutting and weighing. If only fraction p of the N_0 cells enter mitosis, then M_{ave}, as determined above, must be divided by p.

A high degree of synchrony is not required for this method. However, all cells which enter mitosis must divide into two cells. It is not unusual for agents to prevent cytokinesis, leading to a cell with two nuclei, a situation that would lead to an anomalous result.

2. G_1 *Phase.* To obtain a highly synchronized population at the start of G_1, a mitotic population of cells is obtained, usually by selective detachment of cells from monolayer cultures or by use of mitotic inhibitors. Wherever it is possible to use small numbers of cells, a selection method is preferred since it does not lead to unbalanced growth and avoids the toxic side effects of the synchronizing agent. The chemical agent to be tested is added immediately after cells have divided, as monitored by cell count or mitotic index. The length of G_1 phase in treated and untreated cells may be measured by addition of labeled thymidine to the

entire culture followed by removal of aliquots at intervals to determine the percentage of labeled nuclei by autoradiography.[9] The median length of G_1 may be taken as the time required to label 50% of those cell nuclei that will eventually enter S phase. Continuous labeling is preferred over pulse labeling since some cells may finish S phase before others begin.

3. *S Phase.* To obtain the best synchrony in S phase, cells are accumulated at the beginning of S phase using a blocker of DNA synthesis. A double block will give clearer interpretation of the results, since a higher fraction of cells is obtained at the beginning of S phase. The chemical agent to be tested is added at the time when the block is released. Aliquots are removed at frequent intervals, pulsed for 15 minutes with [^{3}H]thymidine, and analyzed autoradiographically[9] for the fraction of labeled nuclei, correcting this fraction for dividing cells if necessary.

The median length for S phase will be the time required for 50% of those cells that were in S phase at the time of release to complete DNA synthesis. Alternatively, the data can be analyzed in a manner similar to that described above for mitosis, i.e., by integrating the area under the curve of S-phase cells versus time to obtain the total cell hours spent in S phase.[18] To obtain the mean length of S phase the total cell hours are divided by the number of cells initially in S phase. While some synchronizing agents are known to affect the length of S phase, this should not mask the difference between control and treated cultures. Clearly, this method will not identify the effects of agents which alter the length of S phase but which execute prior to S phase. To circumvent this problem cells may be synchronized at mitosis instead of at the beginning of S phase. In this case it may be advantageous to use the cell-hour method of analysis.

It is often useful to measure DNA content[19] as cells proceed through S phase, in order to establish that complete replication of DNA has occurred when the autoradiographic labeling of nuclei ceases. In the case where agents block completely, this measurement will give an indication of the point in S phase where the block occurred.

4. *G_2 Phase.* The best synchrony in G_2 is usually obtained by synchronizing cells at the beginning of S phase. The chemical agent may be added during S or at the start of G_2 and the effect on G_2 be observed.

Two measurements are required to determine the length of G_2—the time at which cells end S and the time at which cells enter M. A median time for the end of S is determined by treating aliquots taken at appropriate intervals with short pulses of labeled thymidine and determining the

---

[18] T. Terasima and L. J. Tolmach, *Exp. Cell Res.* **30**, 344 (1963).
[19] See this series, Vol. 3 [99].

fraction of labeled nuclei. The time at which 50% of the synchronized cells enter M can be obtained by following the mitotic index on Colcemid-blocked cells, or, in the absence of mitotic blockers, by obtaining a curve of cells entering mitosis by adding the corrected fraction of mitotic cells to the fraction of divided cells (see Appendix I). The length of G_2 is taken as the time difference between the end of S and the beginning of M.

If one simply wishes to know whether a chemical agent acts on G_2 without knowing the length of G_2 in control cultures, one could add a DNA synthesis inhibitor along with the agent at some time after a significant fraction of cells had finished S phase. Any observed lengthening of time required for cells to enter mitosis over the control culture could then be attributed solely to G_2 effects. (See also Section 4 below.)

Under Steady-State Cell Cycle Conditions

Under the proper conditions, hormones or other chemical agents may produce a cell cycle change that is stable for as long as the agent is present. Thus the culture may establish a new generation time and new phase lengths that will last for many cycles, a situation that may be detected by observing exponential growth with a new exponent (a new slope when graphed on semilog paper). It follows that the same methods that are used to determine the length of the stages of the CDC in untreated cells may be used in this situation. Selected methods for making these determinations will be discussed for each of the stages.

1. *Mitosis.* The most commonly used method to determine the length of mitosis (M) in exponential cells is by the formula[20,21]

$$M = [T' \ln (1 + I_M)]/\ln 2 \qquad (4)$$

where I_M is the mitotic index and T' is the population doubling time. T' and I_M are obtained before and after a new steady state has been achieved. (See Consideration of Doubling Times, below, for a discussion of population doubling time.) Since the derivation of this equation assumes completely asynchronous exponential growth and no variation in length of mitosis, the absolute values obtained may be only approximations of M_{ave}. While other methods for measuring length of mitosis in logarithmic populations have been suggested, this method remains the easiest and most straightforward. An analysis has also been worked out for the case where partial synchrony is present.[20]

2. *G_1 Phase.* The duration of G_1 phase in exponential cultures is usually obtained by subtraction of the length of S, G_2, and M from the

[20] C. P. Stanners and J. E. Till, *Biochim. Biophys. Acta* **37**, 406 (1960).
[21] O. Sherbaum and G. Rasch, *Acta Pathol. Microbiol. Scand.* **41**, 161 (1957).

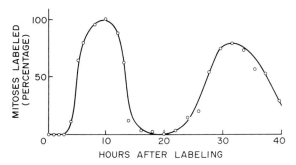

FIG. 3. The appearance of labeled mitotic cells after labeling with [³H]thymidine for 20 minutes. A human kidney cell line grown in monolayer cultures was used. From D. Bootsma, *Exp. Cell Res.* **38**, 429 (1965).

average generation time (T_{ave}). A more direct method is to add Colcemid and [³H]thymidine to cultures and to monitor the accumulation of labeled nuclei as under 3 in the section Determination of Execution Point. By graphing log $(1 + I_L/K)$ versus time one obtains a linear curve which eventually plateaus (Fig. 2C). Extrapolation of the plateau portion of the curve back to an extension of the exponential portion will give $G_{1,\text{ave}} + M_{\text{ave}}$[8] (assuming Colcemid executes at the beginning of mitosis); subtraction of M_{ave}, gives the length of G_1. An alternative analysis based on the fraction of unlabeled cells has also been presented.[22] In either case, the possibility that Colcemid affects parameters other than length of mitosis must be considered. However, a lengthening of G_1 by Colcemid would not be expected to mask the difference between control cultures and cultures treated with chemical agent.

3. S Phase

METHOD 1. FRACTION OF LABELED MITOSES (FLM).[23] Cells are pulsed for a short period of time (about 15 minutes) with [³H]thymidine, which is then washed out thoroughly, replacing the media and adding a low level of unlabeled thymidine (high concentrations inhibit DNA replication). At short time intervals thereafter, samples are prepared for autoradiography.[9] The slides are examined carefully for mitotic cells, and the fraction of mitotic cells which is labeled is determined. The results should show that, after an interval during which there are no labeled mitotic cells (the length of this interval will provide an estimate of the minimum length of G_2), the fraction of labeled cells will rise rapidly (Fig. 3). The fraction will remain high for a time and then fall. If the

[22] T. Maekawa and J. Tsuchiya, *Exp. Cell Res.* **53**, 55 (1968).
[23] H. Quastler and F. G. Sherman, *Exp. Cell Res.* **17**, 420 (1959).

curve reaches 100% and then drops to zero, the time between the 50% marks in the rising and falling portions of the curve will give an estimate of the length of S phase. However, generally the curve will not reach 100% or fall completely to 0%, so that arbitrary points on the curve must be used. Several recent discussions have dealt extensively with the uses and misuses of the FLM method.[24-27]

METHOD 2. While it would seem that the fraction of labeled cells $I_L(t)$ after pulsing a cell population with [^{3}H]thymidine should be a measure of S, the relationship is direct only in certain unique situations. In exponentially growing cells, the length of G_2 and the I_M must also be known in order to calculate the length of S from I_L. By using a generalized form of Eq. (4)[28]

$$X_{ave} = (T'/\ln 2) \ln (1 + I_X) \qquad (5)$$

where $I_X = I_M + I_{G_2} + I_S$ and $X_{ave} = M_{ave} + G_{2,ave} + S_{ave}$, one can obtain X_{ave}. [I_{G_2} can be obtained from the same formula, knowing I_M, M_{ave}, and $G_{2,ave}$; M_{ave} is obtained from I_M by Eq. (4).] S_{ave} is then easily obtained by $S_{ave} = X_{ave} - M_{ave} - G_{2_{ave}}$.

Other analyses have been described which use the same general principles but make multiple determinations of $I_L(t)$ with time for $t < G_1$[20] or with added colchicine for longer times.[8] These methods allow a slightly more precise estimation of the I_L than a single point determination.

4. G_2. The most widely used method for measuring the length of G_2 is that of the FLM[23] (see Section 3, above). Cells in exponential growth are pulse labeled with [^{3}H]thymidine, and the fraction of labeled mitoses is then followed with time. The time after pulsing at which 50% of the mitotic cells become labeled has been designated as $G_2 + M/2$. If the percentage of mitotic cells which are labeled fails to reach 100% (due to cells in G_1 at $t = 0$ passing quickly through S and into M before the slowest unlabeled cells leave M) then the half-maximal value is chosen for $G_2 + M/2$. Length of G_2 may be obtained after an independent mea-

---

[24] M. L. Mendensohn and M. Takahaski, in "The Cell Cycle and Cancer" (R. Baserga, ed.), p. 55. Dekker, New York, 1971.

[25] G. G. Steel and S. Hanes, *Cell Tissue Kinet.* **4**, 93 (1971).

[26] G. G. Steel, *Cell Tissue Kinet.* **5**, 87 (1972).

[27] D. S. Nachtwey and I. L. Cameron, *Methods Cell Physiol.* **3**, 213 (1968).

[28] At time t the number of cells within X minutes of dividing is $N(t + X) - N(t)$; thus, utilizing Eq. (6), $I_X = [N(t + x) - N(t)]/N(t) = [N_0 2^{(t+x)/T'} - N_0 2^{t/T'}]/N_0 2^{t/T'} = 2^{x/T'} - 1$. Solving this equation for X gives $X = (T'/\ln 2)\ln(1 + I_X)$. Equation (5) is similar but makes the additional assumption that the fraction of cells within X minutes of dividing is equal to the fraction of cells that accomplished a growth event which occurs on the average X minutes before cell division.

sure of length of M. It has been pointed out that this analysis need not lead to a correct mean length of G_2.[24]

A logical extension of the FLM method of measuring G_2 might be to block DNA synthesis completely immediately after a 15-minute pulse with [^{3}H]thymidine. This would produce a cohort of labeled cells which have just finished S. The entrance of these cells into mitosis could be monitored in the presence of a mitotic inhibitor.

Two other methods of measuring G_2 after adding mitotic inhibitor and [^{3}H]thymidine to exponentially growing cells have been described.[8] The first is to measure I_{G_2} as the difference between total and labeled fraction of mitotic cells and to calculate G_2 by use of Eq. (5). The second is to measure G_2 directly by the lag time between the graphs of total mitotic and labeled mitotic cells.

Consideration of Doubling Times

In asynchronous, exponentially growing cultures, two doubling time parameters must be distinguished: The averaged individual cell doubling time, T_{ave}, and an approximation, the population doubling time, T'. T_{ave} would equal T' if there were no cell death or variation in T, the individual cell doubling time. Variation is always present, however, and causes T' to differ from T_{ave}.[29,30] The cell number can be expressed either as a function of T' and t

$$N = N_0 2^{t/T'} \quad (6)$$

or as a function of T_{ave} and t

$$N = N_0 g^{t/T_{ave}} \quad (7)$$

where g is a constant determined by the variation in T.

T' is obtained by measuring the time it takes a population to double in number or by measuring the slope of log N versus time and applying the following modification of Eq. (6):

$$\log N = \log N_0 + (\log 2/T')t \quad (8)$$

It may also be predicted from the rate of mitotic cell accumulation or labeled cell accumulation, both methods applied in the presence of mitotic block. This follows from the proof that in the presence of a mitotic inhibitor, the mitotic cell accumulation function of Puck and Steffen,[8] log $(1 + I_M)$, rises at the same rate as log cell number curve in exponentially growing cells in absence of a mitotic block (see Appendix II).

[29] T. E. Harris, *in* "The Kinetics of Cellular Proliferation" (F. Stahlman, Jr., ed.), p. 368. Grune & Stratton, New York, 1959.
[30] J. E. Sisken, *Methods in Cell Physiol.* **1**, 387 (1964).

There are various methods for measuring T_{ave}. The most direct method is microscopic observation of individual attached cells[30] to determine the average time from one cytokinesis to the next. Other methods have been described which can be applied to suspension cultures as well as to attached cells. A synchronized population of cells may be obtained and the time be measured between peaks of mitosis[31] or peaks of cell division.[32] The FLM method also has been widely applied to this determination.[24] A population of exponential cells is pulsed with tritiated thymidine, and the fraction of labeled mitoses followed as described in Section 3 above. The time between peaks of this fraction is a measure of T_{ave}. The precision of these methods is limited by the difficulty in accurately locating one or both of the peaks. In the FLM method the validity of representing T_{ave} by the peak to peak distance has been questioned,[24] and in fact this distance appears to represent the median T rather than the average T.[25]

Appendix I. A Method for Determining the Fraction of Cells Entering Mitosis in the Absence of Mitotic Inhibitors

The fraction of cells from the original population at $t = 0$ which is in mitosis at time t is described by Eq. (1). The fraction of original cells which have divided by time t is $[N(t) - N_0]/N_0$. Thus the fraction F of original cells which have entered mitosis is

$$F = I_{\text{M}_{\text{ob}}}(t)N(t)/N_0 + [N(t) - N_0]/N_0 - I_{\text{M}_{\text{ob}}}(0) \tag{9}$$

This analysis can be used only if the agent does not cause abnormal division, e.g., nuclear division without cytokinesis.

Appendix II. Proof That the Mitotic Cell Accumulation Function Increases with the Same Slope as Log Cell Number

An essential part of this proof is the assumption that I_{Me}, the mitotic fraction in an exponentially growing population, is constant with time. At $t = 0$, let us arbitrarily decide that all cells in a population have already passed through M and entered G_1 once and that all mitotic cells have entered M a second time. Then the number of G_1 entries, $N_{G_1}(t)$ is always equal to the number of cells (assuming no disappearance of cells), and the number of mitotic cells is always the difference between the number of entries into mitosis, $N_M(t)$, and the number of entries into G_1.

---

[31] D. M. Prescott and M. A. Bender, *Exp. Cell Res.* **29**, 430 (1963).
[32] J. Engelberg, *Exp. Cell Res.* **23**, 218 (1961).

By these definitions, for any population of cells,

$$I_M(t) = \frac{N_M(t) - N_{G_1}(t)}{N(t)} = \frac{N_M(t)}{N(t)} - 1 \qquad (10)$$

In the case of an exponentially growing population of cells where $N(t) = N_0 2^{t/T'}$ [Eq. (6)],

$$N_M(t) = N_0(1 + I_{Mc})2^{t/T'} \qquad (11)$$

Thus assuming I_{Mc} is constant, the number of entries into mitosis rises exponentially with the same base and exponent as the cell number [compare Eq. (6)].

Assume a mitotic inhibitor affects cells only during mitosis and produces a complete and lasting block. If addition of inhibitor at $t = 0$ immediately blocks all cell division, cells will continue to enter mitosis at the same rate as in the absence of inhibitor [Eq. (11)]. Substitution of $N(t) = N_0$ and Eq. (11) into Eq. (10) gives

$$I_M(t) = 2^{t/T'}(1 + I_{Mc}) - 1 \qquad (12)$$

The accumulation function is thus

$$\log[1 + I_M(t)] = \log(1 + I_{Mc}) + (\log 2/T')t \qquad (13)$$

This derivation allows for variation in T and M. If the variation in M is small enough that Eq. (5) applies, then

$$I_M(t) = 2^{(t + M_{ave})/T'} - 1 \qquad (14)$$

Equation (14) is substantially the same as that derived by Puck and Steffen for this type of inhibitor.

In the case where a mitotic inhibitor blocks only cells which enter mitosis after addition of inhibitor and not those already in mitosis, then, after all mitotic cells have left mitosis, the total cell number will be $N_0 + I_{Mc}N_0$, while the number of mitotic cells will still be the difference between entries into M and G_1. Thus

$$I_M(t) = 2^{t/T'} - 1 \qquad (15)$$

which is nearly the same equation as derived by Puck and Steffen for this case. The accumulation function is

$$\log[1 + I_M(t)] = (\log 2/T')t \qquad (16)$$

It can be thus seen that the accumulation function will exactly parallel the log of the cell number in an uninhibited exponential population [compare Eqs. (8), (13), (16)]. The only assumptions made have been that of constant I_M in exponentially growing cells and no cell loss.

[4] Methods for Analysis of Cell Cycles *in Vivo*

By S. H. Socher

How cells divide to produce two similar daughter cells is a question that has attracted and challenged biologists for the past century. The formulation of the concept of a mitotic cell cycle by Howard and Pelc[1] has had a tremendous effect on the field of cell biology. The mitotic cell cycle can be defined as a regular sequence of synthetic events which results in a doubling of cell contents and which shows a regular alternation with mitosis. The cell cycle is composed of 4 phases: (1) mitosis, (2) G_1 (presynthetic) phase, (3) S (DNA synthetic) phase, and (4) G_2 (postsynthetic) phase. For kinetic analysis, the cell cycle is the time interval between the formation of a cell by the division of its mother cell and the time at which the cell divides to give rise to its two daughter cells.

Current kinetic analyses of the mitotic cell cycle are based on the model of Howard and Pelc[1] and the theoretical analysis of this model by Quastler and Sherman.[2] Precise methods for the analysis of cell cycles and proliferating populations *in vivo* are almost as numerous as there are investigators in the field. These can be classified into several general approaches: (1) direct observation, (2) isotopic labeling, (3) antimetabolite treatments, and (4) quantitative cytophotometry.

Direct Observation

Analysis of cell cycles *in vivo* by direct observation in vertebrate systems is limited since the mitotic behavior of individual cells cannot be followed. Yet, this approach is important particularly for determination of the level of mitotic activity. Commonly the level of mitotic activity is expressed as the mitotic index (MI), that is, the percentage number of cells in mitosis. The MI within a particular tissue or tissue cell type is determined by counting the number of mitotic figures in a particular fraction of the cell population being examined. Estimations of MI can provide information on (1) basal levels of division in a population, (2) the existence of natural mitotic rhythms, (3) changes in the frequency of division during development, and (4) the effects of hormones on the ability of cells to divide. This information is a necessary prerequisite for any detailed kinetic analysis of proliferating cell populations.

---

[1] A. Howard and S. R. Pelc, *Heredity, Suppl.* **6**, 261 (1953).
[2] H. Quastler and F. G. Sherman, *Exp. Cell Res.* **17**, 420 (1959).

Isotopic Labeling

The study of mitotic cell cycle durations and other characteristics of cell populations has been simplified by the use of radioisotopes and autoradiography. If an isotope is bound firmly within cells, the progress of these marked cells through one or more mitotic cycles may be followed by autoradiographic methods. In order for an isotope to be useful in an analysis of dividing cells, the isotope (1) must mark cells at a particular point in the cell cycle, (2) must mark the cells instantaneously, and (3) must not disturb cellular processes. The most common and useful radioisotope for cell cycle analysis is [^{3}H]thymidine which is used as a specific precursor of newly synthesised DNA. Many of the isotopic labeling methods for obtaining cell cycle data are familiar and have been discussed in both theoretical and practical terms.[2-12]

The most widely used method for analysis of the duration of the cell cycle and its component phases (G_1, S, G_2, and mitosis) is to pulse label a population of cells with [^{3}H]thymidine. The pulsing method used for labeling animal cells *in vivo* is to inject [^{3}H]thymidine into the animal intraperitoneally. The standard dose of [^{3}H]thymidine is 0.5–1.0 μCi per gram body weight.[5,6] After such treatment, the isotope appears to be evenly distributed throughout the body and only incorporated by cells in the S phase. The isotope is firmly bound within the nucleus and is diluted by subsequent division.[13,14] In most systems, about 50% of the

---

[3] H. Quastler, *in* "Cellular Proliferation" (L. F. Lamerton and R. J. M. Fry, eds.), p. 18. Davis, Philadelphia, 1963.

[4] D. E. Wimber, *in* "Cell Proliferation" (L. F. Lamerton and R. J. M. Fry, eds.), p. 1. Blackwell, Oxford, 1963.

[5] J. D. Trasher, *in* "Methods in Cell Physiology" (D. M. Prescott ed.), Vol. II, p. 324. Academic Press, New York, 1966.

[6] S. Gelfant, *in* "Methods in Cell Physiology" (D. M. Prescott, ed.), Vol. II, p. 359, Academic Press, New York, 1966.

[7] H. Rothstein, *in* "Methods in Cell Physiology" (D. M. Prescott, ed.), Vol. III, p. 45. Academic Press, New York, 1968.

[8] J. Van't Hof, *in* "Methods in Cell Physiology" (D. M. Prescott, ed.), Vol. III, p. 95. Academic Press, New York, 1968.

[9] D. S. Nachtwey and I. L. Cameron, *in* "Methods in Cell Physiology" (D. M. Prescott, ed.), Vol. III, p. 214. Academic Press, New York, 1968.

[10] I. L. Cameron, *in* "Methods in Cell Physiology" (D. M. Prescott, ed.), Vol. III, p. 261. Academic Press, New York, 1968.

[11] J. E. Cleaver, *in* "Frontiers of Biology" (A. Neuberger and E. L. Tatum, eds.), Vol. 16. North-Holland, Publ., Amsterdam, 1967.

[12] M. L. Mendelsohn and M. Takahashi, *in* "The Cell Cycle and Cancer" (R. Baserga, ed.), p. 58. Dekker, New York, 1971.

[13] C. P. Leblond and B. Messier, *Anat. Rec.* **132**, 247 (1958).

[14] R. B. Painter, F. Forro, and W. L. Hughes, *Nature (London)* **181**, 328 (1958).

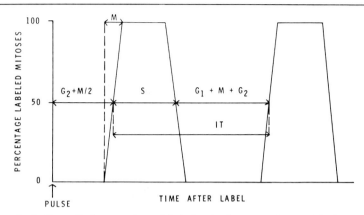

Fig. 1. A theoretical percentage labeled mitosis curve for an ideal population of cells. The curve records the percentage of labeled mitotic figures as a function of time following the pulse label with [^{3}H]thymidine.

[^{3}H]thymidine is incorporated into DNA and the unincorporated precursor is rapidly catabolized by the liver in 20–30 minutes.[5,9] However, as a word of caution, each system must be tested for the fate of the injected isotope.[11,15,16,17]

The pulse labeling method can be used to generate a percentage labeled mitosis curve which was devised by Quastler and Sherman.[2] This method was based on the behavior of a theoretically ideal cell population in which (1) all the cells divide, (2) all cells divide at the same rate, and (3) the cells are evenly distributed throughout the cell cycle. After a pulse label, samples of treated tissue are taken at regular time intervals. Autoradiographs are prepared, and the percentage of labeled mitotic cells is determined for each time sample. The data is plotted as illustrated in Fig. 1. This method involves tagging a group of cells, namely those in S at the time of the pulse label, and following these cells as they go through mitosis. The duration of the cell cycle, intermitotic time (IT), and the durations of its component phases are determined by the percentage labeled mitosis (PLM) curve.

After a short exposure to [^{3}H]thymidine, the first cells to appear at mitosis are those that were in G_2 at the time of the pulse. They will not be labeled. The next cells to reach division will be those that were in S at the time of the radioactive exposure. These cells have incorporated

[15] V. R. Potter, in "Kinetics of Cellular Proliferation" (F. Stohlman, ed.), p. 104. Grune & Stratton, New York, 1959.
[16] J. W. Steiner, Z. M. Pery, and L. B. Taechman, Exp. Mol. Pathol. 5, 146 (1966).
[17] E. Stocher and A. Pfeifer, Z. Zellforsch. Mikrosk. Anat. 79, 374 (1967).

[³H]thymidine into their nuclear DNA, and thus they are labeled. An unlabeled population of cells will then arrive at mitosis. These are cells that were in G_1 at the time of the [³H]thymidine pulse. The next labeled population of cells to appear in division will be composed of those cells that were in S at the time of the pulse label and which have reached their second mitosis after incorporation.

A PLM curve for an ideal population of cells is illustrated in Fig. 1 and is analyzed in the following manner to determine the phase durations. The time interval between the 50% intercepts of successive ascending portions of the curve is used to estimate the cell cycle duration or IT. The duration of G_2 + mitosis/2 is obtained from the interval between the time of the administration of the label and the 50% intercept of the first ascending portion of the curve. The length of mitosis is derived by drawing a line parallel to the y axis at the intersection of the first rise of the PLM curve with the x axis. The time interval between this line and the time at which 100% of the mitosis first appear labeled is equal to the mitotic duration. The G_2 duration is then estimated by subtraction. The duration of the S period is obtained by measuring the time interval between the 50% intercepts of the first ascending and descending portions of the curve. The duration of G_1 is calculated by subtraction; that is: $IT - (S + M + G_2) = G_1$.

In vivo few cell populations are ideal proliferating populations as defined by the theoretical model of Quastler and Sherman, and yet PLM curves can be used to estimate cell cycle parameters of nonideal populations in nature. In fact, the extent of deviation of the experimental PLM curve from the theoretical curve can provide information about both cell behavior and population heterogeneity.[12,18]

Inhibitor Treatments

The use of metabolic inhibitors for cell cycle analysis *in vivo* is limited owing to perturbation of cellular processes throughout the animal. Colchicine and Colcemid both inhibit anaphase and lead to metaphase accumulation. This effect of colchicine treatment can be used to determine the duration of the cell cycle[19] and in combination with other marking techniques to estimate other cell cycle parameters.[9] Caution must be exercised in using colchicine for kinetic analyses since the block of anaphase is not complete and the drug can cause alterations in the cell cycle. For *in vitro* studies of cell cycles and mitotic synchrony, inhibitors

---

[18] P. L. Webster and D. Davidson, *J. Cell Biol.* **39**, 332 (1968).

[19] C. P. Leblond, in "Kinetics of Cellular Proliferation" (F. Stohlman, ed.), p. 31. Grune & Stratton, New York, 1959.

of DNA syntheses have been exploited.[20] The use of DNA synthetic inhibitors for cell cycle analysis *in vivo* in vertebrates is plagued with problems that limit their usefulness.[21]

Quantitative Cytochemistry

The formulation of the hypothesis of DNA constancy in nondividing cells[22] and early measurements of DNA content of individual cells with different tissues[23,24] laid the foundation for cytophotometric and histochemical analyses of the cell cycle. The theoretical basis and practical application of these approaches have been discussed.[25-28] In practice these methods have limited application, principally owing to difficulties in the practical aspect of the methods and in analysis and interpretation of the data obtained.

The importance of cell division in biology and medicine is evident to all investigators. At the present time our knowledge of cell cycle and cell population kinetics in both target and nontarget tissues is in an embryonic state and the field is in need of studies of cell cycle kinetics in cell populations *in vivo*. It is only through these studies, coupled with biochemical analyses of hormone action, that we will be able to understand the hormonal regulation of the division of cells in the growth, differentiation, and maintenance of target tissues.

---

[20] E. Stubblefield, *in* "Methods in Cell Physiology" (D. M. Prescott, ed.), Vol. III, p. 25. Academic Press, New York, 1968.
[21] S. H. Socher and B. W. O'Malley. Unpublished observation.
[22] A. Boivin, R. Vendrely, and C. Vendrely, *C. R. Acad. Sci.* **226**, 1061 (1948).
[23] A. E. Mirsky and H. Ris, *Nature (London)* **163**, 666 (1949).
[24] H. H. Swift, *Physiol. Zool.* **23**, 169 (1950).
[25] C. Vendrely, *in* "The Cell Cycle and Cancer" (R. Baserga, ed.), p. 227. Dekker, New York, 1971.
[26] A. J. Hale, *in* "Introduction to Quantitative Cytochemistry" (G. L. Wied, ed.), p. 183. Academic Press, New York, 1966.
[27] M. L. Mendelsohn, *in* "Introduction to Quantitative Cytochemistry" (G. L. Wied, ed.), p. 202. Academic Press, New York, 1966.
[28] A. M. Garcia and R. Iorio, *in* "Introduction to Quantitative Cytochemistry" (G. L. Wied, ed.), pp. 216 and 239. Academic Press, New York, 1966.

[5] Electron Microscopy and Autoradiography of Chromosomes

By ELTON STUBBLEFIELD

Investigations of the mechanisms controlling cellular processes ultimately reach a point where possible interaction is suspected between some cellular or hormonal agent and the cell genome. One may wish to examine directly the chromosomes of the cells involved to determine whether the agent can be localized in the karyotype. The experiments of Price et al.[1,2] serve as examples of the usefulness of this approach. However, some of their results have been challenged on technical grounds by Bishop and Jones[3] and Prensky and Holmquist.[4] Nevertheless, since it seems likely that repeated attempts with such an approach will be utilized in the future, this description of autoradiographic and electron microscopic techniques for handling chromosomes is included in this volume to serve as an indication of the possibilities and problems for those who may be contemplating such studies.

General Considerations

Most cellular metabolic phenomena occur in interphase cells, where the chromosomes are dispersed in the cell nucleus as chromatin, and it is probable that most interactions of the genome with controlling agents would occur in this state. It is not yet possible to isolate the interphase chromosomes as recognizable entities, however, so metaphase chromosomes must be used instead. The studies of Rao and Johnson[5] have demonstrated that interphase chromosomes can be condensed by fusion of an interphase cell with a metaphase cell, so this restriction may eventually be overcome. However, current technology requires that any agent be linked to isolated chromosomes from metaphase cells or be linked to the genome *in vivo* during interphase and the chromosomes isolated from the next metaphase stage of the cell cycle. In the latter case it is quite apparent that one can only use a growing system where metaphase cells can be obtained in quantity for the isolation of the chromosomes.

[1] P. M. Price, J. H. Conover, and K. Hirschhorn, *Nature (London)* **237**, 340 (1972).
[2] P. M. Price, K. Hirschhorn, N. Gabelman, and S. Waxman, *Proc. Nat. Acad. Sci. U.S.* **70**, 11 (1973).
[3] J. O. Bishop and K. W. Jones, *Nature (London)* **240**, 149 (1972).
[4] W. Prensky and G. Holmquist, *Nature (London)* **241**, 44 (1973).
[5] P. N. Rao and R. T. Johnson, *Advan. Cell Mol. Biol.* **3**, in press.

The chromosomes of an organism provide recognizable packages, each containing a specific fraction of the genome. The chromosomes of different organisms are not equally useful in this respect, however. In the Chinese hamster most of the chromosomes can be identified by simple morphological criteria, whereas in the laboratory mouse all the chromosomes are similar in morphology and form a continuous series from the largest to the smallest. The human karyotype lies somewhere between these two extremes. Recent advances in the methods used to stain chromosomes now permit the identification of all the chromosomes of most karyotypes by their banding pattern,[6] and this may ultimately prove to be useful for the specific identification of isolated chromosomes. Although this is not yet practical, the chromosome still provides a convenient package of a genome fraction which can be exploited in many ways.

Chromosome Isolation

A primary requisite for the isolation of chromosomes is a population enriched in mitotic cells. Ideally, all the cells should be in metaphase, and this is not practical with many cell cultures. However, chromosomes can be isolated from cultures containing as few as 10% mitotic cells; only the yield is reduced. The procedures which isolate chromosomes from mitotic cells usually isolate intact nuclei from interphase cells. Nuclei are much larger than chromosomes and can be removed by differential centrifugation or filtration through Nuclepore filters. Thus, the more mitotic cells, the better the yields will be.

Mitotic cells can be accumulated in growing cultures or tissues by treatment with Colcemid, colchicine, vinblastine (Velban), or other mitotic inhibitors. Treatment of some cells must be limited to about 5 hours, since after this time the cells may return to interphase without dividing, and in doing so may form micronuclei that are difficult to separate from the chromosomes. Some cells may be blocked indefinitely, but in other cases the chromatids may eventually separate, producing fragments that are difficult to identify because of the lack of an obvious centromere. These effects should be examined carefully in any system under contemplation as a chromosome source.

A variety of procedures have been developed for the isolation of chromosomes in acid,[7] neutral,[8,9] or alkaline[10,11] solutions. The procedure

[6] T. C. Hsu, *Annu. Rev. Genet.* **7**, 153 (1974).
[7] K. P. Cantor and J. E. Hearst, *Proc. Nat. Acad. Sci. U.S.* **55**, 642 (1966).
[8] W. Wray and E. Stubblefield, *Exp. Cell Res.* **59**, 469 (1970).
[9] J. J. Maio and C. L. Schildkraut, *J. Mol. Biol.* **24**, 29 (1967).
[10] W. Wray, E. Stubblefield, and R Humphrey, *Nature (London) New Biol.* **238**, 237 (1972).
[11] P. M. Corry and A. Cole, *Radiat. Res.* **36**, 528 (1968).

chosen will depend on the purpose of the experiment. Isolation in 50% acetic acid is simple, but it may not be useful, since cytoplasmic RNA is adsorbed to the chromosomes in acid solutions. Isolation at high pH serves to control nuclease activity which attacks the chromosomal DNA during the hypotonic treatment common to all these procedures. The chromosomal proteins are probably best retained at near neutral pH, and the procedure detailed below is an example in this category.

The procedures now in use involve breakage of metaphase cells in a hypotonic buffer in which the chromosomes are stable. When the metaphase cells first contact the buffer, they rapidly swell, and the chromosomes become greatly dispersed in the cytoplasm.[12] Once the cell membrane is ruptured, the chromosomes condense to their metaphase state and again become visible to phase contrast microscopy.[8] These effects are seen in the micrographs shown in Figs. 1–3. The difficulties of chromosome studies using thin-sectioned material are readily apparent in these figures.

After a metaphase cell is broken open in the hypotonic buffer solution, the problem becomes one of separation and isolation of the chromosomes from the unwanted cell debris. Various shearing and homogenization procedures have been used successfully. For small volumes of cell suspension, we have found it convenient to rapidly pass the suspension through a small gauge needle. In the procedure detailed below, the pH is quite important; below a pH of 6.5 the cell debris adheres tightly to the chromosomes, and above a pH of 7.0 the chromosomes are not stable unless the divalent ion concentration is increased.[10]

Procedure for the Isolation of Metaphase Chromosomes[8]

1. Suspend population of trypsinized metaphase cells in medium containing serum to inactivate any remaining trypsin. Store at 4°, 30 minutes up to 6–8 hours. Cold storage destroys spindle microtubules remaining after Colcemid treatment.

2. Centrifuge cell suspension, 400 g for 2 minutes. Remove all the medium and suspend the cell pellet in cold chromosome buffer (1.0 M hexylene glycol, 0.5 mM calcium chloride, 0.1 mM PIPES buffer, pH 6.6). (The solution is first buffered and the pH adjusted with sodium hydroxide or hydrochloric acid, 10 mM; the hexylene glycol is added last, since it interferes with the responsiveness of some pH-meters.)

3. Immediately centrifuge the cells again, 400 g for 2 minutes. It is important that this washing be completed before any cells break open.

4. Remove the solution from above the pellet and discard. Carefully resuspend the pellet in cold chromosome buffer, 5.0 ml for every 0.1 ml volume of washed cell pellet. The cells should not yet be broken.

[12] B. R. Brinkley, *J. Cell Biol.* **35** (part 2), 17A (1967).

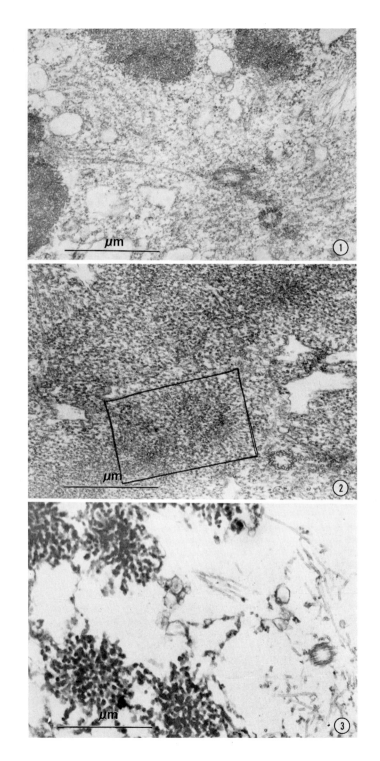

5. Incubate the cell suspension at 37° in a water bath for 15 minutes. During this period the cells will all swell and break open. Confirm this by examining a drop of the solution on a slide by phase microscopy.

6. Shear gently by forcing the preparation through a 22-gauge needle with a syringe, or homogenize with a Dounce homogenizer. Examine the preparation by phase microscopy. Repeat the shearing with increased force as needed to separate and clean the chromosomes. Excessive shearing will distort the chromosomes, leaving long thin fibers. Keep the preparation warm until shearing is completed.

7. Centrifuge the preparation in an angle-head rotor, 400 g for 2 minutes to remove any nuclei and large cell fragments or whole cells. Remove and save the supernatant chromosome suspension.

Preparation of Isolated Chromosomes for Electron Microscopy

Removal of the chromosomes from the broken cell preparation can be accomplished by filtration or differential centrifugation. For direct observation of the whole chromosomes by electron microscopy the procedure which follows has consistently given the best preparations.

A 12 mm circular coverslip (No. 1) is placed in the bottom of a 15-ml round-bottom glass centrifuge tube (Corex, 15 × 100 mm) containing 2.5 ml of 10% sucrose in chromosome isolation buffer. Copper electron microscope grids (200 mesh) with a Formvar film (500 Å thick or less) on one surface are carefully placed beneath the sucrose and arranged in a nonoverlapping pattern on the coverslip. Three grids to a centrifuge tube can be readily managed in this way.

The Formvar films can be made using standard procedures.[13] Solutions of the Formvar powder in ethylene dichloride (0.5% or 1.0%) are

[13] D. C. Pease, "Histological Techniques for Electron Microscopy." Academic Press, New York, 1960.

FIG. 1. Thin section through a mitotic Chinese hamster fibroblast (Don-C) fixed in 3% glutaraldehyde in chromosome buffer, embedded in Epon, and stained after sectioning in uranyl acetate and lead citrate. Parts of three chromosomes and two centrioles are visible. The chromosomes are composed of fine fibrils less than 100 Å in diameter.

FIG. 2. Chinese hamster mitotic cell fixed in 3% glutaraldehyde added to the chromosome buffer after the cells had been incubated in the buffer for 1 minute at 37°. Chromosomal fibers have become greatly dispersed, but the chromosomes can still be found as regions of greater density (rectangle).

FIG. 3. Another mitotic cell fixed in 3% glutaraldehyde (as in Figs. 1 and 2) after 10 minutes of incubation in the chromosome buffer at 37°. The chromosomes are again condensed but now contain fibrils about 200–300 Å in diameter. Of the other cytoplasmic elements, only centrioles, microtubules, and membranes persist.

dropped onto a clean water surface and the grids are placed on the areas of the floating film that appear gray or silver in reflected light. A clean microscope slide is pushed down through the film into the water to trap the grids between the film and the glass surface; it is quickly inverted and brought back through the water surface (without losing the film or grids) and left to air dry. Grids picked up in this manner on the circular coverslips will be held in place by the film and can be placed as a unit in the centrifuge tubes under the sucrose solution.

The broken cell suspension containing the chromosomes is layered over the sucrose solution in the centrifuge tube. The preparation is centrifuged for 15 minutes at 1500 g in a swinging-bucket rotor (Sorvall HB-4). The chromosomes, but not the cell debris, are readily sedimented under these conditions. The supernatant solution is removed by aspiration and the coverslip and grids quickly removed to a beaker containing chromosome isolation buffer (no sucrose). The grids are then gently dislodged from the coverslip. The coverslip is then removed, drained of solution, and air-dried. Examination of the coverslip by phase-contrast microscopy confirms that the chromosome density is not too great or too little; it will be similar to that seen later in the electron microscope grids. If the chromosomes are too tightly packed, then another centrifugation can be done with a more dilute suspension of the chromosomes layered over the sucrose.

Critical Point Drying

Once a usable chromosome density is achieved (Fig. 4), then the grids can be processed further for critical point drying. The purpose of the following procedure is to dry the chromosomes without introducing surface tension artifacts which greatly distort the ultrastructure of air-dried specimens. Several variations have been used,[14,15] but the original procedure devised by Anderson[16] is widely used and generally quite adequate. By repeated addition and draining of absolute ethanol from the beaker containing the grids with the chromosomes sedimented on the Formvar film, the water is removed and the chromosomes are dehydrated. At this point the grids are transferred to a holder made especially for the critical point apparatus (see below) so that many grids can be carried through the remaining solutions together. The transfer is made quickly to the holder immersed in alcohol, so that the chromosomes do not air-dry during the transfer. The holder containing the grids is then carried through

---

[14] A. Cohen, D. P. Marlow, and G. E. Garner, *J. Microsc. (Paris)* **7**, 331 (1968).
[15] T. Koller and W. Bernhard, *J. Microsc. (Paris)* **3**, 589 (1964).
[16] T. F. Anderson, *Trans. N. Y. Acad. Sci. II* **13**, 130 (1951).

Fig. 4. Isolated Chinese hamster chromosomes sedimented onto a coverslip and air-dried as described in the text. Unstained and unmounted, the chromosomes appear as bright objects on a dark background in phase contrast.

a graded series of increasing concentrations of amyl acetate in ethanol (e.g., 25% amyl acetate, 50%, 75%, and two containers of pure amyl acetate). Amyl acetate is used because it is miscible both with ethanol and liquid carbon dioxide, and it does not dissolve the Formvar film. From the last amyl acetate rinse the holder is transferred to the pressure chamber of the critical point apparatus, and liquid carbon dioxide under 800 psi pressure is introduced into the chamber at room temperature. The chamber is slowly flushed with liquid carbon dioxide for 5 minutes to remove all amyl acetate, then the valves are closed and the temperature of the pressure chamber is raised to 60° with a hot water bath. The pressure rises to about 1700 psi, and the carbon dioxide passes through its critical point where the liquid becomes a gas. The hot gas is slowly bled off, and the dry sample is then removed from the chamber.

A critical point apparatus can be purchased from several manufacturers, or the device can be made in the laboratory if modest shop facilities are available.[16] The apparatus marketed by Denton Vacuum, Inc. is well designed and quite satisfactory in our experience. However, we have not been pleased with any of the commercially available grid holders, so we have made one of our own design (Fig. 5) from solid Teflon. The main problem with the commercial holders is that they harbor amyl acetate around bolts and between parts in spaces that do not readily exchange with the liquid carbon dioxide flowing through the holder. The designs that hold the grids in the plane perpendicular to the liquid flow are also bad in that they promote breakage of the fragile Formvar film stretched over the grid.

Chromosomes prepared by this method should have the appearance shown in Fig. 6. They may be photographed directly in the electron microscope without any further treatment or staining, since the natural density of the chromatin fibers is sufficient to interact appreciably with the electron beam. However, additional information can be gained by stereo photography, inasmuch as a three-dimensional image allows a clearer tracing of the fibers through the structure. Many electron microscopes provide a tilting mechanism that allows the specimen to be photographed from two viewpoints 12° apart. The two images differ slightly in information content, and if they can be viewed simultaneously, one by each eye, then the mind is capable of reconstructing the three-dimensional image, as it routinely does in ordinary binocular vision. Interestingly, reversal in position of the two images results in a 180° change of the viewing

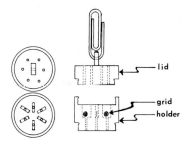

FIG. 5. Grid holder for critical point drying seen in two views. A single lid is attached to as many grid holder units as are needed; the units snap together, one above the next. One grid slot is displaced inward for identification purposes. The units are milled from a solid 1-inch bar of Teflon. The bottom unit must not sit flat on the floor of the pressure chamber, since this would prevent free flow of the liquid carbon dioxide through the holders.

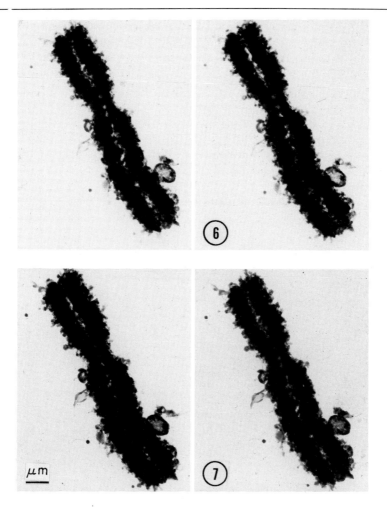

Fig. 6. Stereo electron photomicrograph of critical point dried human chromosome No. 2.

Fig. 7. Same chromosome as that shown in Fig. 6, but with the stereo images reversed so that the chromosome is seen from the other side (through the support film).

angle, so that a chromosome can be viewed from above or below (through the support film) as in Fig. 7.

Autoradiography

A radioactive component added to or incorporated into a chromosome can be detected by autoradiography. For many applications, electron microscopy is not needed, since an agent could be localized to a particular chromosome quite readily at the light microscope level of resolution. The technology associated with light microscope autoradiography has been well developed for many years now, and its direct application to chromosome studies is in wide use. The method of Schmid and Carnes[17] is a good beginning point for anyone wishing to learn these techniques.

A useful modification that is quite helpful in cases where the level of radioactivity is low and one wishes to be sure that a few silver grains found over a particular chromosome are significant, as in the experiments of Price et al.,[1,2] is the technique of repeat autoradiography.[18] If the chromosomes are covered with a thin Formvar film, then a series of autoradiographs can be made using stripping film, examined, photographed, and then peeled off of the slide in succession. Any chromosomal locus that consistently produces an autograph is then obviously radioactive. This approach negates the problem of scattered background grains that always occur over chromosomes at random.

Autoradiography in the Electron Microscope

Techniques already exist to prepare autoradiographs for examination in the electron microscope.[19] Such procedures are used widely in the analysis of isotope distributions in thin-sectioned material. Obviously, consecutive sections through Epon-embedded chromosomes could be made, as in Fig. 8. However, for certain studies, whole isolated chromosomes should be examined intact or after partial disruption,[20] and direct application of the thin-section procedures to whole-mount preparations has certain disadvantages. Direct contact between a thin layer of emulsion and the chromosomes obscures much of the fine detail of chromosome structure. In addition the chromosomes are subjected to drying artifacts after the emulsion is developed, fixed, and washed. We have therefore developed the following procedure for the autoradiographic analysis of critical point dried chromosomes.

Several modifications of the usual procedures are essential to the success of this procedure. The first is the placing of the chromosomes on one side of the Formvar film and the autoradiographic emulsion on the other side. This makes it possible to develop the emulsion without getting the chromosomes wet, but, at the same time, some sacrifice in resolution is made, since the film is not in direct contact with the chromosomes.

[17] W. Schmid and J. D. Carnes, in "Human Chromosome Methodology" (J. Yunis, ed.), p. 91. Academic Press, New York, 1965.
[18] E. Stubblefield, *J. Cell Biol.* **25** (part 2), 137 (1965).
[19] G. C. Budd, *Int. Rev. Cytol.* **31**, 21 (1971).
[20] E. Stubblefield, *Int. Rev. Cytol.* **35**, 1 (1973).

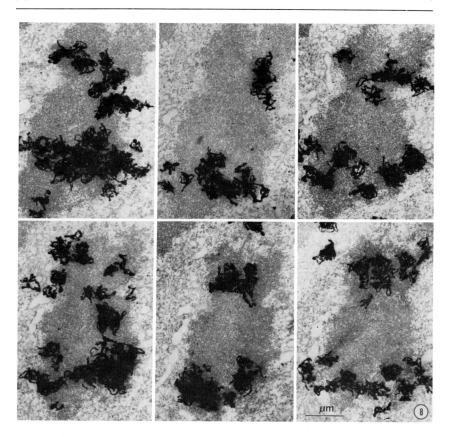

Fig. 8. Chinese hamster chromosome fixed as in Fig. 1. In this case the cell had been labeled with tritiated thymidine (10 μCi/ml; 6.7 Ci/mmole) during the last hour of S phase. The sections were overlaid with a thin film of Ilford L-4 emulsion and exposed for 30 days. Six consecutive serial sections show label over two regions of the chromosome.

A thin coating of carbon is evaporated onto both sides of the Formvar film to make it hydrophobic so that the developer liquids will not creep through occasional holes in the film and wet the chromosomes. Of course, during development only small drops of liquid are carefully applied to the emulsion side of the film.

Procedure for Electron Microscope Autoradiography of Whole Chromosomes

1. Electron microscope specimen grids with a Formvar film are lightly carbon coated on both sides in a vacuum evaporator.[13]

2. Isolated chromosomes are centrifuged though 10% sucrose in isolation buffer onto the support film as described above, except that the grids are placed filmed side down, so that the chromosomes are on the side of the film nearest the grid.

3. The grids and chromosomes are dried by the critical point procedure.

4. The grids are fastened in the center of a glass microscope slide by touching one edge to a small piece of double-surfaced cellophane tape stuck on the slide. The grids are arranged flat with the Formvar film on top of the grids and the chromosomes on the lower surface of the film.

5. A liquid photographic emulsion suitable for electron microscope applications, such as Ilford L-4 or K-5, is prepared.[19] The emulsion is

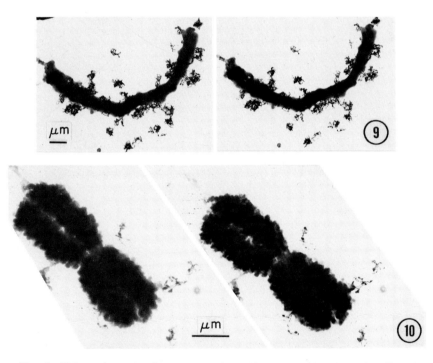

Fig. 9. Chinese hamster chromosome autoradiograph prepared as described in the text, but accidentally wet during development of the emulsion. Chromosome fiber organization is disrupted. The Ilford K-5 silver grains are clearly separated from the chromosome in the stereo view.

Fig. 10. Ilford L-4 autoradiograph properly developed without wetting the chromosome. The stereo view reveals that the radioactive part of the chromosome is elevated from the support film, so that good resolution would not be possible in this case. In Figs. 9 and 10 the chromosomes were labeled as in Fig. 8.

dissolved (under safelight conditions) in 3 parts of water at 45° and then chilled. A thin film of the emulsion is made by dipping a wire loop 2 cm in diameter into the liquid and carefully withdrawing it. The motion of the liquid film in the loop is carefully monitored by observing the reflection of the safelight in it. The emulsion should gell within a minute after the thin film is formed, and any swirling of the reflection pattern ceases.

6. The loop of gelled emulsion is lowered over the grids on the microscope slide and the film is transferred by blowing gently into the loop. Excess emulsion is wiped from around the edges.

7. The grids as stored at 4° in a sealed dry black box for as long as needed to assure sufficient exposure of the emulsion to the radioisotope.

8. Developing solutions are applied to the centers of individual grids as tiny drops (about 1 mm in diameter) on the end of a hypodermic needle (stainless steel) assembled on a syringe. Special developers may be used[19] or a short exposure to Kodak D-19b (20 seconds) is satisfactory. The developer is removed by blotting with the edge of a piece of filter paper or paper towel. A microdrop of Kodak acid fixer is then applied to dissolve out the unreacted silver halide. After 60 seconds it is removed and a drop of distilled water applied and removed to dilute the fixer salts. The preparation is then ready for examination in the electron microscope.

If the chromosomes get wet during the developing of the emulsion, they have the appearance shown in Fig. 9. When properly executed, the procedure yields chromosomes with a crisp image, as in Fig. 10. When photographed and viewed in stereo, the silver grains do not obscure any detail, since they are seen in a plane well separated from the chromosome.

Acknowledgment

I wish to thank Dr. B. R. Brinkley for the use of his excellent electron microscope facilities in the Department of Human Biological Chemistry and Genetics of the University of Texas Medical Branch in Galveston, Texas.

[6] Parallel Isolation Procedures for Metaphase Chromosomes, Mitotic Apparatus, and Nuclei

By WAYNE WRAY

The methods employed for the mass isolation of any cell organelle are limited by the physical and chemical nature of the structure, the adherence of contaminating materials from the cell, and the final desired state of the isolated component. Isolation of subcellular components for experimental study has generally been specific for the cell organelle desired with little or no regard to the slight modifications which might allow nearly identical isolation conditions for a number of cellularly and functionally related structures. The advantage of subjecting biochemical isolations of organelles to similar conditions is that the observations may be compared without concern that differences in experimental isolation parameters may prevent parallel or related observations. This chapter describes a general isolation procedure which is rapid and applicable to the isolation of chromosomes, mitotic apparatus, or nuclei.

System Universality

The methods given are only for the isolation of chromosomes, mitotic apparatus and nuclei from Chinese hamster fibroblasts, but I would like to point out that the buffer system described below may be the nearest approximation to a universal isolation medium yet developed. I have noted that good preparations of many cell organelles may be obtained from all organs, tissues, and tissue culture lines that I have tried. Some require modification (e.g., HeLa chromosomes are easier to isolate at pH 6.7), but most are not that sensitive. Cellular organelles such as ribosomes, polysomes, lysosomes, centrioles, mitochondria, and membranes have been observed and all expected *in vitro* enzymatic activities which have been checked were found, although extensive and necessary biochemistry has not yet been attempted. I hope that laboratories which are routinely isolating a particular cellular organelle from any source would be willing to try these methods and/or buffer system, possibly further modify them, and compare the preparations with those routinely used.

General Methods

Tissue Culture and Cell Line

The experiments presented in this chapter were performed using the cloned strain Don-C[1] which was derived from line Don,[2] a normal diploid lung culture from a male Chinese hamster. The Don-C cells were cultured in monolayers in 600-ml bottles rotating on a wheel at one revolution every 15 minutes. Each bottle contained 50 ml of McCoy's medium 5a supplemented with 0.8 g of lactalbumin hydrolyzate per liter and 20% fetal calf serum. The pH of the medium was buffered by a bicarbonate system requiring 10% CO_2 in the atmosphere. The Don-C strain has a

[1] E. Stubblefield, *J. Nat. Cancer Inst.* **37**, 799 (1966).
[2] T. C. Hsu and M. Zenses, *J. Nat. Cancer Inst.* **32**, 857 (1964).

cell cycle of approximately 12 hours. Cultures were subdivided every 24 hours.

Cell Synchronization

In order to increase the number of cells in mitosis at the time of harvest, mitotic cells were accumulated in the presence of Colcemid (0.06 µg/ml) and selectively trypsinized away from the remaining interphase cells.[3] The mitotic cell populations obtained by this method were routinely 94% to 97% metaphase cells.

Isolation Buffers

The chromosome isolation buffer[4] contains 1.0 M hexylene glycol (2-methyl-2,4-pentanediol), Eastman Organic; 0.5 mM $CaCl_2$; and 0.1 mM piperazine-N,N'-bis(2-ethane sulfonic acid) monosodium monohydrate (PIPES), Calbiochem; at a pH of 6.5. The buffer is most conveniently made by a 1:10 dilution of 5 mM $CaCl_2$, 1 mM PIPES (10 × buffer) which has been adjusted to a pH of 6.5 with 1.0 N sodium hydroxide. Hexylene glycol is then added after dilution. *It is important that the pH be adjusted before the addition of hexylene glycol!* Apparently some pH electrodes respond sluggishly if the solution contains an appreciable amount of this organic liquid.

Chromosomes, mitotic apparatus, and nuclei are all stable in the chromosome isolation buffer, but it was found that for Chinese hamster cells slight modifications made in the buffer system for mitotic apparatus and for nuclei augmented cell breakage and decreased the amount of contamination of the product. For the mitotic apparatus buffer, the PIPES concentration was changed to 50 µM. In the nuclei buffer the hexylene glycol was reduced to 0.5 M, the PIPES was also 50 µM, and the $CaCl_2$ was increased to 1 mM. In these isolation buffers the pH is nearly neutral, heavy amphoteric metals like zinc are absent, and the chromosomes, mitotic apparatus, and nuclei are stabilized but not fixed.

Other Methods

DNA,[5] RNA,[6] and protein[7] were measured colorimetrically. Whole chromosomes could be studied using electron microscopy after air-drying

[3] E. Stubblefield and R. Klevecz, *Exp. Cell Res.* **40**, 660 (1965).
[4] W. Wray and E. Stubblefield, *Exp. Cell Res.* **59**, 469 (1970).
[5] K. Burton, *J. Biochem.* **62**, 315 (1956).
[6] R. B. Hurlbert, H. Schmitz, A. F. Brumm, and V. R. Potter, *J. Biol. Chem.* **209**, 23 (1954).
[7] O. H. Lowry, N. J. Rosebrough, A. L. Farr, and R. Randall, *J. Biol. Chem.* **193**, 265 (1951).

or critical point drying unstained preparations on a collodion or Formvar-coated grid.

Isolation Methods and Properties

A flow chart which outlines in parallel all steps in these isolation procedures and indicates pertinent explanations for some of these steps is shown in Table I.

The reader is urged to examine the morphology of these preparations from light and electron microscopy presented in Wray and Stubblefield[4] and Wray.[8] Only the methods are presented in this chapter.

Chromosomes

Mammalian metaphase chromosome isolation procedures may be catalogued by the pH of the isolation medium. Most of the methods[9-16] utilize low pH, generally between 3.0 and 3.7. According to Huberman and Attardi,[17] lowering the pH has the effect of increasing the contraction of the chromosomes. This could increase resistance of the isolated chromosomes to mechanical damage at low pH or the contraction could be the result of denaturation and precipitation of some chromosomal proteins. The use of low pH may introduce the possibility of undesirable side effects such as the extraction of histones. Huberman and Attardi[17] found that most histones are not extracted by the conditions of the various chromosome isolations, but some lysine-rich histones in HeLa chromosomes are extracted. The use of a pH below 6.0 causes the formation of aggregates of materials of high RNA content which are difficult to remove from the final preparation.[18] These aggregates are probably composed of ribosomes which clump at lower pH values and sediment easily at low centrifuge forces.

Chromosome isolations at neutral pH, according to Hearst and

---

[8] W. Wray, *in* "Methods in Cell Biology" (D. M. Prescott, ed.), Vol. VI, p. 283. Academic Press, New York, 1973.
[9] D. M. Prescott and M. A. Bender, *Exp. Cell Res.* **25**, 222 (1961).
[10] C. E. Somers, A. Cole, and T. C. Hsu, *Exp. Cell Res.*, Suppl. 9, 220 (1963).
[11] M. Chorazy, A. Bendich, E. Borenfreund, and D. J. Hutchison, *J. Cell Biol.* **19**, 59 (1963).
[12] K. P. Cantor and J. E. Hearst, *Proc. Nat. Acad. Sci. U.S.* **55**, 642 (1966).
[13] J. A. Huberman and G. J. Attardi, *J. Mol. Biol.* **29**, 487 (1967).
[14] P. Franceschini and D. Giacomoni, *Atti Ass. Genet. Ital.* **12**, 248 (1967).
[15] J. Mendelsohn, D. E. Moore, and N. P. Salzman, *J. Mol. Biol.* **32**, 101 (1968).
[16] G. D. Burkholder and B. R. Mukherjee, *Exp. Cell Res.* **61**, 413 (1970).
[17] J. A. Huberman and G. J. Attardi, *J. Cell Biol.* **31**, 95 (1966).
[18] M. G. Hamilton and M. L. Peterman, *J. Biol. Chem.* **234**, 1441 (1959).

TABLE I
Flow Chart of Procedures for Isolating Chromosomes, Mitotic Apparatus, and Nuclei from Exponential Fibroblast Culture of Cells Blocked with Colcemid (0.06 μg/ml; 3 hr), with or without Differential Trypsinization[a]

Metaphase cells	Metaphase cells	Interphase cells
1. Centrifuge 1000 rpm, 2 min	1. Centrifuge 1000 rpm, 2 min	1. Centrifuge 1000 rpm, 2 min
2. Suspend in media to inactivate trypsin	2. Suspend in media to inactivate trypsin	2. Suspend in media to inactivate trypsin
3. Cool 20 min or longer at 4° to dissolve mitotic apparatus	3. Incubate 20 min at 37° to allow formation of mitotic apparatus	3. Centrifuge 1000 rpm, 2 min
4. Centrifuge 1000 rpm, 2 min	4. Centrifuge 1000 rpm, 2 min	4. Wash in 25° isolation buffer, pH 6.5 0.5 M hexylene glycol 1.0 mM $CaCl_2$ 0.05 mM PIPES
5. Wash in 4° isolation buffer, pH 6.5 1.0 M hexylene glycol 0.5 mM $CaCl_2$ 0.1 mM PIPES	5. Wash in 25° isolation buffer, pH 6.5 1.0 M hexylene glycol 0.5 mM $CaCl_2$ 0.05 mM PIPES	5. Centrifuge 2000 rpm, 3 min
6. Centrifuge 2000 rpm, 3 min	6. Centrifuge 2000 rpm, 3 min	6. Suspend gently in nuclei buffer
7. Suspend gently in cold chromosome buffer and incubate in a water bath at 37°, 10 min	7. Suspend gently in 25° mitotic apparatus buffer	7. Syringe through 22-gauge needle to lyse cells
8. Syringe gently through 22-gauge needle to lyse cells	8. Swirl on Vortex mixer at high speed or syringe gently to lyse cells	8. Centrifuge 2000 rpm, 3 min
9. Centrifuge 3000 rpm, 5–10 min	9. Centrifuge 3000 rpm, 5 min	ISOLATED NUCLEI
ISOLATED CHROMOSOMES	ISOLATED MITOTIC APPARATUS	

[a] Procedures developed during the course of this investigation. Starting with an exponentially growing culture, either chromosomes or mitotic apparatus may be isolated from the metaphase population after blocking the cells with Colcemid to increase the yield of mitotic cells. Nuclei may be isolated either directly from the exponential culture or from the interphase cells remaining after differential trypsinization to remove the metaphase cells. If an exponential culture is chosen for nuclei isolation, the cells may be removed without the use of trypsin. The culture flask is purged with air to remove the carbon dioxide, which would influence the pH of the isolation buffer. This is followed by removing the media, washing the culture with nuclei isolation buffer twice, and allowing the cells to stand at room temperature in a minimal volume (5 ml) of buffer until they detach from the glass.

Botchan,[19] should be viewed as a major advantage over acidic extraction conditions. Stabilization with 1 mM $ZnCl_2$ has been used for isolation of chromosomes at a neutral pH.[20] In neutral solutions, $ZnCl_2$ forms amphoteric ions which seem to stabilize the chromosomes, but the ions also extract nuclear ribosomes and nonchromosomal proteins, and dissolve the nuclear membrane.[21] The method of Wray and Stubblefield[4] presented here utilizes hexylene glycol as a stabilizing agent at the nearly neutral pH of 6.5. The rationale of using hexylene glycol to stabilize metaphase chromosomes is attributed to Kane,[22] who stated, ". . . the addition of glycol to water reduces the dielectric constant and increases the electrostatic free energy of the protein molecules, thus reducing their solubility. The addition of glycol to water also influences the hydrophobic interactions between the nonpolar groups of the protein molecules, which play an important role in controlling the conformation of protein molecules in solution."

Chromosome isolation procedures at high pH[23,24] also may have unique advantages over the procedures at low and neutral pH in that the isolated chromosomes contain high molecular weight DNA. For a complete discussion of these techniques and their relative importance see Wray.[25]

Chromosome Isolation

The starting material for chromosome isolation is a 95–99% pure mitotic cell population. Cooling to 4° in fresh media inactivates trypsin, dilutes any remaining Colcemid, and causes many microtubules remaining after Colcemid treatment to spontaneously disassemble,[26] thereby helping to minimize aggregation of chromosomes and contamination. The cells are then centrifuged, the medium is removed with a Pasteur pipette and the cells washed rapidly in 4° chromosome isolation buffer. After centrifugation, the pellet is gently resuspended in about 50 volumes of the cold buffer and incubated in a 37° water bath for 10–15 minutes.

[19] J. E. Hearst and M. Botchan, *Annu. Rev. Biochem.* **39**, 151 (1970).
[20] J. J. Maio and C. L. Schildkraut, *J. Mol. Biol.* **24**, 29 (1967).
[21] J. H. Frenster, *Exp. Cell Res.*, Suppl. 9, 235 (1963).
[22] R. E. Kane, *J. Cell Biol.* **25**, 136 (1965).
[23] P. M. Corry and A. Cole, *Radiat. Res.* **36**, 528 (1968).
[24] W. Wray, E. Stubblefield, and R. Humphrey, *Nature (London) New Biol.* **238**, 237 (1972).
[25] W. Wray, *in* "Methods in Cell Biology" (D. M. Prescott, ed.), Vol. VI, p. 307. Academic Press, New York, 1973.
[26] S. Inoue, *in* "Primitive Motile Systems in Cell Biology" (R. D. Allen and N. Kamiya, eds.), p. 549. Academic Press, New York, 1964.

After incubation, the cells may be broken by homogenization. However, for small volumes of cells, gentle syringing through a 22-gauge needle breaks the cell membrane and frees the chromosomes into the buffer. The number of times one passes the solution through the needle and the amount of force necessary vary with cell line, cell incubation time, and thoroughness of the washing. All steps of these procedures should be monitored with phase contrast microscopy. It is better to syringe too gently than too vigorously, since the latter damages chromosome morphology and is irrevocable. Do not allow the temperature to drop until after the desired breakage is accomplished. After the cells are broken and the chromosomes liberated, further operations may be done in the cold. The most convenient equipment for these isolation procedures with a cell pellet of 0.1 ml volume are 15-ml Pyrex or plastic conical centrifugation tubes, 5-ml plastic B-D syringes with 1.5-inch 22-gauge needles and a clinical centrifuge. For large volumes of cells, comparable results may be obtained with a Dounce glass homogenizer from Kontes. To facilitate removal of nuclei and unbroken cells, Nucleopore membrane filters (8 and 5 μm pore sizes) from Wailabs, Inc., San Rafael, California, may be used. The entire procedure is conveniently done within 1 hour.

Chromosome Fractionation. The isolation of metaphase chromosomes leads to the next logical step, the fractionation of these chromosomes into groups according to their size. On a 24-ml linear 10% to 40% w/v buffered sucrose gradient with a bottom cushion of 85% sucrose buffer, is layered 3 ml of an isolated chromosome suspension. The 85% sucrose buffer cushion prevents a pellet of whole cells and nuclei which would occlude the puncture needle. Coating with Siliclad (Clay-Adams) minimizes the adherence of chromosomes to the nitrocellulose centrifuge tubes and all glassware. The gradient is centrifuged 90 minutes at 1000 rpm, in a HB-4 swinging-bucket rotor in a Sorvall RC2-B centrifuge, after which the tube may be punctured and absorbance monitored by a recording spectrophotometer. The contaminating whole cells and nuclei sediment to the 85% sucrose-buffer cushion, and mitotic apparatus and aggregates of chromosomes also sediment far down the gradient. The broad absorbance peak that occurs from the middle of the gradient to near the top is a rough distribution of chromosomes according to size. For Chinese hamster cells the large metacentric chromosomes (A group) which include the 1's and 2's are sedimented the greatest distance into the gradient. The medium-sized metacentric chromosomes (B group) are the X, Y, 4's, and 5's. The majority of these are located above the A group and below the acrocentric chromosomes (6's, 7's, and 8's) which comprise the C group. The D group chromosomes which are the small metacentric chromosomes (9's, 10's, and 11's) are sedimented the least distance into the gradient and run only slightly ahead of a large peak of ultraviolet-absorbing material comprised of light cellular debris. If the gradient is monitored closely by phase contrast microscopy, four fractions corresponding to the four groups may be obtained which include the majority of the chromosomes in the appropriate group, but are contaminated with those from adjoining groups. Analysis of the centrifugation pattern shows that chromosomes that are separated by more than one group are not contaminants of each other (i.e., group B and group D). Further purification is accomplished by repeating the velocity sedimentation centrifugation on each of the obtained fractions.

Chromosome Composition. Hearst and Botchan[19] reviewed the gross chemical composition of chromosomes isolated by a variety of reported methods. RNA is reported on an almost equal weight basis with the DNA of the metaphase chromosome. It has been shown that this RNA present in isolated chromosome preparations is primarily the 18 S and 28 S rRNA,[17,20] although it has not been demonstrated that these RNA's are on the chromosome as ribosomes. Electron microscopy of these chromosome isolation "products" from other methods has not been reported. Therefore, visual comparisons of methods may be evaluated for contamination only at the grossly inadequate light microscope level.

TABLE II
RELATIVE CONTENT OF ISOLATED CHROMOSOMES, MITOTIC APPARATUS, AND NUCLEI[a,b]

Preparation	DNA	RNA	Protein	Investigator
Chromosome fraction				
ABCD	100	0	209	Wray and Stubblefield[a]
ABCD	100	17	220	Wray and Stubblefield[a]
AB	100	18	240	Mendelsohn, Moore, and Salzman[c]
C	100	79	540	
D	100	140	850	
Mitotic apparatus	100	35	1502	Wray and Stubblefield[a]
Nuclei				
Chinese hamster	100	31	272	Wray and Stubblefield[a]
Rat liver	100	27	400	Busch[d]

[a] From W. Wray and E. Stubblefield, *Exp. Cell Res.* **59**, 469 (1970).
[b] The amount of DNA in each preparation was taken to be 100, and the relative amounts of RNA and protein were calculated. DNA values are not related from one preparation to another. Analyses were measured colorimetrically as referenced in the text.
[c] J. Mendolsohn, D. E. Moore, and N. P. Salzman, *J. Mol. Biol.* **32**, 101 (1968).
[d] H. Busch, this series, Vol. 12A, p. 421.

Mendelsohn et al.[15] showed that contaminant RNA and protein remained near the top of a sucrose velocity gradient with the small chromosomes. The fractions enriched for larger chromosomes sedimented farther down the gradient and contained much less RNA and protein. The reduced amount of RNA in this A + B (large chromosome) fraction supports their previous finding[27] that RNA is not an intrinsic component of metaphase chromosomes.

Table II shows the relative amount of DNA, RNA, and protein present in total heterogeneous chromosome fractions isolated by this method compared to partially purified fractions prepared by Mendelsohn et al.[15] It may be seen that the results obtained were very similar to those of the A + B fraction of Mendelsohn. In one instance RNA could not be detected in the preparation while both the DNA and protein compared favorably. Apparently there is little, if any, adventitiously adsorbed RNA or protein in these chromosomes which can be shown biochemically. Morphologically they are excellent.[4,8]

Mitotic Apparatus

Mitotic apparatus from sea urchin eggs were isolated by Kane[22] using a 1 M solution of hexylene glycol buffered in a pH range of 6.2 to 6.4. The cells cytolyze hypotonically, and the intact mitotic apparatus are released by very mild agitation. The fact that the cells tolerate 0.1 M hexylene glycol during division and development suggested that there was no direct chemical effect on the mitotic apparatus.

Mammalian mitotic apparatus have been isolated by Sisken et al.[28] using a technique based on the Kane procedure. The modification consisted of omitting the phosphate buffer, and adding 0.1 mM $CaCl_2$ which tended to enhance the stability of the mitotic apparatus. The solution was left unbuffered because the addition of phosphate or Tris prevented lysis of the cells.

Both Kane and Sisken noted that the chromosomes were stable in the isolated mitotic apparatus, and Sisken observed that nuclei could also be isolated by the same procedure used for isolation of mitotic apparatus. Wray and Stubblefield[4] confirmed and greatly extended these observations. They noted that, under conditions where spindle microtubules are not stable, free chromosomes can be isolated from mitotic cells. Nuclei may also be obtained from the interphase cells.

[27] N. P. Salzman, D. E. Moore, and J. Mendelsohn, *Proc. Nat. Acad. Sci. U.S.* **56**, 1449 (1966).
[28] J. E. Sisken, E. Wilkes, G. M. Donnelly, and T. Kakefuda, *J. Cell Biol.* **32**, 212 (1967).

Mitotic Apparatus Isolation

The isolation of mitotic apparatus differs slightly from the isolation of chromosomes in that the Colcemid-treated mitotic cells are not refrigerated to dissolve the microtubules. Instead, after trypsinization they are incubated in fresh medium at 37° for 20 minutes to allow complete formation of the mitotic spindle. Monitoring by phase contrast microscopy is essential.

The intact mitotic apparatus preparation may be concentrated and purified by differential centrifugation or sedimentation velocity gradients similar to those for chromosomes. Most of the cell contents are extracted, leaving chromosomes, microtubules, centrioles, and in many cases adhering membranes. The membranes are more difficult to remove if the PIPES concentration is higher than 50 μM. The microtubules in this cell line are unstable in $CaCl_2$ concentrations below 0.5 mM. Table II shows the relative amount of DNA, RNA, and protein present in a mitotic apparatus preparation.

Nuclei

The common nuclei isolation methods use either aqueous or nonaqueous homogenization media, the choice of which depends upon the nature of the end products desired. The nonaqueous organic solvent techniques unavoidably damage the nuclear membrane, whereas the aqueous methods allow a leakage of soluble nuclear components. A theoretically perfect nuclei isolation technique should yield a product which contains no cytoplasmic contamination and has lost no nuclear components during the isolation. As yet, no isolation method described meets all the theoretical considerations. Excellent reviews by Wang,[29] Busch,[30] Muramatsu,[31] and Roodyn[32] have been published which discuss and critique in detail many isolation procedures and their relative advantages. Muramatsu[31] notes that nuclear isolation procedures may be specific for the tissue or cell type. The method presented here has been successful for all tissues and cell types in which it has been tried and, owing to the buffer's combined aqueous and nonaqueous properties, may well minimize the bad aspects and maximize the advantages of both systems.

[29] T. Y. Wang, this series, Vol. 12A, p. 417.
[30] H. Busch, this series, Vol. 12A, p. 421.
[31] M. Muramatsu, in "Methods in Cell Physiology" (D. M. Prescott, ed.), Vol. IV, p. 195. Academic Press, New York, 1970.
[32] D. B. Roodyn, in "Subcellular Components: Preparation and Fractionation" (G. D. Birnie, ed.), 2nd ed., p. 15. Butterworth, London, 1972.

Nuclei Isolation

Rupture of contaminating interphase cells in the isolation of chromosomes or mitotic apparatus releases the nuclei as contaminants of these preparations. As is the case with mitotic apparatus the interphase cells lyse more readily if the PIPES concentration is 50 μM or lower. However, raising the $CaCl_2$ concentration to 1 mM seems to stabilize the nuclei against breakage during shearing. The hexylene glycol is reduced to 0.5 M in most preparations; this also seems to reduce cytoplasmic contamination.

Nuclei can be isolated directly from tissue culture cell monolayers. It is necessary to purge the CO_2 from the culture atmosphere in order to prevent a drop in pH which prevents cell lysis. Fibroblasts are washed with the nuclei isolation buffer. A second wash is left on the cells for about 10 minutes at room temperature, after which the cells spontaneously detach from the glass. The suspension is then sheared by syringing or homogenizing as in the case of chromosomes. Preparations should be routinely checked by phase contrast microscopy. The analogous procedure also works for cell suspensions after trypsinization so that it is possible to isolate chromosomes or mitotic apparatus from the metaphase cells, and nuclei from the interphase cells of the same culture. To test the procedure on freshly excised tissue routinely used in many biochemical laboratories, rat liver was chosen, and the liver nuclei were found to be cleanly isolated after homogenization in the nuclei buffer.

When isolating nuclei from freshly excised tissue, the tissue may be minced with scissors and should be kept cold on ice until the worker is ready to break the cells. It is very important that the only solution to contact the tissue should be the isolation buffer. If the tissue is washed with any other solutions, the efficiency of isolation and the quality of the product are greatly diminished.

In isolated nuclei, most of the chromatin appears in the form of 250 Å fibers very similar to those comprising isolated chromosomes. Isolated Don-C nuclei appear to have a composition comparing favorably with nuclei from other sources and techniques[30] shown in Table II. Although the protein values are lower than those reported, enzymatic studies reveal that the isolated nuclei contain DNA polymerase, RNA polymerase, and adenylate kinase. Table III shows these activities along with numerous cytoplasmic enzymes which are found associated only with the cytoplasmic fraction. It is obvious, also, that exposure to the buffer solution used does not destroy the activity of most enzymes. Determination of enzyme kinetic rates after isolation has not been attempted.

Studies of DNase activity in the cytoplasmic lysate indicate that this

TABLE III
ENZYMES ASSOCIATED WITH CELL FRACTIONS

Enzyme	Nuclei	Cytoplasm	Chromosome
DNA polymerase	+	+	*
RNA polymerase	+		*
Adenylate kinase	+	+	
Catalase	−	+	
Alkaline phosphatase	−	+	
Acid phosphatase	−	+	
Lactic dehydrogenase	−	+	
Alcohol dehydrogenase	−	+	
Phosphoglucomutase	−	+	
Creatine kinase	−	+	
6-Phosphogluconate dehydrogenase	−	+	
Glucose-6-P dehydrogenase	−	+	
Hexokinase	−	+	
Isocitric dehydrogenase	−	+	
α-Glycerolphosphate dehydrogenase	−	+	

^a From W. Wray and E. Stubblefield, *Exp. Cell Res.* **59,** 469 (1970).

^b After isolation of nuclei, starch gel electrophoresis was performed on samples of nuclei and cytoplasm. Enzyme activity was then determined by appropriate staining reactions. [C. E. Shaw and A. L. Koen, *in* "Chromatography and Electrophoresis" (I. Smith, ed.), 2nd ed., p. 325, Wiley (Interscience), New York, 1968.] The presence (+) or absence (−) of enzyme is noted in the table. Both DNA polymerase and RNA polymerase activities (*) were found in isolated chromosomes without added template material in *in vitro* systems.

enzyme is not released when the cell is broken open (Table IV). The DNA in chromosomes and nuclei prelabeled with [2-^{14}C]thymidine appears to remain stable when incubated at 37° in the chromosome isolation buffer. Added Mg^{2+} (1.6 mM) enhances stability, whereas in the presence of added DNase (670 µg/ml) it greatly accelerates degradation. The nuclei are more resistant to added DNase, and the kinetics suggest that about two-thirds of the nuclei may be totally resistant, perhaps because of an intact nuclear membrane. At any rate, it seems clear that the isolated preparations do not contain appreciable DNase activity.

Rigorous studies have not been attempted to determine the minimum concentrations of the buffer components. It has been determined, however, that the presence of hexylene glycol and calcium are essential to the stability of both chromosomes and mitotic apparatus while nuclei demonstrate a requirement for calcium.

The relative nontoxicity of the buffer has been demonstrated by growing Chinese hamster fibroblasts in media containing 10% chromosome

TABLE IV
DNase Digestion of Chromosomes and Nuclei[a,b]

Substrate	Additions	% DNA hydrolyzed (minutes)		
		0	15	30
Chromosomes		0.2	6.1	6.2
Chromosomes	$+Mg^{2+}$		0.4	1.1
Chromosomes	$+$DNase		43.1	65.5
Chromosomes	$+Mg^{2+} +$ DNase		85.3	94.6
Nuclei		0.3	1.1	1.5
Nuclei	$+Mg^{2+}$		0.5	0.6
Nuclei	$+$DNase		27.5	26.0
Nuclei	$+Mg^{2+} +$ DNase		37.8	36.5

[a] From W. Wray and E. Stubblefield, *Exp. Cell Res.* **59**, 469 (1970).
[b] The reaction system contained in 0.3 ml either ^{14}C-labeled chromosomes or ^{14}C-labeled nuclei in buffer. Where indicated $MgCl_2$ was added to a concentration of 1.6 mM and DNase was added to a concentration of 670 μg/ml. The hydrolysis of DNA was measured as increase of TCA nonsedimentable isotope into the supernatant. After centrifuging for 10 minutes at 10,000 rpm at 4°, 0.1-ml samples were counted in a NCS-toluene scintillating fluid in a Beckman Liquid Scintillation Counter.

isolation buffer.[4] Since this dilution would contain 0.1 M hexylene glycol, analogous results to those of Kane[22] may be inferred.

Discussion

The isolation system of Wray and Stubblefield[4] is derived from that used by Sisken *et al.* for the isolation of mitotic apparatus. The selection of PIPES for the buffer, however, was due to the statements of Good *et al.*[33] in which it was emphasized that PIPES was a nonmetal-binding buffer with a pK_a of 6.8 at 20°. Subsequently Ris[34] demonstrated that the morphology of chromatin in electron microscope preparations depends upon the buffer chosen. His pictures suggest buffer artifacts when chelating buffers like phosphate, Veronal acetate, cacodylate, and *s*-collidine are used; but when nonmetal-binding buffers are used the normal configurations of nucleohistone fibers are exhibited.

It is logical to assume that the stability of isolated components should depend to some extent upon the approximation of the intracellular envi-

[33] N. E. Good, G. D. Winget, W. Winter, T. N. Connolly, S. Izawa, and R. M. M. Singh, *Biochemistry* **5**, 467 (1966).
[34] H. Ris, *J. Cell Biol.* **39**, 158a (1968).

ronment *in vitro*. This particular system does not mimic the intracellular conditions. However, using the rigorous guides of localization of enzyme activity, enzyme functionality, and morphological stability of components, it appears to act primarily as a support media which preserves biologic activity and prevents subcellular degradation of most of the cellular organelles.

Let it again be emphasized that the chromosomes, mitotic apparatus, and nuclei are all stable in the chromosome isolation buffer, and the only reason for the buffer modification in the isolation procedures for mitotic apparatus and nuclei in this cell line is the ease of isolation and decontamination of the component chosen as the end point.

Chromosomes isolated are stable morphologically for periods of several months at 4°. They may also be quick frozen in 10% glycerol, stored, and thawed with no apparent harm. The appearance of chromosomes when isolated by gentle shearing forces compares favorably to that demonstrated by many other laboratories. If the shearing force is too vigorous, however, stretched chromosomes may result.[35] The reasons for preincubating the whole cells at 37° are that the cells break easier and that this incubation time allows the chromosomes to equilibrate with the buffer and not be shocked or stretched by the difference in environment upon breakage. If the cells are broken in a buffer lacking hexylene glycol, the chromosomes are unstable and gradually disintegrate over several hours' time. At Ca^{2+} concentrations below 0.5 mM, both chromosomes and nuclei rapidly dissolve.

The composition of the heterogeneous chromosome population isolated compares favorably with the most highly purified fraction of Mendelsohn *et al.*[15] and appears to fulfill the prediction by Salzman and Mendelsohn[36] that preparations of chromosomes will be ultimately isolated which contain no contaminant RNA.

When isolating mitotic apparatus from Colcemid-blocked cells one must allow the premetaphase cells, which consist of condensed chromosomes in a spherical arrangement around the unseparated parent and daughter centrioles,[37] to reverse from the mitotic inhibitor. Incubation at 37° for 20 minutes allows a normal mitotic spindle to form before the isolation is attempted. By carefully controlling the timing of the incubation one may select for the isolation of metaphase or anaphase configurations of the mitotic apparatus.

The isolation of mitotic apparatus has as its biggest problem the limi-

---

[35] E. Stubblefield and W. Wray, *Chromosoma* **32**, 262 (1971).

[36] N. P. Salzman, and J. Mendelsohn, *in* "Methods in Cell Physiology" (D. M. Prescott, ed.), Vol. III, p. 277. Academic Press, New York, 1968.

[37] B. R. Brinkley, E. Stubblefield, and T. C. Hsu, *J. Ultrastruct. Res.* **19**, 1 (1967).

tation imposed by the physical nature of its composition. This large, unwieldy, and rather fragile structure is more subject to physical disruption by shear forces than either chromosomes or nuclei. Consequently the isolation must be more gentle; lowering the amount of PIPES tends to facilitate removal of the membranes which tends to contaminate the isolated mitotic apparatus.

In this system the microtubules are well defined and appear to be free of contaminating cellular material. The centrioles are demonstrated to be stable and to be physically attached to the ends of the spindle.

The nuclei when isolated were found to be free from cytoplasmic tags. There was one predominant chromatin fiber type of 250 Å material, heterochromatic and euchromatic areas were observed, and RNA content was found to be similar to the amount found in other cells. Differences in the amount of protein found in these nuclei could be attributed to contamination in other preparations, cell type deviations, or loss of intranuclear protein from these nuclei. The latter is questioned by enzymatic data which indicate that cytoplasmic enzymes are excluded from the nuclei as expected and that both DNA and RNA polymerase are found in the nuclear fraction. It would be expected that damaged nuclei would exchange proteins with the cytoplasm.

The advantages these procedures have over classical homogenization methods are: (1) the ready adaptation to as few as a million cells; (2) applicability to all tissue culture cell lines, organisms, and tissues; (3) the speed, simplicity, and minimum of equipment needed for the isolation techniques; and (4) the essentially identical isolation conditions for the functionally related organelles.

Acknowledgments

This work was supported by grants from the National Science Foundation, The National Cancer Institute, The American Cancer Society, and The National Institute of Health.

Section II

The Cell Nucleus and Chromatin Proteins

[7] Fractionation of Chromatin

By DeLill S. Nasser and Brian J. McCarthy

The first reported attempts at chromatin fractionation were made by Frenster et al.[1] Thymocyte chromatin, fragmented by sonication, was separated by differential centrifugation into two fractions corresponding by biochemical and ultrastructural criteria to eu- and heterochromatin. Modifications of this differential centrifugation methodology have been reported by Yasmineh and Yunis,[2] Chalkley and Jensen,[3] and Duerksen and McCarthy[4] with varying degrees of separation and evidence for physical, chemical, and functional differences. Over the past few years several other approaches have been utilized although the basis for fractionation is often obscure. For example, Wilt and Ekenberg[5] reported that a small fraction of sea urchin embryo chromatin containing nascent RNA chains banded at a lighter density in Cs_2SO_4 equilibrium gradients than the bulk of the chromatin. Although they first interpreted this in terms of fractionation of functionally different chromatin, this possibility was later discounted in view of the failure of DNA/RNA hybridization experiments to reveal differences in the DNA sequences present in the two fractions.[6] Other methods, which have been proposed but will not be discussed here in detail, include nonrandom attack of chromatin by deoxyribonuclease resulting in precipitation of resistant portions[7] and differential precipitation by Mg^{2+} or Hg^+ ions.[8]

In the ensuing pages we summarize four simple convenient procedures where one or more obvious criteria for separation of functionally different fractions have been applied. These include association of nascent RNA chains, detection of transcribed DNA sequences by molecular hybridization, *in vitro* template activity, and thermal denaturation behavior.

Methods

Methods for isolation and purification of chromatin are described in Volume 12B of this series.

[1] J. Frenster, V. G. Allfrey, and A. E. Mirsky, *Proc. Nat. Acad. Sci. U.S.* **50**, 1026 (1963).
[2] W. G. Yasmineh and J. J. Yunis, *Biochem. Biophys. Res. Commun.* **35**, 779 (1969).
[3] G. R. Chalkley and R. Jensen, *Biochemistry* **7**, 4380 (1968).
[4] J. D. Duerksen and B. J. McCarthy, *Biochemistry* **10**, 471 (1971).
[5] F. H. Wilt and E. Ekenberg, *Biochem. Biophys. Res. Commun.* **44**, 831 (1971).
[6] F. H. Wilt, in press (1973).
[7] R. J. Billing and J. Bonner, *Biochim. Biophys. Acta* **28**, 453 (1972).
[8] H. W. Dickermann, B. C. Smith, and M. A. Isaacs, *Fed. Proc., Fed. Amer. Soc. Exp. Biol.* **32**, 1793 (1973).

Differential Centrifugation

The method described is that of Nishiura and McCarthy.[9] This sucrose gradient method is modified from earlier methodology[3] by employing a very steep sucrose gradient which facilitates clean separation of the two fractions. The method was optimized for *Drosophila melanogaster* cultured cell chromatin in which approximately one-third of the DNA is transcribed, but the general approach appears to be valid for mammalian cells.

Chromatin is isolated and purified by standard methods, sheared in 10 mM Tris buffer, pH 8.0 and layered on a 0.17 M to 1.7 M linear gradient of sucrose dissolved in the same buffer, in a Beckmann SW 40 tube (13 ml). The degree of separation is influenced by the shearing method and the centrifugation time. Two procedures have been most successful for a variety of chromatin preparations. If the chromatin is sheared in a Virtis homogenizer at 40 V for 5 minutes, optimal separation is obtained by centrifuging at 113,000 g for 6 hours at 4°. Alternatively chromatin sheared in a French pressure cell at 3000 psi is centrifuged for 14 hours at 113,000 g at 4°. In the case of *Drosophila* chromatin, both procedures result in two peaks; roughly one-third sedimenting slowly, corresponding by the above criteria to active chromatin, and two-thirds sedimenting rapidly.

Thermal Elution from Hydroxyapatite (HA)

The method described is that of McConaughy and McCarthy.[10] The procedure is based upon that described for fractionation of DNA according to G + C content.[11,12] However, in the case of chromatin the fractionation results from differential stabilization of sections of the DNA against thermal denaturation by associated chromosomal proteins. Since active chromatin fractions display lower thermal denaturation profiles, the corresponding DNA sequences are eluted first from the column.[10]

Either Bio-Rad or Clarkson HA may be used, although the latter product has a higher capacity and leads to a higher T_m. In both cases the HA is prepared for use by boiling for 5 minutes in 10 mM sodium or potassium phosphate pH 6.8. The HA (about 1 g) is packed into water-jacketed columns and equilibrated with 0.12 M phosphate, pH 6.8, at 60°. Potassium phosphate is recommended over sodium phosphate

[9] B. J. McCarthy, J. T. Nishiura, D. Doenecke, D. S. Nasser, and C. B. Johnson, *Cold Spring Harbor Symp. Quant. Biol.* **38**, 763 (1973).
[10] B. L. McConaughy and B. J. McCarthy, *Biochemistry* **11**, 998 (1972).
[11] G. Bernardi, *Nature (London)* **206**, 779 (1965).
[12] Y. Miyazawa and C. A. Thomas Jr., *J. Mol. Biol.* **11**, 223 (1965).

since the elution of DNA is shifted to lower temperatures. Chromatin sheared at 12,000 psi in a French pressure cell is then absorbed to the column. The column is heated in 5° increments to 100° by means of an attached circulating water bath. At each temperature, 5 ml buffer in 2 equal successive increments is forced through the column under air pressure to elute single-stranded DNA. Elution of any DNA remaining on the column at 100° is accomplished by washing with 8 M urea, 0.24 M sodium phosphate, pH 6.8, 10 mM EDTA.[13] In attempting to reduce the exposure to high temperatures, we have recently found that thermal elution occurs at considerably lower temperatures if 10 mM cesium phosphate is used throughout as the eluting buffer.

This method is not useful for obtaining chromosomal proteins for obvious reasons. However, it may be used to fractionate the DNA moiety of chromatin according to the nature of its association with chromosomal proteins. Transcribed DNA sequences are specifically eluted at lower temperatures.

Chromatography on ECTHAM-Cellulose

The procedure is that of Reeck et al.[14] using ECTHAM cellulose.[15] Chromatin in 1 mM Tris, pH 8.0, is sheared in Waring Blendor for 90 seconds at 90 V. After centrifugation at 10,000 g for 30 minutes, the chromatin is sonicated in a Branson Sonifier Model 185 for 2 minutes at 2° at a concentration of about 0.5 mg DNA per milliliter. The microtip used is extended 4.5 cm into the fluted sample cell. ECTHAM cellulose is prepared for use by extensive washing in 0.5 N HCl, 0.5 N NaOH, and finally 10 mM Tris pH 7.2. Ten milliliters of sonicated chromatin is added with pressure at 3 ml per hour to a 1.2 × 15 cm column containing 2 g of ECTHAM cellulose. Elution is achieved with 10 mM Tris buffer, pH 7.2, containing 10 mM NaCl flowing at 2 ml per hour and collecting 1-ml fractions.

The published elution patterns reveal no structure suggestive of a clear separation. However, the leading edge of the peak displayed a T_m of about 85° in contrast to 77° for the trailing edge. Later fractions are enriched in chains of nascent RNA and contain some different nonhistone proteins.[14,16,17]

[13] R. J. Britten, M. Pavich, and J. Smith, *Carnegie Inst. Wash. Yearb.* **68**, 400 (1970).
[14] G. R. Reeck, R. T. Simpson, and H. A. Sober, *Proc. Nat. Acad. Sci. U.S.* **69**, 2317 (1972).
[15] E. A. Peterson and E. L. Kuff, *Biochemistry* **7**, 2916 (1969).
[16] I. Palacow and R. T. Simpson, *Fed. Proc., Fed. Amer. Soc. Exp. Biol.* **32**, 1798 (1973).
[17] G. R. Reeck, *Fed. Proc., Fed. Amer. Soc. Exp. Biol.* **32**, 1799 (1973).

Fractionation by Agarose Gel Filtration

The method is that of Janowski, Nasser, and McCarthy.[18] The basis of the fractionation appears to depend on the differential tendency of active and inactive chromatin to aggregate and differences in hydrodynamic behavior.

Bio-Rad agarose beads A-50 or A-150 are packed under low pressure (<50 cm) in a column 2.6 × 40 cm fitted with upward flow plungers. The column is equilibrated with either 10 mM Tris pH 8.0, 1 mM mercaptoethanol, or 10 mM Tris pH 8.0 containing 0.15 M KCl, 0.1 M MgCl$_2$ and 1 mM mercaptoethanol. Fractionation may be achieved in either buffer representing the two extreme conditions of ionic environment in which chromatin is soluble. Under low salt conditions the active chromatin appears at the end of the elution pattern while in high salt it appears at the front.

The chromatin, 0.7 mg to 1 mg DNA per milliliter in 10 mM Tris is sheared as for the sucrose gradient method either in the French pressure cell at 3000 psi or in a Virtis homogenizer for 5 minutes at 40 V. The sample is applied in 1–3 ml of the chosen eluting buffer, which is then passed through the column at 15 ml per hour. Collection and analysis of 3-ml fractions reveals two peaks. Even though histone F1 is dissociated under the high salt conditions, this eluting buffer is recommended since the active peak, representing 5–20% of the total depending on the tissue, is clearly resolved from the remainder. Criteria for activity which have been applied in this case include *in vivo* and *in vitro* template activity and association of hormone-receptor complexes.[18]

Prospects

The methods available at the present time are crude and empirical, and the basis for success is often not understood. Other methods such as electrophoresis and equilibrium banding in density gradients have not yet proved to be successful. Another weakness of present approaches is the practice of mechanical shearing to obtain fragments of the size that can be handled with available techniques. Since these fragments containing DNA of 1 to 3 × 10^6 daltons are small compared to transcription units in many eukaryotes, a major effort should be directed to developing methods for fractionating larger chromatin segments as well as gentler enzymatic procedures to obtain them.

[18] M. Janowski, D. S. Nasser, and B. J. McCarthy, *in* "Karolinska Symposia on Research Methods in Reproductive Endocrinology" (E. Diczqalusy, ed.), 5th Symposium, p. 112. Karolinska Institutet, Stockholm, 1972.

More recent evidence suggests that although chromatin can be fractionated as described the basis of fractionation may be the nonrandom distribution of proteins along the chromatin fiber rather than the differences in the overall structure of active and inactive regions of chromatin.

[8] Isolation of Template Active and Inactive Regions of Chromatin

By JAMES BONNER, JOEL GOTTESFELD, WILLIAM GARRARD, RONALD BILLING, and LYNDA UPHOUSE

In the differentiated eukaryotic cell, only a portion of the nuclear DNA is transcribed into RNA while the majority of the genetic material is repressed. Part of this transcriptional heterogeneity appears to be determined by the protein complement of chromosomes. To better understand the role of chromosomal proteins in chromatin function and structure, we and other investigators have developed techniques to fractionate chromatin into template active and inactive regions. Such an approach encounters two main technical problems. First, chromatin must be sheared to a size small enough to allow the subsequent separation of active and inactive regions. Second, a means of selectivity fractionating the material is required.

To minimize cross-contamination of active and inactive regions, chromatin should be sheared to less than the size of an average unit of transcription. Most investigators have used mechanical methods (including sonication[1] and French pressure cell passage[2,3]) to shear chromatin. Marushige and Bonner[4] have introduced the use of DNase II for this purpose. Nucleolytic cleavage of chromatin DNA is gentle and does not lead to detectable levels of protein rearrangement during shearing or fractionation. We have adopted this technique and describe appropriate conditions on p. 98.

Sheared chromatin has been separated into active and inactive regions

---

[1] J. Frenster, V. G. Allfrey, and A. E. Mirsky, *Proc. Nat. Acad. Sci. U.S.* **50**, 1026 (1963).

[2] B. L. McConaughy and B. J. McCarthy, *Biochemistry* **11**, 998 (1972).

[3] M. Janowski, D. S. Nasser, and B. J. McCarthy, *in* "Karolinska Symposia on Research Methods in Reproductive Endocrinology" (E. Diczfalusy, ed.), 5th Symposium, p. 112. Karolinska Institutet, Stockholm, 1972.

[4] K. Marushige and J. Bonner, *Proc. Nat. Acad. Sci. U.S.* **68**, 2941 (1971).

by differential centrifugation,[1,5] and by chromatography on hydroxyapatite,[2] agarose,[3] and anion-exchange resin.[6] Marushige and Bonner[4] have developed a simple and rapid fractionation technique which involves selective precipitation of the inactive region with standard saline-citrate. This approach is based on the fact that chromatin is highly aggregated at physiological ionic strength. Under such conditions, the limited portions of chromatin available to RNA polymerase would be predicted to be less aggregated and perhaps soluble in sheared chromatin preparations. Our method, based on the same principle, uses $MgCl_2$ to precipitate the inactive region. The minor portion of chromatin soluble in 2.0 mM $MgCl_2$ is the template active region.

Procedure for Fractionation

Sucrose-purified chromatin[7] is homogenized at 4° in 10 volumes of 10 mM Tris·Cl (pH 8) using a glass-Teflon homogenizer (10 strokes by hand followed by 2 minutes of gentle motor homogenization). Unless otherwise stated, all subsequent operations are performed at 4°. The resulting chromatin solution is dialyzed overnight against 200 volumes of 25 mM sodium acetate (pH 6.6) and pelleted by centrifugation at 25,000 g for 20 minutes. The resulting material is homogenized as above in 25 mM sodium acetate (pH 6.6) to yield a chromatin solution having an $A_{260}^{1\ cm}$ of 10 (measured in 1 N NaOH). The solution is brought to 24° and DNase II (Worthington, HDAC) is added to a final concentration of 100 units/ml (10 units of enzyme per A_{260} unit of chromatin). After incubation at 24°, the pH of the solution is adjusted to 7.5 by the addition of 0.1 M Tris·Cl (pH 11) and the mixture is cooled to 4°. Unsheared chromatin is removed by centrifugation at 25,000 g for 20 minutes. The amount of chromatin DNA remaining in the supernatant is naturally dependent upon the duration of incubation with DNase. After 5 minutes of enzyme treatment, 15% of the chromatin DNA is found in the supernatant, while after prolonged incubation (15 minutes or more), 70 to 85% is found.

The sheared chromatin is now fractionated into template active and inactive portions by $MgCl_2$ precipitation. One ninety-ninth volume of 0.20 M $MgCl_2$ is added dropwise with rapid stirring. After 30 minutes of further stirring, the turbid suspension is centrifuged at 25,000 g for

[5] G. R. Chalkley and R. Jensen, *Biochemistry* **7**, 4380 (1968).
[6] G. R. Reeck, R. T. Simpson, and H. A. Sober, *Proc. Nat. Acad. Sci. U.S.* **69**, 2317 (1972).
[7] J. Bonner, G. R. Chalkley, M. Dahmus, D. Fambrough, G. Fujimura, R.-C. Huang, J. Huberman, R. Jensen, K. Marushige, H. Ohlenbusch, B. Olivera, and J. Widholm, this series, Vol. 12B, p. 3.

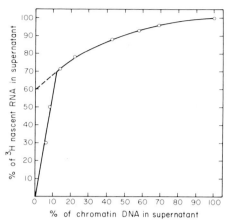

FIG. 1. Localization of ^{3}H-labeled nascent RNA. Sheared chromatin was prepared from ascites cells, pulse labeled for 5 minutes with [^{3}H]uridine as described by R. J. Billing and J. Bonner [*Biochim. Biophys. Acta* **28**, 453 (1972)]. Aliquots of chromatin were precipitated with various concentrations of MgCl$_2$. Soluble DNA and radioactivity were determined. Circles along curve represent successively decreasing MgCl$_2$ concentrations used for precipitation of template inactive portions.

20 minutes. The resulting supernatant contains the template active chromatin fraction, while the pellet contains the template inactive material. The length of the DNA in the template active fraction is dependent upon the conditions of nuclease treatment. A 5 minute exposure produces fragments of 700–850 base pairs (500 nucleotides single stranded).[8] Longer times of incubation produce much shorter fragments.[9]

Evidence for Fractionation

Evidence for the separation of the template active and inactive regions by MgCl$_2$ precipitation is based on the cofractionation of nascent RNA with the active region.[9,10] Chromatin was prepared from Novikoff ascites cells[11] which were pulse labeled for 5 minutes with [^{3}H]uridine. Under such conditions, the majority of the label would be predicted to be associated with the template active region as growing RNA chains. The chromatin was then sheared with DNase II, and aliquots were precipitated with decreasing concentrations of MgCl$_2$. Figure 1 shows the

---

[8] The size of the sheared DNA was estimated by neutral and alkaline sedimentation [W. F. Studier, *J. Mol. Biol.* **11**, 373 (1964)].

[9] R. J. Billing and J. Bonner, *Biochim. Biophys. Acta* **28**, 453 (1972).

[10] J. Bonner, W. T. Garrard, J. Gottesfeld, D. S. Holmes, J. S. Sevall, and M. Wilkes, *Cold Spring Harbor Symp. Quant. Biol.* **38**, 303 (1973).

[11] M. E. Dahmus and D. J. McConnell, *Biochemistry* **8**, 1524 (1969).

CHEMICAL COMPOSITION OF THE TEMPLATE ACTIVE AND
INACTIVE REGIONS OF CHROMATIN[a]

Sample	Novikoff ascites[b]			Rat liver[c]		
	% Total DNA	Acid-extractable protein: DNA	Nonhistone: DNA	% Total DNA	Acid-extractable protein: DNA	Non-histone: DNA
Sheared chromatin	80 ± 6	1.14 ± 0.02	0.55 ± 0.20	78 ± 3	1.12	0.36
Template active	12 ± 1	0.43 ± 0.19	0.85 ± 0.20	24 ± 2	0.60	0.46
Template inactive	67 ± 6	1.24 ± 0.04	0.50 ± 0.15	53 ± 4	1.43	0.29

[a] Chemical composition was determined as follows. DNA was estimated by absorbance at 260 nm in 1 N NaOH (1.0 A_{260} = 34 µg/ml). Chromatin fractions were extracted twice with 0.4 N H_2SO_4. The nonhistone residue was dissolved in 1 N NaOH. Protein was estimated by a modification of the method of H. Kuno and H. K. Kihara [*Nature (London)* **215**, 974 (1967)], using 25% TCA for precipitation. Purified histone and bovine serum albumin were used as standards.
[b] Average of five preparations plus or minus standard deviation.
[c] For DNA, average of three preparations plus or minus standard deviation. For protein, average of two preparations. Chromatin from both rat liver and ascites cells was incubated with DNase II for 15 minutes prior to $MgCl_2$ fractionation.

amounts of chromatin DNA and nascent RNA remaining soluble at various $MgCl_2$ concentrations. The bulk of labeled RNA remains in solution with a small portion of the DNA until the $MgCl_2$ concentration becomes sufficiently high to precipitate the active region. From the initial slope of the fractionation curve, it is estimated that the active region has been enriched 5.5-fold over unfractionated chromatin. (This value is estimated by assuming that all the labeled RNA is associated with the template active region. The enrichment of template active DNA would be greater if any redistribution of the labeled RNA had occurred.) More direct evidence for fractionation is based on DNA renaturation, RNA–DNA hybridization, and template activity measurements, and is reported elsewhere.[12]

Properties of the Chromatin Fractions

The table shows the chemical compositions of chromatin fractions isolated from Novikoff ascites and rat liver. The template active regions

[12] J. M. Gottesfeld, W. T. Garrard, G. Bagi, R. F. Wilson, and J. Bonner, *Proc. Nat. Acad. Sci. U.S.* **71**, 2193 (1974).

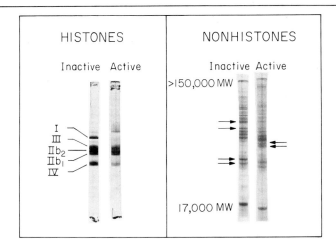

FIG. 2. Disc electrophoresis of chromosomal proteins from the template active and inactive regions. Histone gels: Electrophoresis and staining was performed as described by S. Panyim and G. R. Chalkley [*Arch. Biochem. Biophys.* **130**, 337 (1969)]. Each gel was loaded with 30 μg of acid-extracted, ethanol-purified protein. Nonhistone gels: Electrophoresis and staining was performed as described by J. King and U. K. Laemmli [*J. Mol. Biol.* **62**, 465 (1972)]. Acid-extracted chromatin was homogenized in and dialyzed against 2.5% sodium dodecyl sulfate, 65 mM Tris · Cl (pH 6.8). Prior to electrophoresis, metastable aggregates were disrupted by adding 5% β-mercaptoethanol followed by boiling for 1 minute. Gels were each loaded with 100 μg of protein.

from both sources are impoverished in acid-extractable protein, while the template inactive fractions are enriched in this material. Fractionation of chromatin also yields an unequal partitioning of the nonhistone proteins. The active region contains about twice the amount of nonhistone protein per milligram of DNA as the inactive region. About 80% of the chromatin of both ascites and liver is sheared by prolonged (15 minute) enzyme treatment. The amount of template active DNA found in the supernatant after $MgCl_2$ fractionation is strikingly different, however, in the two cases. The proportion of rat liver template active DNA is 2-fold higher than that of ascites, 24% versus 12%. These values are in accordance with the previously reported template activities of the two chromatins as measured with exogenous RNA polymerase.[13]

Disc electrophoretic patterns of the acid-extractable proteins from the active and inactive fractions of rat liver chromatin are shown in Fig. 2. The most pronounced difference in the banding patterns is the complete absence of histone I in the active region. Furthermore, although over 95%

[13] J. Bonner, M. E. Dahmus, D. Fambrough, R.-C. Huang, K. Marushige, and D. Y. H. Tuan, *Science* **159**, 47 (1968).

of the acid-extractable protein from sheared chromatin and inactive region fractions are histone proteins, we estimate that only 60% of the acid-extractable protein from the active region is histone (from the amount of protein that is applied to the gels and the intensity of the histone staining bands). On this basis the true histone:DNA ratio of the template active fraction is less than 0.36, while the nonhistone:DNA ratio of this fraction is greater than 0.70. It is clear that the nonhistone proteins are concentrated in the template active fraction and impoverished in the template inactive fraction. Separation of nonhistone polypeptides on the basis of molecular size by sodium dodecyl sulfate disc electrophoresis is also shown in Fig. 2. The inactive region has a higher proportion of high molecular weight bands. In addition, quantitative differences in the concentration of individual bands exist (marked with arrows).

Comment

Sheared chromatin can be fractionated into template active and inactive portions on the basis of predicted differences in the physical properties between the two regions. If separation without gross cross contamination is to be achieved, the size of the fragments is critical. Mild digestion with DNase II has the obvious advantage over other reported techniques in that it yields substantially smaller fragments. Fractionation of chromatin with $MgCl_2$ provides a rapid and reliable means of separating active and inactive regions.

Acknowledgments

Report of work supported in part by U.S. Public Health Service Grant GM 13762, by a U.S. Public Health Service postdoctoral fellowship (L.U.), by a Damon Runyon Memorial Fund for Cancer Research postdoctoral fellowship (W.G.), by U.S. Public Health Service Training Grant GM 86 (J.G.) and by the California Institute of Technology Gosney Fellowship (R.B.).

[9] Methods for Analysis of Histones

By LUBOMIR S. HNILICA

The Composition and Heterogeneity of Histones

Although histones were discovered almost a century ago,[1] their limited heterogeneity and lack of substantial tissue and species specificity were

---

[1] A. Kossel, "The Protamines and Histones." Longmans, Green, New York, 1928.

not fully realized until recently. In definition, histones can be characterized as acid soluble, basic, low molecular weight proteins (11,000–22,000 daltons) found associated with DNA in chromatin of most eukaryotes. They lack tryptophan and, with the exception of the arginine-rich, ARE (F3 or III), histones in higher organisms,[2-4] and the GRK (F2a1 or IV) histone in sea urchins,[5,6] they do not contain cystine or cysteine. Because of their excess of lysine, arginine, and histidine over glutamic and aspartic acids, histones carry a positive charge and complex readily with acids. While the free histone bases are only sparingly soluble in water or neutral buffers, histone salts with strong acids, such as HCl, H_2SO_4, are solubilized with ease. Perhaps the most characteristic feature of histones distinguishing them from other proteins of similar amino acid composition (in ribosomes, nervous tissue, etc.) is their association with nuclear DNA in chromatin.

In higher animals and many lower eukaryotes, there are five main types of histones. Two of them contain an excess of arginine over lysine and are classified as the arginine-rich histone fractions. The remaining three histones contain more lysine than arginine and are known as the lysine-rich histones. In some specialized cells, such as nucleated erythrocytes in birds, spermatozoa of various animals, additional histone fractions were found, differing in their amino acid composition from the five main histone species.[7-10]

Despite the extensive interest in histones of numerous investigators which is reflected in the ever-increasing numbers of publications concerned with these proteins, there is no unified histone nomenclature. Two of the most widely accepted enumerations are the Roman numeral nomenclature initiated by Luck et al.[11] and the F nomenclature of Butler and his associates.[12] Both of these nomenclatures reflect the respective

[2] S. Panyim, K. R. Sommer, and R. Chalkley, *Biochemistry* **10**, 3911 (1971).
[3] R. J. DeLange, J. Hooper, and E. L. Smith, *Proc. Nat. Acad. Sci. U.S.* **69**, 882 (1972).
[4] W. F. Brandt and C. Von Holt, *FEBS Lett.* **23**, 357 (1972).
[5] J. A. Subirana, *FEBS Lett.* **16**, 133 (1971).
[6] A. W. Johnson, J. A. Wilhelm, D. N. Ward, and L. S. Hnilica, *Biochim. Biophys. Acta* **295**, 140 (1973).
[7] L. S. Hnilica, "The Structure and Biological Functions of Histones." Chem. Rubber Publ. Co., Cleveland, Ohio, 1972.
[8] R. J. DeLange and E. L. Smith, *Annu. Rev. Biochem.* **40**, 279 (1971).
[9] R. H. Stellwagen and R. D. Cole, *Annu. Rev. Biochem.* **38**, 951 (1969).
[10] D. T. Wigle and G. H. Dixon, *J. Biol. Chem.* **246**, 5636 (1971).
[11] J. M. Luck, P. S. Rasmussen, K. Satake, and A. N. Tsvetikov, *J. Biol. Chem.* **233**, 1407 (1958).
[12] J. A. V. Butler, E. W. Johns, and D. M. P. Phillips, *Progr. Biophys. Mol. Biol.* **18**, 209 (1968).

sequential elution of the histone fractions during ion exchange chromatography on Amberlite IRC 50 (or CG 50) or carboxymethyl cellulose columns. A third system resulting from attempts to find a nomenclature acceptable to most concerned scientists emerged from the first Gordon Research Conference on Nuclear Proteins, Chromatin Structure and Gene Regulation in Beaver Dam, Wisconsin (July 3–7, 1972). A proposal was accepted by the conferees to base the new nomenclature on the amino acid composition of the individual histones. One-letter abbreviations of the three amino acids most abundant or more characteristic for the fraction determine its name. Since the one-letter codes for lysine, alanine, proline, leucine, glutamic acid, glycine, and serine are K, A, R, P, L, E, G, and S, respectively, the symbols of the five common histones are

KAP (rich in lysine, alanine, and proline), also F1, I, or the very lysine-rich histones

KAS (rich in lysine, alanine and serine), also F2b, IIb2, or the moderately lysine-rich histone

LAK (rich in leucine, alanine, and lysine), also F2a2, IIb1, or the slightly lysine-rich histone

ARE (rich in alanine, arginine, and glutamic acid), also F3, III, or arginine-rich histone

GRK (rich in glycine, arginine, and lysine), also F2a1, IV, or glycine-arginine-rich histone

At the Ciba Foundation Symposium on Structure and Function of chromatin (London, April 1974) a compromise was reached regarding the nomenclature of the common histone fractions. According to this new nomenclature the five main histone fractions are labeled as follows:

1 = KAP or F1 or I
2A = LAK or ALK or F2a2 or IIb1
2B = KAS or KSA or F2b or IIb2
3 = ARE or ARK or F3 or III
4 = GRK or F2a1 or IV

The amino acid composition of these fractions from calf thymus substantiating this nomenclature is shown in the table on p. 105.

Additional information concerning the new nomenclature of histones can be obtained from Professor James Bonner, Division of Biology, California Institute for Technology, Pasadena, California 91109. In this chapter, the new histone nomenclature will be used systematically with the corresponding F and Roman numeral symbols in parentheses. The three-letter symbol for the lysine-rich histone specific for nucleated erythrocytes in birds, and possibly in other lower vertebrates, is KSA (rich in lysine, serine, and alanine, F2c or histone V). Another tissue-specific histone fraction discovered by Wigle and Dixon[10] is AKP histone (or histone

Amino Acid Composition of the Main Histone Fractions from Calf Thymus[a]

Amino acid	KAP[b] F1 (I)	KAS F2b (IIb2)	LAK F2a2 (IIb1)	GRK F2a1 (IV)	ARE F3 (III)
Lysine	**28.7**	16.7	12.5	9.8	10.1
Histidine	0.0	2.3	2.8	1.9	2.4
Arginine	1.7	6.4	9.3	**13.9**	**13.6**
ϵ-N-Methyllysine	0.0	0.0	0.7	1.1	0.6
Aspartic acid	2.0	4.9	5.5	5.0	4.4
Threonine	5.4	6.2	4.9	6.6	6.5
Serine	6.7	**10.9**	5.0	2.5	3.8
Glutamic acid	3.4	7.6	8.7	6.2	**10.2**
Proline	**10.1**	4.8	4.1	1.3	4.4
Glycine	6.9	5.6	9.0	**15.9**	5.8
Alanine	**25.1**	10.2	13.2	7.5	**13.5**
Half-cystine	0.0	0.0	0.0	0.0	0.3
Valine	4.1	7.1	6.0	7.8	4.6
Methionine	0.0	1.6	0.3	1.0	1.5
Isoleucine	0.8	4.9	4.2	5.6	4.8
Leucine	4.1	4.9	**10.1**	8.0	9.0
Tyrosine	0.5	3.8	2.3	3.5	2.1
Phenylalanine	0.5	1.6	1.0	2.2	2.0
Basic/acidic	5.6	2.0	1.8	2.4	1.8

[a] All values are expressed as percent of total moles of amino acids recovered. The serine values were corrected (10%) for hydrolytic losses. The main features distinguishing each histone fraction are printed in bold figures.
[b] Unfractionated.

T), which is rich in alanine, lysine, and proline. Several less common nomenclatures for histones are compared in recent reviews by Johns[13] and Hnilica.[7]

Isolation of Histones

Since histones are primitive proteins without enzymatic activity or other features which would allow for their selective assay in protein mixtures, they must be first dissociated from the DNA and extracted with an appropriate solvent. It is customary to extract the analyzed material with dilute mineral acid (H_2SO_4 or HCl) and express the acid-soluble protein concentration as the histone content of nuclei or chromatin. Because there are other acid-soluble nonhistone proteins present in chro-

[13] E. W. Johns, in "Histones and Nucleohistones" (D. M. P. Phillips, ed.), p. 1. Plenum, New York, 1971.

matin and especially in nuclei, estimates of the total acid-soluble protein in isolated chromatin or nuclei are therefore higher than their actual histone contents. Cytoplasmic microsomes and mitochondria as well as nucleoli and nucleoplasm all contain acid-soluble proteins, some of which resemble histones in their amino acid composition. Consequently, isolated chromatin should be used for quantitative studies on DNA-bound histones. The isolation of nuclei and chromatin can be achieved by numerous procedures, description of which is beyond the scope of this chapter. Critical review articles concerned with the isolation of uncontaminated nuclei, chromatin, and other organelles appeared in previous volumes of this series[14-16] and in other monographs.[17-20] It should be remembered, however, that the same isolation procedure can be only rarely applied to a variety of tissues of cellular types. In other words, a technique yielding exceptionally clean nuclei from liver may be utterly unsuitable when applied to a hepatoma or tissue-cultured cells.

If isolated nuclei are used for extraction and quantitation of histones, it is advisable to remove the bulk of acid-soluble nonhistone proteins by repeated homogenization and washing of the disrupted nuclei with saline citrate solution [standard saline-citrate (SSC) contains 0.15 M NaCl and 15 mM sodium citrate[21,22]] and then once or twice with 0.1 M Tris·HCl buffer, pH 7.4.[23,24] The Tris·HCl buffer wash can be either complemented or substituted by one extraction with 0.3 M NaCl as recommended by Johns and Forrester.[25] Homogenization of the SSC and Tris·HCl buffer-washed nuclei in 80% aqueous ethanol is said to preferentially decrease the solubility of nonhistone proteins in acid and facilitate more selective solubilization of histones.[26] Although the volumes of

[14] H. Busch, this series, Vol. 12A, p. 421.
[15] T. Y. Wang, this series, Vol. 12A, p. 417.
[16] J. Bonner, R. Chalkley, M. Dahmus, D. Fambrough, F. Fujimura, R. C. C. Huang, J. Huberman, R. Jensen, K. Marushige, H. Ohlenbusch, B. M. Olivera, and J. Widholm, this series Vol. 12B [96], p. 3.
[17] A. Dounce, *Exp. Cell Res.*, Suppl. 9, 126 (1963).
[18] H. Busch and K. Smetana, "The Nucleolus." Academic Press, New York, 1970.
[19] G. Siebert, *Methods Cancer Res.* 2, 287 (1967).
[20] J. Bonner, M. E. Dahmus, D. Fambrough, R. C. C. Huang, K. Marushige, and D. Y. H. Tuan, *Science* 159, 47 (1968).
[21] J. A. V. Butler, P. F. Davison, D. W. F. James, and K. V. Shooter, *Biochim. Biophys. Acta* 13, 224 (1954).
[22] J. Widholm and J. Bonner, *Biochemistry* 5, 1753 (1966).
[23] V. G. Allfrey, *Exp. Cell Res.*, Suppl. 9, 183 (1963).
[24] L. S. Hnilica, L. J. Edwards, and A. E. Hey, *Biochim. Biophys. Acta* 124, 109 (1966).
[25] E. W. Johns and S. Forrester, *Eur. J. Biochem.* 8, 547 (1969).
[26] E. W. Johns, D. M. P. Phillips, P. Simson, and J. A. V. Butler, *Biochem. J.* 77, 631 (1960).

solutions used for the processing of isolated nuclei are not critical within a reasonable range, 10- to 20-fold excess (v/v) of liquid over the sedimented material is advisable.

Care must be exerted when using the 0.3 M NaCl step for removal of some of the acid-soluble nonhistone proteins. Although this salt concentration was shown not to alter the chromatin preparations from calf thymus, rat liver, and other tissues of higher animals,[7,25,27] there is a possibility of selective extraction of the KAP (F1 or I) histones by 0.3 or 0.35 M NaCl from partially degraded chromatin preparations or from nuclei of animal species containing less stabilized chromatin.

To obtain nearly all the DNA-bound histones, it is necessary to repeat the extraction of nuclear material at least twice, each time with a fresh portion of solvent. Essentially, the speed of extraction is determined by the diffusion of histones from acid-precipitated chromatin material. Obviously, clumpy or aggregated chromatin or nuclei will release their histones slower than finely dispersed samples. When a more concentrated acid (2.5 M HCl or 2 M H_2SO_4) is added to a salt-dissociated chromatin, the acid precipitates DNA and nonhistone proteins on contact, forming membranous material which encloses significant quantities of the solubilized chromatin inside small spheres. Attempts to disrupt such spheres by homogenization produce a rubbery substance which is quite difficult to extract quantitatively. Shearing such material in a blender followed by an overnight extraction may be necessary to recover most of its histone contents.

The extent to which histones were extracted from chromatin can be estimated by dissolving the acid-insoluble residue in buffered 1% sodium dodecyl sulfate, removing the DNA by centrifugation at 105,000 g for 24–36 hours and subjecting the protein-containing supernatant, after reducing its volume, to electrophoresis in polyacrylamide gels.[27,28] It should be noted, however, that there are acid-insoluble nonhistone proteins in chromatin which are firmly associated with the DNA and migrate in polyacrylamide gel electrophoresis (in the presence of 0.1% sodium dodecyl sulfate) with mobilities similar to those of the ARE (F3 or III) and GRK (F2a1 or IV) histones.[29,30]

The concentration of acid employed for histone extraction influences the efficiency of the procedure as well as the fractional composition of the histones. Much diluted HCl or H_2SO_4 (0.01–0.1 M) extracts prefer-

[27] J. A. Wilhelm, A. T. Ansevin, A. W. Johnson, and L. S. Hnilica, *Biochim. Biophys. Acta* **272**, 220 (1972).
[28] S. C. R. Elgin and J. Bonner, *Biochemistry* **9**, 4440 (1970).
[29] G. L. Patel, *Life Sci.* **11**, 1135 (1972).
[30] K. Wakabayashi, S. Wang, G. Hord, and L. S. Hnilica, *FEBS Lett.* **32**, 46 (1973).

entially the lysine-rich histone fractions and higher acid concentrations (0.2–0.25 M) are necessary for complete solubilization of the arginine-rich histones. However, further increase in the concentration of acid, especially HCl, results in considerable contamination of solubilized histones by nonhistone proteins, some of which appear to be preferentially soluble in 1–2 M HCl. To keep the nonhistone protein contamination of histones minimal, some investigators recommend the use of 0.2 M H_2SO_4 which solubilizes most of the histones without a significant solubilization of other proteins.[16] Unless a partial fractionation of histones is the objective, weak organic or inorganic acids are not suitable for quantitative extraction of histones.

Extraction with 0.25 M HCl of nuclei or chromatin prepared from tissues rich in nuclear nonhistone proteins frequently yields an acid-soluble protein ("histone") fraction which represents twice or more the DNA content of the nuclear material. Dialysis and lyophilization of such "histones" renders them partially insoluble in water, SSC, or neutral buffer solutions. Amino acid analysis of the insoluble material shows that it consists of mostly acidic proteins which became denatured by the lyophilization. Most of the histones proper can be recovered from such a mixture by centrifugation, dialysis, and relyophilization of the supernatant.

Histones solubilized with acid can be recovered by dialysis against deionized water and lyophilization, or they can be precipitated with various reagents.[7] Since most histones are relatively small proteins (12,000–15,000 daltons), losses of histones during prolonged dialysis were reported in the literature.[7] However, because histones solubilized by most common procedures contain detectable quantities of proteolytic enzymes,[31,32] losses attributed to their passage through the dialysis membrane may actually reflect the proteolytic degradation of these proteins. Obviously, the dialysis time should be minimized by frequent changes of the deionized water or by employing systems for rapid dialysis, such as the hollow fiber dialysis device (Dow Chemical Co.). Some investigators prefer to dialyze the histones first against diluted (10–20 mM) HCl or 1% CH_3COOH. The acid is then removed rapidly by a short dialysis against deionized water, or by lyophilization.

Acid-extracted histones can be almost quantitatively precipitated with 10 or more volumes of cold acetone. Presence of ions (mineral acids) is essential for good precipitation of histones with acetone and, especially, ethanol. Histone hydrochlorides are soluble in most aliphatic alcohols. Large excess of ethanol precipitates sulfates of the KAP (F1 or I) histone fraction. Complexes of histones with trichloroacetic or citric acid are partially soluble in acetone and mineral acid (H_2SO_4 or HCl) must be

[31] D. M. P. Phillips and E. W. Johns, *Biochem. J.* **72**, 538 (1959).
[32] D. M. P. Phillips, *Prog. Biophys. Biophys. Chem.* **12**, 211 (1961).

added to the solvent to precipitate all the histone fractions. Usually, acetone is mixed with 5 N HCl (99:1, v/v) and the histones complexed to organic acid are converted to hydrochlorides. The precipitate should be washed several times with pure acetone and can then be dried *in vacuo* to yield a fine powder.[13,26]

Lindh and Brantmark[33,34] developed a technique for virtually quantitative recovery of even small quantities of all histone fractions. In the presence of an excess of Reinecke salt, $NH_4[Cr(NH_3)_2(SCN)_4] \cdot H_2O$, the reineckate complex binds stoichiometrically to positively charged amino acid residues. This renders most proteins insoluble in aqueous solvents. For selective quantitative precipitation of histones, the precipitation is performed between pH 1 and 2 with ice-cold, saturated solution of Reinecke salt. As little as 10 µg of histones can be recovered quantitatively. For reclaiming, the precipitate is suspended in cold acetone containing 2% of 1 M HCl. This treatment releases free Reinecke acid, which is soluble in acetone leaving the histone hydrochloride insoluble. The latter can be recovered by washing with pure, ice-cold acetone, and drying *in vacuo*.

The recovery of individual histones, their mixtures as well as their quantitative determinations which will be discussed later can be profoundly affected by proteolytic degradation. Histones are structurally simple proteins and are readily degraded by a variety of proteolytic enzymes, including some proteases said to be specific for these proteins.[31,35-40] The vulnerability of histones to proteolysis increases once they become separated from the DNA in chromatin. Since many proteolytic enzymes are especially active in the pH range 7–9, histones can be rapidly damaged during the dissociation of chromatin within these pH limits.[35] The presence of concentrated urea or guanidine·HCl in buffered NaCl solutions, i.e., solvents frequently used in chromatin dissociation and reconstitution experiments, does not abolish all the proteolytic activity directed against the histones.[40] Addition of protease inhibitors such as 5 mM NaHSO$_3$, 1 mM diisopropyl fluorophosphate, etc., to the media used for dissociation of chromatin and isolation of nuclei, chromatin or histones has been used and recommended by several investigators.[38-40] Although the use of protease inhibitors is definitely advantageous for quantitative

[33] N. O. Lindh and B. L. Brantmark, *Anal. Biochem.* **10**, 415 (1955).
[34] N. O. Lindh and B. L. Brantmark, *Methods Biochem. Anal.* **14**, 79 (1966).
[35] A. L. Dounce and R. Umana, *Biochemistry* **1**, 811 (1962).
[36] M. Furlan and M. Jericijo, *Biochim. Biophys. Acta* **147**, 135 (1967).
[37] M. Furlan and M. Jericijo, *Biochim. Biophys. Acta* **147**, 145 (1967).
[38] S. Panyim, R. Jensen, and R. Chalkley, *Biochim. Biophys. Acta* **160**, 252 (1968).
[39] J. Bartley and R. Chalkley, *J. Biol. Chem.* **245**, 4286 (1970).
[40] T. C. Spelsberg, L. S. Hnilica, and A. T. Ansevin, *Biochim. Biophys. Acta* **228**, 550 (1971).

extraction of undegraded histones at neutral pH, it is not clear whether other components of chromatin are not irreversibly modified by the highly reactive inhibitor reagents (modification of seryl and threonyl residues, sulfhydryl groups, DNA, etc.). It was reported that $NaHSO_3$ interacts with pyrimidines in nucleic acids forming sulfonates and deaminating cytosine.[41,42]

Quantitative Determination of Histones

Although it is possible to estimate the lyophilized or precipitated histones gravimetrically, this is obviously not a method of choice for quantitation of histones, especially when numerous small samples or fractions must be assayed. Perhaps the most widely accepted method for quantitative determination of proteins was described by Lowry et al.[43] This method utilizes the biuret reaction of proteins complemented by the Folin–Ciocalteau reagent. The Lowry et al.[43] method is only little dependent on the aromatic amino acid content of individual proteins, much less than the direct absorbancy reading at 280 nm. It is very sensitive, about 10–20 times more sensitive than the absorbancy measurement at 280 nm, about 100 times more sensitive than the ninhydrin reaction for proteins. Conversely to ninhydrin, the Lowry et al.[43] procedure is only little affected by the presence of free amino acids or small peptides. As was already mentioned, one of the main advantages of the Lowry et al.[43] procedure is its little dependency on the amino acid composition of proteins. However, the intensity of color does vary slightly with various proteins, especially if they differ greatly in their contents of aromatic amino acids. Therefore, if a great accuracy is required, known amounts of each histone fraction or unfractionated histone should be used for the construction of standard curves. The blue color yield is nearly linear within the absorbancy range between 0 and 0.400 at 700 nm, for higher protein concentrations the color yield decreases progressively.

Reagents

A. 2% Na_2CO_3 in 0.1 M NaOH in water
B. 0.5% $CuSO_4 \cdot 5\ H_2O$ in 1% sodium-potassium tartrate in water. The pH should be adjusted to 7.0 with KOH to keep the copper complex in solution.

[41] R. Shapiro, B. I. Cohen, and R. E. Servis, *Nature (London)* **227**, 1047 (1970).
[42] H. Hayatsu, Y. Wataya, and K. Kai, *J. Amer. Chem. Soc.* **92**, 724 (1970).
[43] O. H. Lowry, N. J. Rosebrough, A. L. Farr, and R. Randall, *J. Biol. Chem.* **193**, 265 (1951).

C. Phenol reagent. Dilute commercial Folin–Ciocalteau reagent with an equal volume of water (the acid concentration should be 1.0 N).

The alkaline copper reagent (biuret) is prepared freshly before use by mixing 50 ml of A with 1.0 ml of B. For protein assay, 0.3 ml of sample (containing 0.02–0.20 g of protein) are mixed well with 3.0 ml of alkaline copper reagent and let stand at room temperature for 10 minutes or longer. The deep blue color is developed by adding 0.3 ml of diluted phenol reagent (C). Each tube should be mixed immediately after addition. Read after 30 minutes or longer at 700 or 800 nm. More concentrated samples and standards can be read at 500 nm.

Other modifications of this procedure are (1) 0.2 ml of sample, 1.0 ml of alkaline copper reagent (A + B), 0.1 ml of phenol reagent (C); or (2) 0.5 ml of sample, 3.0 ml of alkaline copper reagent (A + B), 0.3 ml of phenol reagent (C); or (3) 1.5 ml of sample, 1.5 ml of double-strength alkaline copper reagent (A + B), 0.3 ml of phenol reagent (C). Strong acid or alkaline samples should be neutralized before mixing with the alkaline copper reagent.

According to Lowry et al.[43] and Layne,[44] most purine and pyrimidine bases as well as DNA or RNA did not interfere substantially with the reaction. However guanine, xanthine, uric acid, most phenols, aromatic amino acids, etc., reacted with Folin–Ciocalteau reagent, producing blue color. No interference was observed with the following substances to a final concentration in parentheses: urea (0.5%), guanidine (0.5%), sodium tungstate (0.5%), sodium sulfate or nitrate (1%), neutralized $HClO_4$ or trichloroacetic acid (0.5%), ethanol or ether (5%), acetone (0.5%), $ZnSO_4$ (0.1%), and $Ba(OH)_2$ (0.5%). Glycine and $(NH_2)_2SO_4$ decreased the color yield of the Lowry et al.[43] reaction with proteins.

Reagents containing sulfhydryl or disulfide groups give false color complexes and interfere with the determination of proteins.[45,46] Most of this interference can be abolished by adding small amounts of H_2O_2 to the alkaline copper reagent containing the protein sample and heating it to 50° for 10 minutes before the addition of phenol reagent.[45] Small amounts of sucrose (less than 5%) in samples do not seriously affect the assay. In higher concentrations, sucrose either increases or decreases the color yields of proteins, depending on the protein concentration. According to Gerhardt and Beevers,[47] doubling the copper content in the reagent and lowering the sucrose concentration below 10% results in es-

---

[44] E. Layne, this series, Vol. 3 [73], p. 447.
[45] P. J. Geiger and S. P. Bessman, *Anal. Biochem.* **49**, 467 (1972).
[46] C. G. Vallejo and R. Lagunas, *Anal. Biochem.* **36**, 207 (1970).
[47] B. Gerhardt and H. Beevers, *Anal. Biochem.* **24**, 337 (1968).

sentially normal color values for protein concentrations over 50 μg/ml. More diluted proteins cannot be assayed accurately in the presence of sucrose.

Hexosamines,[48] various monosaccharides,[49] penicillin and some of its semi-synthetic derivatives,[50] EDTA,[51] oxidized lipids,[52] and many organic buffers[53,54] react strongly with the Folin–Ciocalteau reagent and their presence prevents the use of Lowry et al.[43] protein assay. The following buffers interfere strongly[53-55]:

> HEPPS, or N-(2-hydroxyethylpiperazine-N'-2-propane)sulfonic acid
> HEPES, or N-(2-hydroxyethylpiperazine-N'-2-ethane)sulfonic acid
> Bicine, or N,N-bis(2-hydroxyethyl)glycine

Lesser interference was reported with the following:

> BES, or N,N-bis(2-hydroxyethyl)-2-aminoethanesulfonic acid
> PIPES, or piperazine-N,N'-bis(2-ethane)sulfonic acid
> MOPS, or 2-(n-morpholino)ethanesulfonic acid
> ADA, or N-(2-acetamido)-2-aminodiacetic acid
> ACES, or N-(acetamido)-2-aminoethanesulfonic acid
> TAPS, or tris(hydroxymethyl)methylaminopropanesulfonic acid
> CAPS, or cyclohexylaminopropanesulfonic acid and glycine amide

In addition to giving color reaction with the phenol reagent, the following buffers were reported to inhibit the color reaction of proteins:

> ADA (see above)
> Tris, or tris(hydroxymethyl)aminomethane
> Tricine, or N-tris(hydroxymethyl)glycine
> TAPS (see above)

Innocuous, at least to the concentration of 50 μg in assay mixture are the following:

> MES or 2-(N-morpholino)ethanesulfonic acid
> TES, or N-tris(hydroxymethyl)methyl-2-aminoethanesulfonic acid

Obviously, it is advisable to use the solvent in which the assayed protein is dissolved as a control (blank) to eliminate erroneous results origi-

[48] J. K. Herd, Anal. Biochem. **44**, 404 (1971).
[49] J. O'Sullivan and G. E. Mathison, Anal. Biochem. **36**, 540 (1970).
[50] D. J. Silverman, Anal. Biochem. **27**, 189 (1969).
[51] M. F. Gellert, P. H. von Hippel, H. K. Schachman, and M. F. Morales, J. Amer. Chem. Soc. **81**, 1384 (1959).
[52] J. Eichberg and L. C. Makrasch, Anal. Biochem. **30**, 386 (1969).
[53] M. A. Peters and J. R. Fouts, Anal. Biochem. **30**, 299 (1969).
[54] J. D. Gregory and S. W. Sajdera, Science **169**, 97 (1970).
[55] L. V. Turner and K. L. Manchester, Science **170**, 649 (1970).

nating from the solvent interactions with the Lowry et al.[43] reagent.

Little dependent on amino acid composition of the histone fractions are turbidimetric procedures. The turbidity (precipitate) formed when a protein solution is mixed with a low concentration of most common protein precipitants can be used to determine the protein concentration. Procedures employing diluted trichloroacetic acid, potassium ferrocyanide, and sulfosalicylic acid were described in more detail by Layne.[44] A modification of the trichloroacetic acid procedure of Kunitz[56] was adapted for quantitative assay of histones by Luck et al.[11] Aqueous solution of histones is rapidly and thoroughly mixed with 3 M solution of trichloroacetic acid (2:1, v/v). The turbidity is allowed to develop at room temperature, and the absorbancy of each sample is determined at 400 nm, 15 ± 2 minutes after the addition of trichloroacetic acid. The authors reported that the turbidity readings are strictly proportional to the protein concentration up to the absorbancy of 0.350 (1 cm light path). This reading corresponds to about 150 µg of unfractionated histone in the sample. The turbidity per unit weight of histone is approximately the same for all the fractions. This technique which was first used to measure the concentration of chromatographic histone fractions in the presence of relatively large amounts of guanidine hydrochloride was used by many investigators. In a minor modification which employs 3.3 M (instead of 3.0 M) trichloroacetic acid, 10 µg of histones per milliliter of the final acidified solution have the OD_{400} of 0.083. In the presence of guanidinium hydrochloride, the original sample must be diluted with H_2O (at least 1:3, v/v) because of the rather limited solubility of guanidinium hydrochloride in 1.0 or 1.1 M trichloroacetic acid. The turbidimetric procedure cannot be used in the presence of other macromolecules insoluble in trichloroacetic acid (e.g., DNA or RNA).

The low content of tyrosine and phenylalanine in most histone fractions (especially in the KAP or F1, I histones) limits the use of direct absorbancy measurements for quantitative assay of histones at 275 nm (i.e., at their aromatic amino acid maximum). However, like any other proteins, histones absorb strongly at the short end of the UV spectrum and, in the absence of interfering substances, their concentration can be determined with considerable accuracy from the absorbancy readings between 220 and 240 nm. Although shorter wavelengths (180–200 nm) offer much higher sensitivity, unfortunately most other materials, especially buffers, absorb strongly over this range and less sensitive readings at longer wavelengths are the only alternative. Unfractionated calf thymus histone gives the following readings in 1 cm light path: OD_{275}, 0.445

---

[56] M. Kunitz, J. Gen. Physiol. **35**, 423 (1952).

(1.0 mg); OD_{240}, 0.580 (1.0 mg); OD_{230}, 0.330 (0.1 mg); OD_{220} 0.780 (0.1 mg); OD_{210}, 1.700 (0.1 mg).

Another procedure for direct quantitative assay of histones takes the advantage of the fluorescence of tyrosine. Since histones do not contain any tryptophan, and phenylalanine does not fluoresce, the fluorescence of histones is proportional to their tyrosine content. Bonner et al.[16] recommend the excitation wavelength of 280 nm for histones; the fluorescence is measured between 305 and 308 nm. Any standard fluorometer is suitable. However, it should be standardized before each use with a sealed fluorescence standard, e.g., solution of 9-aminoacridine. The histone samples can be dissolved in any aqueous solvent which does not fluoresce or absorb at the excitation wavelength. The use of 0.1 N HCl is recommended. The fluorescence intensity is a function of the tyrosine content of each histone fraction.

Because tryptophan has the excitation maximum at 290 nm and fluorescence maximum at 345 nm, contamination of histones by tryptophan-containing proteins can be easily detected by fluorometry. The very weak fluorescence of nucleic acids at 305–308 nm permits the measurement of protein content of chromatin directly, without previous dissociation or other treatment. The fluorescence of histones in nucleohistone remains unaffected by the salt concentration between 0.01 and 5.0 M NaCl. This indicates that the fluorescence of histones is not altered by their binding to the DNA. As stated by Bonner et al.,[16] the direct fluorometric determination of histones is rapid and accurate over a range of 25–600 µg/ml. It is not damaging to the sample, which can be easily recovered by dialysis and lyophilization.

Another method for quantitative determination of histones also takes advantage of their tyrosine content. It is based on the Millon reaction for proteins and in the modification of Mirsky and Pollister,[57] the reagent precipitates all other proteins but the histones. For an assay, 2 ml of a nuclear or chromatin suspension are mixed well with an equal volume of 0.34 M $HgSO_4$ in 1.88 M H_2SO_4 (i.e., 10% solution of $HgSO_4$ in 10% H_2SO_4) and heated to 60° for 15 minutes. The cherry red color is developed by addition of 0.2 ml of 1% $NaNO_2$ solution in water and heating at 60°C for 10 minutes. If the solution is cloudy or contains a precipitate, it should be centrifuged. Histone concentration is determined from the absorbancy at 354 nm which for 1 mg/ml of calf thymus histone in 1 cm cell is approximately 0.80. Other proteins than histones form red precipitates or reddish turbid solutions which can be sedimented by centrifugation at approximately 10,000 g. Since the color in this reaction is produced

---

[57] A. E. Mirsky and A. W. Pollister, *J. Gen. Physiol.* **30**, 117 (1946).

mainly by the diazotation of tyrosine, the color yield of calf histone fractions is proportional to their tyrosine contents.

Interactions of histones with certain dyes and the subsequent fluorometric determination of such complexes can be also used for histone quantitation. Dyes such as 8-anilino-1-naphthalenesulfonic acid (ANSA) and 1-fluorodinitrobenzene (FDNB) are relatively nonfluorescent in acid solutions but fluoresce very strongly when complexed with proteins. According to Shepherd and Noland,[58] histone samples are mixed with at least 2-fold excess of ANSA (w/w) at pH 2.0 and the fluorescence of the complex is determined at an excitation maximum of 375 nm and emission (fluorescence) maximum of 500 nm. At the excess of ANSA to histone, there is a linear fluorescence intensity to histone relationship over a range of 1–300 µg of protein per milliliter. However, as was shown by Laurence,[59] individual histone fractions differ considerably in their ANSA binding and the binding is greatly favored by histone aggregation. Although at pH 2.0 the aggregation of histones is minimal, this phenomenon decreases the practical value of this method.

Another very sensitive fluorometric technique utilizes the interaction of histones with FDNB.[60] Since FDNB reacts principally with primary amines, the lysine content of histone fractions will determine intensity of the fluorescence. Using the ratio of native protein fluorescence to the DNP-protein fluorescence, it is possible to distinguish between the histones (low ratio of less than 2.0) and the nonhistone proteins (high ratio over 10.0).[60]

A fluorometric method first introduced for quantitative determination of serum phenylalanine which is based on the interaction of phenylacetaldehyde formed by reacting with ninhydrin in the presence of primary amines[61-63] can be adapted for the assay of picomole quantities of histones. The fluorescent product of the ninhydrin reaction with amino acids and primary amines was characterized and synthesized by Wiegele *et al.*[64-65] This reagent (4-phenylspiro[furan-2(3H),1'-phthalan]3,3'-dione) is available under the name of Fluorescamine (Hoffmann-La Roche, Inc., Nutley, New Jersey, 07110). It reacts directly with primary amines form-

[58] G. R. Shepherd and B. J. Noland, *Anal. Biochem.* **11**, 443 (1965).
[59] D. J. R. Laurence, *Biochem. J.* **99**, 419 (1966).
[60] K. Sheth and L. H. Cohen, *Anal. Biochem.* **37**, 142 (1970).
[61] M. W. McCaman and E. Robins, *J. Lab. Clin. Med.* **59**, 885 (1962).
[62] K. Samejima, W. Dairman, and S. Udenfriend, *Anal. Biochem.* **42**, 222 (1971).
[63] K. Samejima, W. Dairman, J. Stone, and S. Udenfriend, *Anal. Biochem.* **42**, 237 (1971).
[64] M. Weigele, J. F. Blount, J. P. Tengi, R. C. Czaijkowski, and W. Leimgruber, *J. Amer. Chem. Soc.* **94**, 4052 (1972).
[65] M. Weigele, S. L. De Bernardo, J. P. Tengi, and W. Leimgruber, *J. Amer. Chem. Soc.* **94**, 5927 (1972).

ing the same fluorescent products as are generated in the ninhydrin-phenylacetaldehyde reaction.[66] In excess of fluorescamine, the reaction with primary amines is rapid and essentially quantitative. Excess reagent is destroyed in a few seconds, and free fluorescamine as well as its hydrolytic products are not fluorescent. The fluorescent complex with amino acids, peptides, or proteins is assayed at 390 nm excitation and 475 nm emission wavelength. As little as 0.05 µg of protein can be determined quantitatively.[66]

Although the fluorometric methods for assaying proteins are quite sensitive, simple, and economical, it must be realized that all buffers or reagents containing groups reactive in the proteins will also react with the fluorogen, creating unwanted interference. Conversely, other chemicals can quench the fluorescence, decreasing the sensitivity of the method.

Quantitative Determination of Histone Fractions

Among the numerous methods described for the fractionation of histones, only a few are suitable for quantitative estimation of the individual fractions. These are, in the order of increasing applicability and accuracy: chemical fractionation, chromatography on ion exchange resins, gel filtration, cellulose acetate (Cellogel) or starch gel electrophoresis, polyacrylamide gel electrophoresis.

Chemical Fractionation

Although chemical fractionation of histones is not very practical for quantitative estimation of the fractions, it can be used for crude estimates. The differences in amino acid composition between the KAP (F1 or I) histones and the rest of the fractions are sufficient to permit their clean separation. Since the KAP (F1 or I) histones are soluble in 5% trichloroacetic acid or 5% $HClO_4$, precipitation of a histone mixture or extraction of isolated chromatin with one of these two acids can be used to obtain a fairly accurate information about the quantities of KAP (F1 or I) histones in various tissues.[67-69] As a rule, 5% $HClO_4$, is preferred since it does not solubilize many other macromolecules. The KAP (F1 or I) histones can be precipitated at the final trichloroacetic concentration of 15–20%.

Another fractionation schedule occasionally used to determine the

[66] S. Udenfriend, S. Stein, P. Böhlen, W. Dairman, W. Leimgruber, and M. Weigele, *Science* **178**, 871 (1972).
[67] E. H. DeNooij and H. G. K. Westenbrink, *Biochim. Biophys. Acta* **62**, 608 (1962).
[68] E. W. Johns and J. A. V. Butler, *Biochem. J.* **82**, 15 (1962).
[69] E. W. Johns, *Biochem. J.* **92**, 55 (1964).

amounts of histone fractions is the selective extraction of arginine-rich ARE (F3 or III) and GRK (F2a1 or IV) histones with a mixture of absolute ethanol and 1.25 M HCl (4:1, v/v), followed by the extraction of lysine-rich KAP (F1 or I), KAS (F2b or IIb2), and LAK (F2a2 or IIb1) histones with 0.25 M HCl.[26,68,69] These two groups of histones can be further fractionated according to the scheme in Fig. 1, which follows essentially the method of Johns.[69] The individual histone fractions resulting from this scheme are reasonably pure and their yields can be determined gravimetrically or by colorimetric assay of appropriate aliquots. Although the scheme in Fig. 1 is quite useful for the isolation of quantities of individual histones, it is very cumbersome and inaccurate for quantitative studies.

Chromatographic Procedures

The discovery of ion exchange resins introduced a new and powerful tool for fractionation of proteins, including histones. Despite the initial disappointments caused by an excessive retention of histones by the resins, techniques were developed permitting the fractionation of the whole histone into its individual fractions. Presently, synthetic organic resins based on styrene of methacrylic acid polymers which are either sulfonated or contain carboxylic acid residues are available for the fractionation of histones in addition to several cellulose-based ion exchange

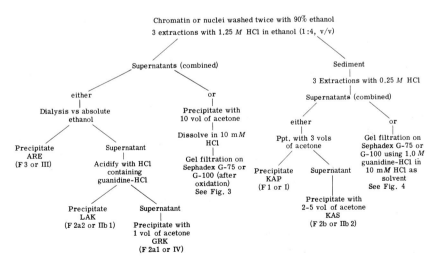

Fig. 1. Chemical fractionation of histones from higher animals. Adapted from E. W. Johns, in "Histones and Nucleohistones" (D. M. P. Phillips, ed.), p. 1, Plenum, New York, 1971.

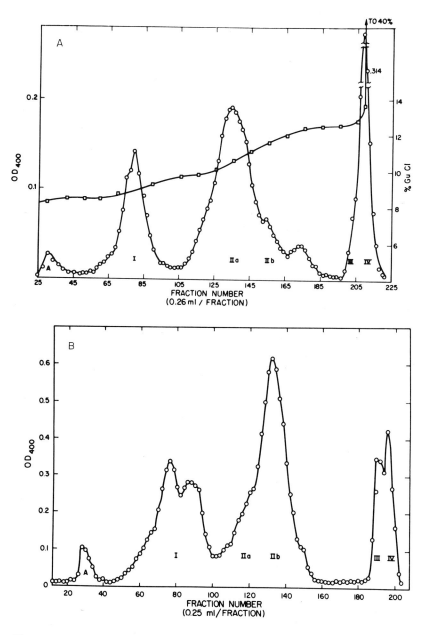

Fig. 2. Chromatography of acid-extracted pea bud (A) and calf thymus (B) chromatin histones on Amberlite IRC-50 (CG-50). Protein load was 3.0 mg per each column. Column size was 0.6 × 55 cm. □——□, concentration of guanidine-HCl

media. Although quite useful for histone fractionation, the cellulose-based cation exchange resins are not suitable for quantitative studies on histones. The method of choice uses weakly acidic Amberlite IRC 50 (or CG-50) resin and was developed Luck et al.[11]

Commercially available chromatographic grade of Amberlite CG-50 (200–400 mesh) is freed of fine particles by settling from an aqueous suspension. The wet resin is cycled several times using a sequence of washes with 2 M HCl, distilled water, 2 M NaOH, distilled water, 2 M HCl, etc. After the final wash with 2 M HCl, the resin is converted to the sodium form with 2.0 M NaCl, and its pH is adjusted to 7.0 with NaOH. The neutralized resin is suspended in the initial chromatographic medium consisting of 8% guanidine-HCl in 0.1 M sodium phosphate buffer, pH 6.8 and packed in the desired column (2.5 cm $\times$ 60 cm for preparative or 0.6 cm $\times$ 60 cm for analytical work). Details of the packing procedure as well as the methods for purification of commercial guanidine-HCl were described by Luck et al.[11] and by Bonner et al.[16] After packing, the column is washed with several volumes of buffered 8% guanidine-HCl solution. Histone sample dissolved in the same solution (clarified by centrifugation, if necessary) is applied to the column and washed with small quantities of the buffered 8% guanidine-HCl solution; the elution of histone fractions is accomplished by the application of a linear gradient of guanidine-HCl in 0.1 M sodium phosphate buffer, pH 7.6. A gradient from 8% to 13% elutes the KAP (F1 or I) histones followed by a broad peak of the LAK (Fb2 or IIb1) and KAS (F2b or IIb2) fractions. Finally, a steep gradient of guanidine-HCl from 13% to 40% brings out first the ARE (F3 or III) histone followed by the GRK (F2a1 or IV) fraction. The column can be regenerated by washing with 8% guanidine-HCl in 0.1 M sodium phosphate buffer. For a 2.5 cm $\times$ 60 cm column, at least 700 ml of the first gradient solution followed by 100 ml of 40% guanidine-HCl and 150 ml of the 8% guanidine-HCl rinse is recommended to obtain a good separation of the histones (Fig. 2).[16]

To assay the histone content in the effluent, 0.2-ml aliquots are taken from the collected fractions, diluted with 0.6 ml of H_2O and mixed with 0.4 ml of 3.3 M trichloroacetic acid. The turbidity of the precipitated

---

gradient; ○——○, protein concentration determined by turbidimetry of the trichloroacetic acid precipitates at 400 nm. The eluted protein fractions are A: nonhistone protein contamination; I: KAP (F1 or I) histones; IIa: LAK (F2a2 or IIb1) histone; IIb: KAS (F2b or IIb2) histone; III: ARE (F3 or III) histone; IV: GRK (F2a1 or IV) histone. From J. Bonner, G. R. Chalkley, M. Dahmus, D. Fambrough, F. Fujimura, R. C. C. Huang, J. Huberman, R. Jensen, K. Marushige, H. Ohlenbusch, B. M. Olivera, and J. Widholm, this series, Vol. 12B [96], p. 3.

histones is determined after 13–16 minutes at 400 nm.[16] For radioactivity measurements, fraction aliquots (50–100 μl) can be counted either directly after mixing with a fluor containing commercial solubilizer solution or after drying on small paper or glass filter disks. There is essentially no quenching of ^{14}C-labeled compounds. The quenching of ^{3}H increases with the concentration of guanidine-HCl.

For quantitative estimates, the recovery of histones is about 80–90% if measured directly in the effluent. After histone recovery, e.g., by dialysis and lyophilization, the total yield of all the fractions rarely exceeds 70–75% of the original sample. Among the disadvantages of this method is its poor resolution of LAK F2a2 or IIb1) and KAS (F2b or IIb2) histones and a significant contamination of the arginine-rich ARE (F3 or III) and GRK (F2a1 or IV) fractions with nonhistone proteins coeluted with the steep gradient of 40% guanidine-HCl.[70] Presence of a large "runoff" peak at the beginning of the elution sequence, usually heralds such as a contamination.

Modifications of the Amberlite IRC 50 (CG-50) chromatographic procedure using a shallow gradient of guanidine-HCl were successfully applied to the subfractionation of the lysine rich KAP (F1 or I) histones by Cole and his co-workers.[71,72] The LAK (F2a2 or IIb1) and KAS (F2b and IIb2) histones can also be separated by a similar modification.

Gel Filtration. The feasibility of histone fractionation by gel filtration was first demonstrated by Cruft,[73] who separated calf thymus histone into several peaks on Sephadex G-75. The resolution of individual peaks was poor, however, and 20 mM HCl had to be used for elution to decrease the aggregation of histones and their adsorption to Sephadex. Like other chromatographic methods, gel filtration cannot resolve all the histone fractions in a one-step procedure. Its main advantage is the better than 90% recovery of histones.

In combination with selective extraction of arginine-rich histones with absolute ethanol–1.25 M HCl mixture (4:1), the oxidized ARE (F3 or III) histone can be separated from a mixture of GRK (F2a1 or IV) and LAK (F2a2 or IIb1) fractions by gel filtration on Sephadex G-75 or G-100.[74,75] By recycling the second peak, a good resolution of the GRK (F2a1 or IV) and LAK (F2a2 or IIb1) histones can be obtained.

[70] R. H. Stellwagen and R. D. Cole, *J. Biol. Chem.* **243**, 4452 (1968).
[71] J. M. Kinkade and R. D. Cole, *J. Biol. Chem.* **241**, 5790 (1966).
[72] M. Bustin and R. D. Cole, *J. Biol. Chem.* **244**, 5286 (1969).
[73] H. R. Cruft, *Biochim. Biophys. Acta* **54**, 611 (1961).
[74] L. S. Hnilica, *Experientia* **21**, 124 (1965).
[75] L. S. Hnilica and L. G. Bess, *Anal. Biochem.* **12**, 421 (1965).

A long (2.5 cm × 180 cm) chromatographic column is filled with a suspension of G-100 or G-75 Sephadex in 10 mM HCl saturated with chloroform. Arginine-rich histones dissolved in the same solvent are allowed to pass through the column at an approximate flow rate of 36–40 ml per hour. The histone concentration in the effluent can be determined directly from the absorbancy of individual fractions at 275 nm, and aliquots can be counted for their radioactivity either after mixing with an appropriate scintillation cocktail or after drying on small filter paper or glass disks (Fig. 3).

The lysine-rich histones, obtained by extraction of the ethanolic-HCl residue with 0.25 M HCl can be fractionated also by gel chromatography. Columns of Sephadex G-75 and G-100 (2.5 cm × 180 cm) are equilibrated with a solvent consisting of 1.0 M solution of guanidine-HCl in 10 mM HCl. A three-peak pattern is obtained for lysine-rich histones from most animal tissues (Fig. 4). The first peak represents high molecular weight aggregates, the second peak is the KAP (F1 or I) histones and the KAS (F2b or IIb2) histone is eluted in the third peak. Owing to the low tyrosine and phenylalanine content of KAP histones, the absorbancy of the fractions should be determined at 230 or 235 nm.

Because of the tendency of the acid-soluble nonhistone proteins to form aggregates with both themselves and the histones, the first peak

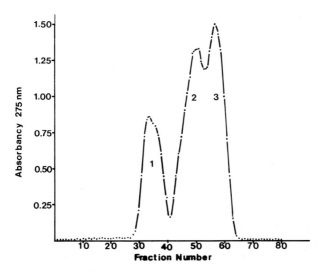

Fig. 3. Gel filtration on Sephadex G-75 of rat liver arginine-rich histones (ethanol–1.25 M HCl, 4:1, v/v, extract). Column size: 25 × 1800 mm. Solvent: 10 mM HCl saturated with chloroform. Peak 1: oxidized ARE (F3 or III) dimer. Peak 2: LAK (F2a2 or IIb1) histone; the ARE monomer (reduced) will also elute in this position. Peak 3: GRK (F2a1 or IV) histone.

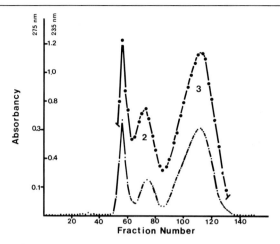

FIG. 4. Gel filtration on Sephadex G-75 of rat liver lysine-rich histones (0.25 M HCl extract of the ethanol-HCl residue). Column size: 25 × 1800 mm. Solvent: 1.0 M guanidine hydrochloride in 10 mM HCl. Peak 2: KAP (F1 or I) histones. Peak 3: KAS (F2b or IIb2) histones. ·-·-, Absorbancy at 275 nm; ●——●, absorbancy at 235 nm.

to emerge from the column will contain such aggregates. Hence the oxidized ARE (F3 or III) and the KAP (F1 or I) histones will be affected by this phenomenon. Addition of urea or guanidine-HCl to the solvents for gel filtration greatly reduces the aggregation.

Electrophoresis

Superior to any other method for quantitative and qualitative analysis of histones is their electrophoresis in gels. Among the most common techniques are the polyacrylamide and starch gel, and Cellogel electrophoresis.

Electrophoresis in Cellogel. Cellogel, a modified form of cellulose acetate which contains structurally bound water, and swells up to its total water content of 75–80%, can be used for electrophoretic separation of most histones. Machicao and Sonnenbichler[76] described a method capable of resolving all the five main histones from calf thymus. The electrophoresis in Cellogel strips is done in a closed chamber used for paper electrophoresis with 0.6 M ammonium borate buffer pH 10.0 containing 10 mM EDTA, 10 mM 2-mercaptoethanol, and 6 M urea as the electrolyte. Calf thymus histones separated by this method and stained in amido

[76] F. Machicao and J. Sonnenbichler, *Biochim. Biophys. Acta* **236**, 360 (1971).

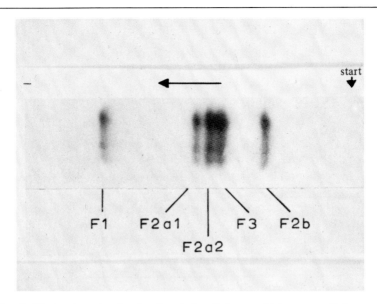

FIG. 5. Electrophoresis of unfractionated calf thymus histone on cellulose acetate (Cellogel) in 0.6 M ammonium borate buffer, pH 10.0. From F. Machicao and J. Sonnenbichler, *Biochim. Biophys. Acta* **236**, 360 (1971).

black solution in methanol–water–glacial acetic acid mixture (45:45:10, v/v/v) are shown in Fig. 5.

A similar technique, developed independently by Creighton and Trevithick[77] uses 25 mM HCl as electrophoretic medium. The Cellogel strips are first soaked briefly in a solution containing 2 M urea, 1% Nonidet P-40 (a nonionic detergent, similar to Triton X-100) and 25 mM HCl. The electrophoresis is performed in 25 mM HCl as the electrolyte. Although the conditions of electrophoresis are quite different from those described by Machicao and Sonnenbichler,[76] the individual histone fractions separate in the same sequence as shown in Fig. 5, except that the KAP (F1 or I) histones are very slow and the fractions GRK (F2a1 or IV) and LAK (F2a2 or IIb1) are not resolved. After electrophoresis, the Cellogel strips can be stained either in amido black solution, or, as recommended by the authors, in a solution of Procion blue in acidified methanol. The sensitivity of this procedure can be considerably increased by allowing the electrophoretically separated histones to react with dansyl chloride (1-dimethylaminonaphthalene-5-sulfonyl chloride) and scanning the strips in a suitable spectrofluorometer. In this way, as little as 0.1 μg of histone can be measured accurately.

[77] M. O. Creighton and F. R. Trevithick, *Anal. Biochem.* **50**, 255 (1972).

The radioactivity of histones resolved by Cellogel electrophoresis can be either determined directly in an instrument for scanning paper chromatograms, or individual fractions can be cut out and counted in the scintillation spectrometer.

Starch Gel Electrophoresis. Developed by Smithies,[78] the technique of starch gel electrophoresis was first applied to the histones by Neelin and Neelin,[79] who observed 16–22 protein bands in calf thymus histone preparations. Since the number and intensity of some electrophoretic bands in calf thymus histone varied with changing conditions of the electrophoresis it was concluded that at least some of the bands are aggregates of a relatively few histone fractions. Indeed, using a simple version of horizontal starch gel electrophoresis with 10 mM HCl as electrolyte, Johns et al.[80] separated calf thymus histone into only 5–6 bands, all corresponding to the major fractions obtained by chemical fractionation procedures.

Although starch gel electrophoresis is now almost completely overshadowed by the more elegant and sensitive separation in acrylamide polymers, it is still the method of choice for metabolic studies on histones *in vivo*, where low specific activities of the fractions exclude their activity analysis by polyacrylamide gels (one slot of starch will accommodate and resolve 1–2 mg of histone mixture as compared with the 0.025–0.050 mg tolerated by each standard polyacrylamide gel).

The recommended method[24] was adapted from the horizontal system of Johns et al.[80] Starch (hydrolyzed, from Connaught Medical Research Laboratories, Toronto, Canada), 12% suspension in 10 mM HCl containing 0.2 mM AlCl$_3$ is heated to 85–90°, and the hot viscous mixture is poured into a plastic mold permitting the formation of eight slots, each measuring 15 × 6 × 250 mm. The slots connect at both ends through a horizontal cut across all the septa dividing the individual slots. Two rectangular pieces of 3 MM Whatman filter paper (double thickness) inserted into the cross cuts on both ends serve as connectors (bridges) to electrode vessels. They are held in place by solidified starch gel. After cooling, the elevated gel surface is removed by pulling thin molybdenum wire (3 mil) across the gel using the surface of the mold for guidance. The upper starch layer is discarded and histone samples, dissolved in 0.1 M HCl are applied soaked into rectangular pieces (5 × 14 mm) of Whatman No. 17 filter paper by inserting the rectangles into narrow cuts in the individual gels, about 3–4 cm from the anodic end. To prevent evaporation, the gel surface and filter paper bridges connecting the elec-

---

[78] O. Smithies, *Biochem. J.* **61**, 629 (1955).
[79] J. M. Neelin and E. M. Neelin, *Can. J. Biochem. Physiol.* **38**, 355 (1960).
[80] E. W. Johns, D. M. P. Phillips, P. Simson, and J. A. V. Butler, *Biochem. J.* **80**, 189 (1961).

trodes are covered by a sheet of plastic (Saran Wrap). The electrode vessels are 1000-ml beakers containing about 900 ml of 10 mM HCl and 0.2 mM AlCl$_3$. The electrophoresis is performed at 40 mA of constant current for 22 hours (5 mA per slot). At the end of electrophoresis, the filter paper connectors are pulled out and the mold is placed into the refrigerator for about 30 minutes to facilitate transfer of the gels. Approximately 16-cm long pieces of gels are lifted from the slots with a spatula and are placed into test tubes containing 0.1% amido black in a mixture of methanol–water–acetic acid (5:5:2, v/v/v). After staining for about 18 hours, the gels are destained by several changes of the methanol–water–acetic acid solution without the dye. The destained gels are transferred into a cutting frame, where approximately 1-mm portions are sliced off the top and bottom using a thin molybdenum wire (3 mil). The central portions measuring 3 × 15 × ~150 mm contain dark blue-stained histone bands against a pale-blue background (Fig. 6).

For quantitative evaluation and radioactivity measurements, the individual protein bands as well as a piece of gel without any protein (ahead of the fastest GRK, F2a1 or IV fraction) are cut from the gels with a razor blade. The cut pieces of gels are placed into test tubes and dried with acetone (three changes, each 4–6 hours), in a stream of dry, cool air and finally in an oven at 105° for 3–4 hours. The dry pieces are weighed and dissolved each in 7 ml of concentrated formic acid by heating in a boiling water bath for 20 minutes. After rapid cooling, 7 ml of 1 M NaOH are added and the mixture is mixed vigorously. The absorbancy determined at 630 nm is corrected for the background:

$$A_{corr} = A_s[(A_b \times W_s)/W_b]$$

where A_{corr} = corrected absorbance, A_s and A_b are absorbances of the sample and blank gel, respectively; W_s and W_b are dry weights of the sample and blank, respectively. The absorbance remains linear for the amount of histone between 0.05 and 1.0 mg of protein in each band. The amount of dye bound to the protein in starch gel was found to be independent from the composition of histones or the concentration of AlCl$_3$. However, since occasional batches of amido black exhibit differences in their binding to the KAP (F1 or I) histones, it is recommended to calibrate each new batch of this dye with isolated histone fractions.

The radioactivity of histones labeled with [^{14}C]- or [^{3}H]amino acids is determined by combusting the dried gel pieces in oxygen.[81-83] The

---

[51] F. Kalberer and J. Rutschmann, *Helv. Chim. Acta* **44**, 1956 (1961).
[82] H. E. Dobbs, *Anal. Chem.* **35**, 783 (1963).
[83] M. Baggiolini, *Experientia* **21**, 731 (1965).

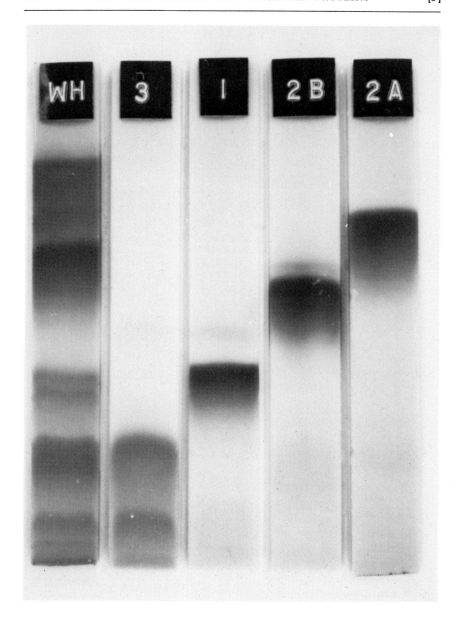

FIG. 6. Electrophoresis in starch gel of calf thymus histones. WH, unfractionated sample; 3, ARE (F3 or III) fraction; 1, KAP (F1 or I) histones; 2B, KAS (F2b or IIb2) histone; 2A, a mixture of LAK (F2a2 or IIb1) and GRK (F2a1 or IV) fractions. Starch gel (12%, w/v) was made in 0.2 mM AlCl$_3$ in 10 mM HCl. The origin of migration is at the bottom of the figure.

method of Kalberer and Rutschmann[81] can be followed with excellent results. Dried gels are wrapped into small strips of Whatman No. 1 filter paper, placed into platinum baskets, and combusted in special flasks filled with oxygen. The radioactive products of combustion are trapped in a mixture of monoethanolamine in methyl Cellosolve (1:8, v/v), aliquots are mixed with a toluene-based scintillation fluid, and their radioactivity is determined in a liquid scintillation spectrometer. A combustion apparatus made by Packard Co. can be used with a great convenience to process large numbers of samples in a single day. Usually one specimen of histone is electrophoresed in all 8 gel slots. One-half (4 gels) is used for quantitative determination of protein, and the other half (4 gels) is combusted to determine the radioactivity. All values are averaged, yielding quite accurate figures of specific activities for each histone band. Histone samples with specific activities of about 100 cpm/mg or more can be analyzed with confidence. It is preferable to use chemically prefractionated histone samples for electrophoresis instead of whole extracts. In a typical determination, chromatin is first washed with 80% ethanol. The arginine-rich histones are extracted with absolute ethanol: 1.25 M HCl (4:1, v/v) followed by extraction of the lysine-rich histones with 0.25 M HCl.[26] Histones recovered from both extracts are electrophoresed separately to obtain a better resolution. A different modification of histone electrophoresis in starch gel was described by Sung and Smithies.[84]

Histone Electrophoresis in Polyacrylamide Gels. The best technique for rapid quantitative separation of all major histone fractions became available through the application of the electrophoretic method in polyacrylamide gels, developed by Ornstein[85] and Davis.[86] It was employed, in various modifications and improvements, by many investigators studying the heterogeneity, tissue specificity, metabolism, chemical modifications, and other properties of histones. The following procedures are most widely used for electrophoretic separation of histones in polyacrylamide gels.

THE METHOD OF BONNER ET AL.[16] Fire-polished 8-cm pieces of 0.6-cm bore glass tubing are washed in chromic acid solution, rinsed well with water, and coated with silicone (the silicone coating is optional and facilitates a smooth removal of gels from the tubes). To prepare for electrophoresis, the tubes are closed on one end with rubber caps and placed into a polymerization rack made of plexiglass or other material. The acrylamide gel is made by the polymerization of a desired concentration of acrylamide monomer in the presence of a cross-linking agent, accelera-

[84] M. Sung and O. Smithies, *Biopolymers* **7**, 39 (1969).
[85] L. Ornstein, *Ann. N.Y. Acad. Sci.* **121**, 321 (1964).
[86] B. J. Davis, *Ann. N.Y. Acad. Sci.* **121**, 404 (1964).

tor, and catalyst. Either 15% or 7.5% gels in respect to polyacrylamide are recommended. The following stock solutions can be stored refrigerated for several months.

 Solution A: To prepare 15% gel, this solution consists of 60 g of acrylamide (electrophoretic grade, Bio-Rad Labs.) and 0.4 g of Bis (N,N'-methylenebisacrylamide, electrophoretic grade, Bio-Rad Labs.) in 100 ml of deionized water. For 7.5% gels, only 30 g of acrylamide and 0.8 g of Bis in 100 ml of deionized water should be used.

 Solution B: This solution contains 48 ml of 1.0 M KOH, 17.2 ml of glacial acetic acid, and 4 ml of TEMED ($N,N,N,',N'$-tetramethylethylenediamine, Eastman Kodak Co.) in 100 ml of deionized water.

 Solution C: The catalyst solution of 0.2% ammonium persulfate in freshly deionized 10.0 M urea must be made fresh daily, immediately before use.

To make 12 gels, 4 ml of A and 2 ml of B are mixed with 10 ml of C. The mixture can be kept on ice for several hours to prevent its polymerization. Chilling the monomer mixture is especially important if higher concentrations of ammonium persulfate (0.6%) in 10.0 M urea are used for accelerated polymerization. The thoroughly mixed solution is filled into the stoppered glass columns to a height of 6 cm. Disposable plastic syringe or a Pasteur pipette can be used conveniently. The filled columns are freed of any entrapped air bubbles and carefully overlaid with 0.1–0.2 ml of 3 M urea to ensure anaerobic polymerization. With 0.2% ammonium persulfate, the polymerization is complete in about 1 hour. If higher concentrations of ammonium persulfate are used (e.g., 0.6%) the gels are first partially polymerized at 10° for 2 hours to eliminate any dissolved air freed by the heat of polymerization. The polymerization is completed at room temperature.

After polymerization, the 3 M urea is carefully removed with a Pasteur pipette and histone samples, dissolved in 10 M urea, are applied to the top of the gels. The sample volume should not exceed 50 μl for best resolution. A buffer consisting of 31.2 g of β-alanine and 8 ml of glacial acetic acid in 1000 ml of deionized water is gently layered over the sample solution and all individual gel tubes are filled to the top. It is also possible to first fill the gel tubings with buffer and layer carefully the dense solution of histones in 10 M urea between the buffer and gel surface. To follow the migration it is recommended to add a tracking dye to at least one sample (about 5 μl). Pyronin Y or benzeneazo-α-naphthylamine are most frequently used. The filled tubes are inserted into plastic retainers built into most of the commercial electrophoretic assemblies, electrode vessels are filled with the β-alanine buffer, and a direct current of 5 mA per tube is applied to the apparatus for 2–3 hours. The electrophoresis is terminated when the band of pyronine Y dye reaches the end of the gels. A regulated power supply providing constant current up to 100 mA is recommended. If the electrophoresis is done at low temperature (4°), the time of electrophoresis must be approximately doubled to compensate for the slower histone migration at low temperatures. Currents greater than 6 mA per gel tube result in excessive heat formation which leads to a deformed protein banding pattern (proteins in central portions of the gel migrate faster than those in peripheral parts). Lower current, on the other hand, would allow excessive diffusion of individual protein bands with subsequent loss of resolution.

After electrophoresis the columns are immediately removed from the apparatus and the gels are carefully reamed free using a stream of water delivered through a blunted 15-gauge hypodermic needle. The gels are stained in 1% solution of amido black (Buffalo black or naphthol blue black) in aqueous 40% ethanol and 7% acetic acid solution. Although about 2 hours of staining are sufficient for qualitative analysis, an overnight staining period is recommended for quantitative evaluation. The gels are destained electrophoretically in 10% aqueous acetic acid, a commercial destaining apparatus and a source of direct current being used. Column retaining devices permitting current flow across the entire length of gel columns are recommended. A 24 V storage battery charger can be used to supply several destaining devices simultaneously. Destained gels can be kept stored in 10% acetic acid for many months without considerable decoloration. However, for quantitative analysis, it is recommended to scan the gels immediately after their destaining.

Destained gels can be scanned intact or a section (slab) can be cut along their long axis to obtain a central part of the column. The gels are scanned in a densitometer or a spectrophotometer with linear transport (Canalco model E microdensitometer, gel scanning attachment to a spectrophotometer Gilford, Beckman Photovolt model 530 densitometer, etc., are the most commonly used instruments) at 600–610 nm. For round gels, it is recommended to scan each gel several times, rotating the gel slightly between the individual scans. There is a linear relationship between the absorbancy at 610 nm and the histone concentration at least up to 10 μg of protein in each band. For most dyes, the dye-protein binding ratios are essentially the same for all the main histone fractions. Calibration curves can be constructed for the individual histones, and their concentration can be determined from the area of the scanned peaks (by integration, planimetry, or gravimetrically). The scan of unfractionated calf thymus histone is shown in Fig. 7.

For radioactivity determination, the scanned gels are sliced into disks

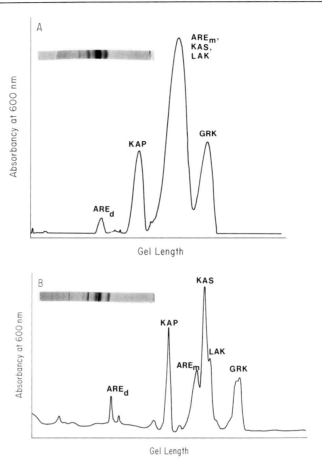

FIG. 7. Electrophoresis in polyacrylamide gel of unfractionated calf thymus histone. (A) The method of J. Bonner et al. (this series, Vol. 12B [96], p. 3). (B) The method of S. Panyim and R. Chalkley [*Arch. Biochem. Biophys.* **130**, 337 (1969)]. The individual fractions are identified by their symbols. ARE_m and ARE_d indicate the reduced (monomer) or oxidized (dimer) form of this fraction, respectively. The direction of migration was from left to right.

of constant thickness (several commercially made slicers are available) and the individual slices are placed into scintillation vials. A small volume (0.1 ml) of 30% H_2O_2 is added and the vials are heated at 50° until the gel slices become depolymerized.[87] The gel slices must not dry completely during their heating. After depolymerization, a small volume (0.1–0.2 ml) of a commercial solubilizer (NCS of Amersham-Searle,

[87] P. V. Tishler and C. J. Epstein, *Anal. Biochem.* **22**, 89 (1968).

Soluene of Packard Instruments, etc.) is added and mixed with 10–15 ml of a toluene-based scintillation fluid. It is also possible to place the individual wet gel slices into 10–15 ml of a commercial solubilizer-fluor solution (e.g., Protosol, New England Nuclear), incubate overnight, and count the eluted protein directly. Specific activity estimates can be obtained by plotting the radioactivities of each slice against the optical scan.

The urea present in the sample prevents aggregation of histones and also inhibits the formation of complexes between histones and some acid soluble nonhistone proteins. As can be seen in Fig. 7 the described method separates histones, in the order of their increasing mobility: oxidized (dimer) ARE (F3 or III), KAP (F1 or I), reduced (monomer) ARE (F3 or III), KAS F2b or IIb1) and GRK (F2a1 or IV). The ARE (F3 or III) monomer, KAS (F2b or IIb2), and LAK (F2a2 or IIb1) histones frequently form one broad band and cannot be clearly separated in a single electrophoresis. Protamines can be resolved by the Bonner et al.[16] procedure from the histones. Their mobility is quite high, similar to that of the tracking dye, pyronine Y. An improved separation of all the main histone fractions was described by Panyim and Chalkley.[88]

THE PROCEDURE OF PANYIM AND CHALKLEY. This modification of the method of Bonner et al.[16] is capable of clearly resolving all the main histone fractions and is rapidly becoming the most used procedure for electrophoretic analysis of histones. It employs long (7.5, 15, or more centimeters) gels and 0.9 M acetic acid as the electrolyte. The following stock solutions, with the exception of 0.2% ammonium persulfate in 10 M urea, can be stored refrigerated for several months.

Solution A: 40% (w/v) of acrylamide (electrophoretic grade or crystallized commercial grade) and 0.4% (w/v) of Bis in deionized water

Solution B: 4% (w/v) of TEMED in 43.2% (v/v) aqueous acetic acid

Solution C: Dissolve 0.2% (w/v) of ammonium persulfate in deionized 10 M urea. Make fresh daily before use.

To make gels, 2 parts of A, 1 part of B, and 5 parts of C are thoroughly mixed and filled into glass or plastic columns of desired length. In the presence of urea, the polymerization is retarded, and about 3 hours should be allowed for complete polymerization.

All polymerized gels are preelectrophoresed to remove the catalyst. The tray buffer for preelectrophoresis is 0.9 M acetic acid. The authors recommend to monitor the preelectrophoresis by applying benzeneazo-α-

---

[88] S. Panyim and R. Chalkley, *Arch. Biochem. Biophys.* **130**, 337 (1969).

naphthylamine to one gel. This dye is red in acid and travels about 1.65 cm per hour at 2 mA per tube. The preelectrophoresis is judged complete when the dye is eluted after crossing the entire length of the gel. Pyronine Y can also be used for tracking.

For analysis, 1-mg samples of histone are dissolved in 1 ml of 0.9 N acetic acid and 15% sucrose (to increase their density). Between 10 and 50 μl of this solution are applied to the gels, depending on the number of fractions in the mixture and desired intensity of the bands. Usually 20 μl of unfractionated histone are sufficient. Since the polyacrylamide gel electrophoresis is a relatively simple and fast procedure, each histone sample can be applied in 2 or 3 different concentrations to select the best concentration range for scanning and radioactivity measurements. The electrophoresis is done in 0.9 M acetic acid as the electrolyte. To keep the ohmic heat to a minimum, the current of 2 mA per 7.5 cm gel is recommended. Pyronine Y can be added to one sample to serve as a migration indicator. The current is disconnected when the dye reaches end of the gel column. The developed gels are stained and processed as was described for the system of Bonner et al.[16]

The separation of individual histones and their subfractions can be manipulated by variations in the pH and urea concentration.[88,89] At pH 2.7, the best separation of histones can be obtained in the presence of 2.5 M urea. In 6.25 M urea at pH 3.2, KAS (F2b or IIb2) and LAK (F2a2) or IIb1) histones migrate together. However, under these conditions, the separation of subfractions of the ARE (F3 or III), KAP (F1 or I), and GRK (F2a1 or IV) histones is at its best. In absence of urea, fractions ARE monomer (F3 or III), KAS (F2b or IIb2), and LAK (F2a2 or IIb1) migrate in this sequence in respect to their increasing mobility. In 6 M urea, the order of KAS (F2b or IIb2) and LAK (F2a2 or IIb1) histones is reversed and they migrate in sequence of their increasing mobilities: ARE monomer (F3 or III), LAK (F2a2 or IIb1), and KAS (F2b or IIb2).

While the electrophoretic heterogeneity of KAP (F1 or I) histones is caused by their true chemical heterogeneity and only to a small extent by chemical modification (phosphorylation),[90,91] the electrophoretic heterogeneities of ARE (F3 or III) and especially of GRK (F2a1 or IV) are caused by differential acetylation of the lysine residues in the parental molecule.[92] The bands in order of their decreasing mobility correspond

---

[89] S. Panyim, D. Bilek, and R. Chalkley, *J. Biol. Chem.* **246**, 4206 (1971).

[90] K. Evans, P. Hohman, and R. D. Cole, *Biochim. Biophys. Acta* **221**, 128 (1970).

[91] D. Sherod, G. Johnson, and R. Chalkley, *Biochemistry* **9**, 4611 (1970).

[92] L. Wangh, A. Ruiz-Carrillo, and V. G. Allfrey, *Arch. Biochem. Biophys.* **150**, 44 (1972).

to 0, 1, 2, and possibly 3 acetylated lysine residues in each histone molecule.

THE METHOD OF JOHNS. Histone separation in polyacrylamide gels without urea at pH 2.4 (about 0.9 M acetic acid) was independently described by Johns.[93] Acetic acid (10 mM) is used for electrolyte. This method resolves clearly all the main histone fractions with the exception of KAS (F2b or IIb2) and ARE monomer (F3 or III) which form one broad band. Using differential staining of parallel gels with amido black and bromophenol blue, the KAS (F2b or IIb2) and ARE (F3 or III) histone bands can be resolved since the bromophenol blue stains only the arginine-rich histones ARE (F3 or III) and GRK (F2a1 or IV).[94]

The KSA (F2c or V) histone which is specific for nucleated erythrocytes in birds and other lower vertebrates, migrates in both the Panyim and Chalkley and the Johns gel systems between the KAP (F1 or I) and ARE monomer (F3 or III) histone fractions.

Polyacrylamide Gel Electrophoresis in the Presence of Sodium Dodecyl Sulfate. A more recent electrophoretic modification introduced by attempts to electrophorese the nonhistone protein species of chromatin is the polyacrylamide gel electrophoresis in the presence of sodium dodecyl sulfate.[95] Because of their interaction with this detergent, most proteins form negatively charged complexes which separate electrophoretically in polyacrylamide gels in order of their molecular weights.[95,96] Indeed the electrophoretic mobility of proteins larger than 10,000 daltons can be used for fairly accurate molecular weight determination.[95,97]

Sodium dodecyl sulfate dissociates most proteins into their subunits (unless covalently linked) and consequently, the number of protein bands observed in polyacrylamide gel electrophoresis performed in the presence of this detergent corresponds to the number of subunits and may be much larger than the true number of the *in vivo* protein species. On the other hand, polypeptide chains of similar or identical molecular weights will form a single protein band. Since histones are simple proteins, their homogeneity is not affected by exposure to sodium dodecyl sulfate. However, their molecular weights just about approach the lowest range of confidence for molecular weight determinations of the sodium dodecyl sulfate–protein complexes.[97]

The electrophoresis in presence of sodium dodecyl sulfate can be per-

[93] E. W. Johns, *Biochem. J.* **104**, 78 (1967).
[94] I. D. Barrett and E. W. Johns, *J. Chromatogr.* **75**, 161 (1973).
[95] A. L. Shapiro, E. Viñuela, and J. V. Maizel, *Biochem. Biophys. Res. Commun.* **28**, 815 (1967).
[96] A. K. Dunker and R. R. Rueckert, *J. Biol. Chem.* **244**, 5074 (1969).
[97] J. G. Williams and W. B. Gratzer, *J. Chromatogr.* **57**, 121 (1971).

formed both with or without urea. Because this detergent eliminates virtually all the protein–protein interactions, the presence of urea is not as beneficial as in the detergent-free systems. Elimination of the positive charges on histones by the detergent makes impossible the electrophoretic separation of differentially acetylated or phosphorylated histones on sodium dodecyl sulfate containing gels.

Currently used methods of polyacrylamide electrophoresis in sodium dodecyl sulfate-containing media are based on the procedure described by Maizel and his co-workers.[95]

A good separation of histones, even in the presence of nonhistone proteins, can be obtained in 10% polyacrylamide gel containing 4 M urea. The gels (10 × 0.6 cm) are prepared by mixing the following stock solutions:

A. Gel buffer: 0.2 M phosphate buffer containing 0.2% sodium dodecyl sulfate and 8 M urea, pH 7.0

B. Acrylamide solution: 24% acrylamide (electrophoretic grade) and 0.6% Bis in deionized water

C. Ammonium persulfate, 90 mg in 10 ml of deionized water (must be made fresh before use)

D. TEMED, undiluted reagent

The solutions are mixed in the ratio A:B:C:D = 6:5:1:0.06. For 12–14 gels (10 × 0.6 cm) 30 ml of A are mixed with 25 ml of B, 5 ml of C and cooled on ice under reduced pressure for 15 minutes. Finally, after the addition of 0.03 ml of TEMED (solution D), the gels are poured immediately and overlaid each with 0.1 ml of solution A diluted 1:1 with deionized water. The gels are polymerized for at least 2 hours at room temperature. In the presence of large excess of nonhistone proteins or DNA, a 3% polyacrylamide stacking gel should be layered over the 10% gel. To prepare the stacking gel, a modified solution B containing 7.5% acrylamide and 1.2% Bis in deionized water is used instead of the regular solution.

Samples containing 10–30 μg of unfractionated histone are dissolved in solution A which was made more dense by the addition of concentrated sucrose or urea. After filling with the electrolyte buffer (solution A) the gels are electrophoresed at 8–10 mA per tube for 8–10 hours. Application of a tracking dye (bromophenol blue) to at least one of the samples helps to monitor the migration. After electrophoresis, the gels are stained in the solution of Coomassie blue or amido black (0.2% or 0.5%, respectively) in methanol–acetic acid–water mixture 5:1:5 or they can be first fixed in 20% aqueous sulfosalicylic acid or 15% trichloroacetic acid. This treatment also removes most of the detergent which may interfere with subsequent staining. Some batches of Coomassie blue are incompatible

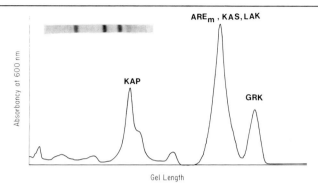

FIG. 8. Electrophoresis of unfractionated calf thymus histone in polyacrylamide gel containing sodium dodecyl sulfate. The histone fractions are identified by their symbols. The direction of migration was from left to right.

with trichloroacetic acid, and it may be necessary first to remove the acid by placing the fixed gels into 20% sulfosalicyclic acid solution for several hours. A mixture of methanol, acetic acid, and water (5:1:5) can be also used to remove the 15% trichloroacetic acid from the gels before their staining with Coomassie blue. The gels can be destained by diffusion in an excess of the methanol–acetic acid–water mixture, or much faster, electrophoretically in 7% acetic acid.

In the presence of sodium dodecyl sulfate, the following histone fractions can be resolved in order of their increasing mobilities: oxidized ARE dimer (F3 or III), the KAP (F1 or I) histones, a mixture of ARE monomer (F3 or III) and the KAS (F2b or IIb2) fraction, the LAK (F2a2 or IIb1), and finally, the GRK (F2a1 or IV) bands (Fig. 8). Because the electrophoresis of histones in the presence of sodium dodecyl sulfate is more time consuming and requires more expensive equipment as compared with the methods of Bonner et al.[16] and of Panyim and Chalkley,[88] it is not recommended for serial analyses of histones, especially since the Coomassie blue-stained gels cannot be directly used for quantitation of the individual fractions. However, this method should be selected when the mixtures of histones and nonhistone proteins are to be analyzed.

Other Methods for Histone Determination

There are several other techniques that can be used to estimate the kinds and quantities of histones in various biological materials. Virtually all are more complicated and frequently not as informative as the electrophoretic analysis of histones. Among several, at least three should be mentioned in more detail.

Cytochemical Measurements

Less specific methods are based on the interactions of histones and other proteins with various dyes and their subsequent determination *in situ* by microspectrophotometry. Quantitative cytochemical measurement of histones after their staining with alkaline fast green was introduced by Alfert and Geschwindt[98] and subsequently modified to increase its selectivity by Bloch.[99] The cytochemical procedures, which were reviewed in several articles,[99–101] offer the advantage of estimating the quantities of histones and other basic proteins in chromosomes, nuclei, chromatin, etc., directly, without disturbing the cellular morphology.

Principally, the material (cells or tissue slices) is fixed and, after removal of the fixative, the cellular structures are stained either directly or after first removing the DNA, usually by acid hydrolysis (saturated hot picric acid solution was found to result only in minimal losses of basic nuclear proteins).[99] Fixation in neutral formaldehyde decreases the stainability of lysine-rich histones, probably through formylation of the ϵ-N-amino residues of lysine.[100] This effect of formaldehyde can be eliminated by exposure to an acid environment. Alkaline fast green solution stains efficiently all basic proteins both *in situ* or when dried on a nonstainable support.[100,102] Variations of the basic staining procedure include specific staining with bromophenol blue (protaminelike proteins),[99] Sakaguchi reaction or fluorescent staining with phenanthrenequinone for arginine-rich proteins,[103] etc. Deamination of proteins or acetylation of the free amino groups can further increase the selectivity of histochemical staining procedures.

Information about the conformational state of chromatin and distribution of nonhistone proteins can be obtained by various combinations of staining with ammoniacal silver,[104] toluidine blue,[105] actinomycin D,[106–108] acridine orange,[108,109] etc.

[98] M. Alfert and I. I. Geschwindt, *Proc. Nat. Acad. Sci. U. S.* **39**, 991 (1953).
[99] D. P. Bloch, *Protoplasmatologia* **5**, Fasc. 3 (1966).
[100] N. R. Ringertz and A. Zetterberg, *Exp. Cell Res.* **42**, 243 (1966).
[101] K. Smetana, *Methods Cancer Res.* **2**, 361 (1967).
[102] L. Berlowitz, D. Pallotta, and P. Pawlowski, *J. Histochem. Cytochem.* **18**, 334 (1970).
[103] B. E. Magun and J. W. Kelly, *J. Histochem. Cytochem.* **17**, 821 (1969).
[104] M. M. Black and H. R. Ansley, *Science* **143**, 693 (1964).
[105] R. Love and R. J. Walch, *J. Histochem. Cytochem.* **11**, 188 (1963).
[106] Z. Darzynkiewicz, L. Bolund, and N. R. Ringertz, *Exp. Cell Res.* **56**, 418 (1969).
[107] N. R. Ringertz, Z. Darzynkiewicz, and L. Bolund, *Exp. Cell Res.* **56**, 411 (1969).
[108] R. Rigler, D. Killander, L. Bolund, and N. R. Ringertz, *Exp. Cell Res.* **55**, 215 (1969).
[109] D. Killander and R. Rigler, *Exp. Cell Res.* **54**, 163 (1969).

Although most histochemical procedures are unable to distinguish between the individual histone fractions, their major advantage is their resolution on subcellular level, permitting accurate localization of basic proteins within the individual cellular structures.

Immunochemical Reactions of Histones

Because of their exceptional sensitivity to proteolytic enzymes and limited species specificity, histones are poor antigens making difficult quantitative immunoassays for the individual fractions. However, the immunochemistry of histones is quite useful for comparative studies on the tissue and species specificity of these proteins.

Since histones alone are not antigenic, antibodies must be elicited against the complexes of histones with other macromolecules. As was first shown by Landsteiner, even a minor chemical modification of the antigenic molecule can alter substantially its immunochemical specificity. It is possible to immunize rabbits with covalently linked complexes of histones with serum albumin or nucleic acids. Using this procedure, Stollar and Ward[110] were able to obtain specific antibodies against each of the main histone fractions and show, with the help of microcomplement fixation technique of Wasserman and Levine,[111] that corresponding homologous histones from various animal species (human, calf, chicken, frog, and lobster) are immunochemically identical. The extreme sensitivity of the complement fixation technique was demonstrated when it was able to clearly distinguish between the antisera from rabbits immunized with the individual subfractions of the KAP (F1 or I) histone from rabbit or calf thymus.[112] It is known from their amino acid sequences that these subfractions are extremely similar, differing in only a few amino acid residues.[113]

Determination of the Lysine–Tryptophan Ratios

This technique, which employs the differential incorporation of lysine (or arginine) and tryptophan into newly synthesized proteins, is too crude for an accurate histone assay. It was used mainly to study histone biosynthesis on isolated polysomes, i.e., a situation where minute quantities of material, together with the interference of some basic ribosomal proteins, make the use of more specific histone assays very difficult. It

---

[110] B. D. Stollar and M. Ward, *J. Biol. Chem.* **245**, 1261 (1970).
[111] E. Wasserman and L. Levine, *J. Immunol.* **87**, 290 (1961).
[112] M. Bustin and B. D. Stollar, *J. Biol. Chem.* **247**, 5716 (1972).
[113] S. C. Rall and R. D. Cole, *J. Biol. Chem.* **246**, 7175 (1971).

was shown by comparing the incorporation of lysine or arginine with that of the differently labeled tryptophan (^{14}C versus ^{3}H), that a class of small polysomes in the cytoplasm can be identified with the site of histone and protamine biosynthesis.[114-116] Within limitations, the lysine or arginine to tryptophan labeling ratios can help to distinguish between the histones and nonhistone proteins in complicated polyacrylamide gel electrophoretograms. However, it must be emphasized that not every protein lacking tryptophan is histone.

Acknowledgments

Supported by the U.S. Public Health Service grants CA-07746 and HD-5803 and by The Robert A. Welch Foundation grant G 138.

[114] E. Robbins and T. W. Borun, *Proc. Nat. Acad. Sci. U.S.* **57**, 409 (1967).
[115] M. Nemer and D. Lindsay, *Biochem. Biophys. Res. Commun.* **35**, 156 (1969).
[116] L. H. Kedes, P. R. Gross, G. Cognetti, and A. L. Hunter, *J. Mol. Biol.* **45**, 337 (1969).

[10] Methods for the Assessment of Selective Histone Phosphorylation

By ROD BALHORN and ROGER CHALKLEY

The cell nucleus contains a great variety of phosphorylated proteins. Accordingly, before selective histone phosphorylation can be analyzed, it is highly advantageous to purify the histones to the greatest degree possible. Methods for isolation and purification of chromatin, the complex of DNA and nuclear proteins, have been adequately defined.[1] However, the ease of chromatin isolation varies from tissue to tissue, and it is important that appropriate criteria be used to guarantee the degree of purity of the material.[1] Histones can be efficiently extracted by the acidification of chromatin using previously documented methods.[1] In general, acidification with sulfuric acid is preferable to the use of hydrochloric acid which introduces adventitious, slightly basic proteins. Extraction using elevated ionic strengths is not recommended as the procedure is lengthy, nuclear proteases may function to give artifacts and aggregation of certain histone fractions may also serve to confuse the unwary.

[1] J. Bonner, G. R. Chalkley, M. Dahmus, D. Fambrough, F. Fujimura, R.-C. Huang, J. Huberman, R. Jensen, K. Marushige, H. Ohlenbusch, B. Olivera, and J. Widholm, this series, Vol. 12B[96], p. 3.

If one wishes to ascertain whether a given histone fraction is phosphorylated, one can employ polyacrylamide gel electrophoresis directly on the whole histone extract (see Section A). If a more detailed analysis of a given histone fraction is required, then it will be necessary to isolate that fraction before proceeding to high resolution gel analysis (see Section B). Methodology is available for the separation and purification of any given histone fraction, and the appropriate literature may be consulted.[2-4] As the lysine-rich (F1) histone is perhaps most frequently encountered in its phosphorylated form, we have confined ourselves to a discussion of the isolation and analysis of this particular fraction, although the methods used for analysis of phosphorylation are generally applicable.

A. Assay for Phosphorylation of a Given Histone Fraction

1. Assays for histone phosphorylation without using ^{32}P are available and will be discussed below. However, the process is greatly simplified if radiolabeled [^{32}P]histone is available. In general the method to be described can detect as little as 500 cpm per milligram of histone, but the conditions for radiolabel incorporation which are presented here will lead to values much in excess of this in replicating cells.

2. Tumor cells, either in tissue culture or carried within the rat are incubated with sodium [^{32}P]phosphate, pH 7.4, in saline. In general, two 300 μCi injections delivered at 0 and 6 hours is an appropriate dose and the cells may be harvested on the ninth hour. Additional doses at later times are probably of little value in increasing specific activity because of the fairly rapid turnover of histone phosphate groups. For tissue culture studies a concentration of 2 μCi/ml is used at a cell concentration of 2×10^5 to 4×10^5 cells per milliliter. Significant histone counts can be detected after 20 minutes of incubation and will reach a maximum after approximately 3 hours.

3. The whole histone material is extracted by standard techniques and should be stored as a solid powder at $-20°$. It is dissolved in 0.9 N acetic acid, 15% sucrose in preparation for polyacrylamide gel electrophoretic analysis. If the material is counted at this time, it will be found to be replete with ^{32}P radioactivity. However, this is of no diagnostic value for information concerning the level of histone phosphorylation. Inorganic phosphate and low molecular weight polyribonucleotides are invariable contaminants of histone preparations, amounting on occasion

[2] E. W. Johns, *Biochem. J.* **92**, 55 (1964).
[3] D. M. P. Phillips and E. W. Johns, *Biochem. J.* **94**, 127 (1965).
[4] D. Oliver, K. R. Sommer, S. Panyim, S. Spiker, and R. Chalkley, *Biochem. J.* **129**, 349 (1972).

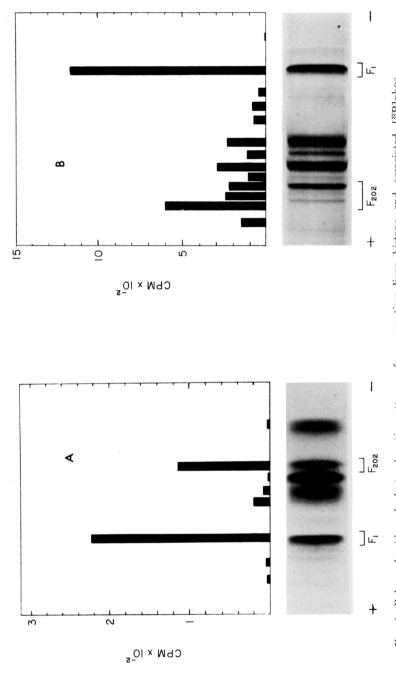

FIG. 1. Polyacrylamide gel electrophoretic patterns of regenerating liver histone and associated [^{32}P]phosphate in acid-urea (A) and Triton (B) gel systems.

to as much as 90% of the radiolabel that may be present.[5] It is for this reason that electrophoresis is of such major value, as the impurities are negatively charged and migrate in the opposite direction to the phosphohistones which have an overall positive charge.

4. Polyacrylamide gels (15% acrylamide, 0.9 N acetic acid, 2.5 M urea) are prepared by mixing solutions A through C in the following proportions: 2 parts solution A, 1 part solution B, and 5 parts solution C.

Solution A: 60% acrylamide monomer (w/v, $R_f = 1.421$) and 0.4% N,N'-methylenebisacrylamide (w/v) in water

Solution B: 43.2% glacial acetic acid (v/v) and 4% N,N,N',N'-tetramethylethylenediamine (v/v) in water

Solution C: 0.2% ammonium persulfate[6] and 4 M urea in water

The resulting mixture is deaerated *in vacuo* (to remove dissolved air that will be released as bubbles during polymerization) and pipetted into glass tubes measuring 0.6 cm (i.d.) by 10 cm. The gels are immediately layered with 0.5 cm of cold 0.9 N acetic acid to ensure the formation of flat gel tops. Polymerization is complete within 60 minutes.

Triton gels[7] are prepared in an identical fashion, except that 0.8% Triton X-100 is added to solution C to give a final gel concentration of 0.5%.

5. Preelectrophoresis is performed at 130 V for 3.5 hours at room temperature in 0.9 N acetic acid (pH 2.8). This step is necessary to remove the ammonium persulfate before electrophoresis of protein.

Prior to the application of samples, the gel tops are gently aspirated with 0.9 N acetic acid to disperse the urea layer that forms just above the gel. Samples are layered on the gels and electrophoresed at 130 V (2 mA/gel) for 3 hours at room temperature.

6. The gels are stained for a minimum of 3 hours in Amido Schwarz (0.1% Amido Schwarz, 7% acetic acid, 30% ethanol). The gels are then destained, either by electrophoresis or diffusion.

After destaining, the gels are scanned at 600 nm in a microdensitometer. The area under each curve is directly proportional to the histone content of the band. The gels are sliced into 3-mm sections, dried, and counted in a gas-flow planchet counter. If liquid scintillation counting is preferred, each gel slice can be digested in 0.20 ml of 30% hydrogen

---

[5] R. Balhorn, W. O. Rieke, and R. Chalkley, *Biochemistry* **10**, 3952 (1971).

[6] To decrease the rate of polymerization, the ammonium persulfate concentration may be lowered to 0.1%.

[7] A. Zweidler and L. H. Cohen, *Fed. Proc., Fed. Amer. Soc. Exp. Biol.* **31**, 4051 (1972).

peroxide at 60° for 1 hour in a tightly covered scintillation vial. After digestion, 10 ml of Bray's solution or BioSolve is added.

7. Identification of phosphorylation either of histone fractions F1 or F2a1 would be conclusive in the acid-urea gel system as these fractions are so totally separated from other histone bands (Fig. 1). However, a conclusive identification of which histone fraction is phosphorylated is more troublesome for the three histone fractions which migrate closely together in this system. Advantage may then be taken of the Triton gel system,[7,8] which depends upon the relative degree of detergent binding and permits a dramatic resolution of the three histone fractions which presented some interpretive problem in the acid–urea gel system.

Thus, by using the two electrophoretic systems in conjunction, there should be no ambiguity in the conclusion of which histone fractions have been phosphorylated. Application of this approach to demonstrate that histone F2a2 (and not histone F2b) is phosphorylated in rapidly dividing tumor cells is shown in Fig. 1.

B. Detailed Analysis of Phosphorylation within a Given Histone Fraction

1. Lysine-rich, ^{32}P-labeled histone can be separated from other histone fractions by treating a solution of whole histones (2 mg/ml in 0.001 N HCl) to a final concentration of 0.5 N perchloric acid. Only the lysine-rich histone remains in solution. After centrifugation, dialysis against 0.4 N sulfuric acid and then against ethanol leads to a precipitate that can be collected, washed with acetone, and dried.

2. The mobility (m_ϵ) of a given histone in the acid-urea system is described by the equation

$$m_\epsilon = k\sigma^+ t/[\eta]^{0.9}\eta_s$$

where k is a constant, σ^+ is the overall positive charge density on the histone, t is the time of electrophoresis, η is the intrinsic viscosity of the histone, and η_s is the viscosity of the solvent system. Thus, since the incorporation of phosphate groups should decrease the overall positive charge density of a given histone, then the phosphohistones should migrate more slowly than the parent proteins, giving rise to a series of discrete bands containing more phosphate groups in the more slowly moving fractions.

[8] R. Balhorn, D. Oliver, P. Hohmann, R. Chalkley, and D. Granner *Biochemistry* **11**, 3915 (1972).

3. Resolution of the phosphoprotein subbands requires the use of high resolution gel electrophoresis. The addition of each phosphate moiety to mammalian lysine-rich histone decreases the overall positive charge on the molecule by approximately 1 part in 65. If electrophoresis is performed in 20–25 cm gels, each species (parental and multiply phosphorylated) may be separated by 3–4 mm. Such separation is obtained by electrophoresis at 200 V in 25-cm gels (15%) at 4° for 60–85 hours (see Fig. 2). The tray buffer (0.9 N acetic acid) must contain urea (2.5 M) to prevent urea loss from the gels. Electrophoresis in the cold markedly decreases diffusion and increases band clarity.

The presence of considerable electrophoretic microheterogeneity in the lysine-rich histone fraction suggests, but does not demand, the presence of phosphorylated species. Most tissues contain at least two electrophoretically distinct parental lysine-rich histones, and it is therefore necessary to distinguish between phosphorylation-induced and sequence-induced microheterogeneity. If the lysine-rich histone can be labeled with [^{32}P]phosphate, phosphorylation-induced species may be identified by

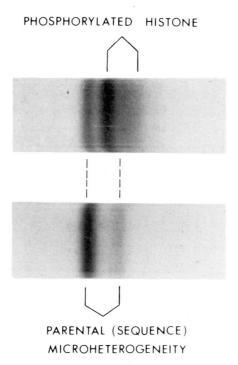

Fig. 2. High resolution polyacrylamide gel electrophoretic patterns of HTC lysine-rich histones before and after phosphatase treatment.

cutting out specific F_1 subbands, drying, and counting them. Assignment can also be made using unlabeled lysine-rich histone by removing phosphate groups enzymatically.

4. Removal of the phosphate groups from phosphohistones removes not only the associated ^{32}P but also the charge-induced microheterogeneity. Phosphate groups can be conveniently removed by treatment with *Escherichia coli* alkaline phosphatase[9] (Fig. 2). Lysine-rich histones are incubated with 1 mole of enzyme per 30–50 moles of histone (F_1) in 10 mM Tris · HCl, pH 8.0, for 24 hours at 37°. The reaction is terminated by exhaustive dialysis against 0.9 N acetic acid, 15% sucrose. Samples can then be directly applied to 25-cm gels and electrophoresed. The electrophoretic patterns thus obtained will be representative of the unphosphorylated, parental species.

[9] The highest quality Worthington enzyme (electrophoretically pure) must be used as the less pure enzymes contain very active proteases.

[11] Methods for Isolation and Characterization of Nonhistone Chromosomal Proteins

By SARAH C. R. ELGIN

The nonhistone chromosomal proteins (NHC proteins) are defined to be those proteins, other than the histones, which isolate in association with nuclear DNA in the purification of chromatin. The methods for the isolation and characterization of NHC proteins given in detail here were developed using chromatin prepared according to the method of Bonner *et al.*[1] Very briefly, this method is as follows. Frozen tissue is disrupted by grinding in a Waring blender, and the crude nuclear pellet is collected by centrifugation. After washing in saline-EDTA, the nuclei are lysed by homogenization in 10 mM Tris, pH 8. The crude chromatin is collected by centrifugation, washed several more times in 10 mM Tris, and purified by centrifugation through 1.7 M sucrose. The purified chromatin is resuspended in 10 mM Tris and solubilized by shearing in a Virtis homog-

[1] J. Bonner, G. R. Chalkley, M. Dahmus, F. Fujimura, R. C. Huang, J. Huberman, R. Jensen, K. Marushige, H. Ohlenbusch, B. Olivera, and J. Widholm, this series, Vol. 12B, p. 3.

enizer. This solubilized chromatin is the starting material for the isolation of NHC proteins. It should be noted that it is important to lyse the nuclei in buffers with low concentrations of salts; lysis in 0.14 N NaCl can apparently lead to adventitious adsorption of additional nuclear and cytoplasmic protein to the chromatin.[2] The purification of chromatin by centrifugation through 1.7 M sucrose is important in eliminating loosely bound protein. That chromatin prepared by this method is equivalent to the complex as it exists *in vivo* is indicated by the fact that its properties as a template for RNA polymerase are the same, to the extent that this can be assessed with present technology.[3–5] The NHC proteins are a complex family of proteins; the set probably overlaps, but is not identical to, the nuclear acidic proteins or nuclear phosphoproteins. The latter are prepared by extracting nuclei rather than chromatin.[6–8]

General Strategies for the Isolation of NHC Proteins

Several methods have been developed for the fractionation of chromatin into its three main constituent components, DNA, histones, and NHC proteins. The problem is not a trivial one; the native complex is difficult to dissociate, and the proteins tend to aggregate with each other when separated from the DNA. The strategy that has been most frequently employed is as follows. The chromatin is dissociated, either with sodium chloride–urea mixtures or with denaturing solvents. The DNA is removed by centrifugation or by exclusion chromatography. The histones and NHC proteins are then separated on the basis of pI; frequently this is done by maintaining the proteins in solution by the use of urea and low concentrations of salt, and fractionating the mixture on cation exchange resin. The tendency of the chromosomal proteins to aggregate forces one to make a choice between methods where clean separations and good recoveries are easily achieved but highly denaturing solvents are used, and methods where separations and recoveries are not so complete but the proteins are exposed only to solvents such that they can reasonably be expected to maintain or recover biological activity. Two

---

[2] E. W. Johns and S. Forrester, *Eur. J. Biochem.* **8**, 547 (1969).
[3] K. Marushige and J. Bonner, *J. Mol. Biol.* **15**, 160 (1966).
[4] J. Paul and R. S. Gilmour, *J. Mol. Biol.* **34**, 305 (1968).
[5] K. D. Smith, R. B. Church, and B. J. McCarthy, *Biochemistry* **8**, 4271 (1969).
[6] T. Y. Wang, *J. Biol. Chem.* **242**, 1220 (1967).
[7] T. A. Langan, "Regulation of Nucleic Acid and Protein Biosynthesis" (V. V. Koningsberger and L. Bosch, eds.), *Biochim. Biophys. Acta Library Ser.* **10**, 233. Elsevier, Amsterdam, 1967.
[8] L. J. Kleinsmith and V. G. Allfrey, *Biochim. Biophys. Acta* **175**, 123 (1969).

protocols that fall into the former category will be presented here; procedures that fall into the latter category are presented in the following two articles of this volume and references cited in footnotes 9–12.

The first procedure discussed here will be referred to as the SDS method. In this procedure the histones are extracted from the chromatin with mineral acid. The remaining complex of DNA and NHC proteins is solubilized in sodium dodecyl sulfate (SDS), and the DNA is removed by centrifugation. The NHC proteins are then characterized by SDS polyacrylamide disc gel electrophoresis. This method is simple, quick, and applicable to small samples. It allows one to obtain a highly reproducible, tissue-specific characterization of the NHC protein fraction based on the molecular weight distribution pattern in the SDS gels. The SDS method is very useful in looking at/for changes in the NHC proteins during changes in template activity of a tissue (during development, following hormone stimulation, etc.). The second procedure discussed will be referred to as the SE method. In this procedure the chromatin is dissociated in salt–formic acid–urea. After removal of the DNA by centrifugation, the proteins are bound quantitatively to Sephadex SE, a strong cation exchange resin, and eluted by a salt gradient. This method is suitable for preparation of large quantities of the NHC proteins for chemical characterization. Four fractions of NHC proteins are separated by the method; furthermore, these fractions now have reasonably favorable solubility characteristics, making further fractionation and the purification of individual NHC proteins possible. Citations of other methods for the isolation of NHC proteins can be found in references cited in footnotes 13 and 14.

Preparation of NHC Proteins Using the SDS Method[15,16]

Solutions

Tris·HCl, 1 M, pH 8
H_2SO_4, 2 N

[9] L. M. J. Shaw and R. C. Huang, *Biochemistry* **9**, 4530 (1970).
[10] T. C. Spelsberg, L. S. Hnilica, and A. T. Ansevin, *Biochim. Biophys. Acta* **228**, 550 (1971).
[11] A. J. MacGillivray, A. Cameron, R. J. Krauze, D. Rickwood, and J. Paul, *Biochim. Biophys. Acta* **277**, 384 (1972).
[12] K. H. Richter and C. E. Sekeris, *Arch. Biochem. Biophys.* **148**, 44 (1972).
[13] T. C. Spelsberg, J. A. Wilhelm, and L. S. Hnilica, *Sub-Cell. Biochem.* **1**, 107 (1972).
[14] S. C. R. Elgin, S. C. Froehner, J. E. Smart, and J. Bonner, *Advan. Cell Mol. Biol.* **1**, 1 (1971).
[15] K. Marushige, D. Brutlag, and J. Bonner, *Biochemistry* **7**, 3149 (1968).
[16] S. C. R. Elgin and J. Bonner, *Biochemistry* **9**, 4440 (1970).

Sodium phosphate buffer, 1 M, pH 7 (0.335 mole $NaH_2PO_4 \cdot H_2O$ and 0.665 mole of Na_2HPO_4 per liter final volume)

Sodium dodecyl sulfate, 20%, w/w. High purity SDS (Sipon WD) can be obtained from Alcolac Chemical Corp., Baltimore, Maryland, or regular SDS can be recrystallized from 80% ethanol. Dissolve 100 g of SDS in 500 ml of boiling 80% ethanol. Filter, using Whatman No. 1 paper in a Büchner funnel with aspirator suction. Cool filtrate to approximately 10°C and refilter in cold room. Wash crystals at room temperature. After most of the alcohol has evaporated, drying can be completed in a vacuum oven.

Buffer I: 1% SDS and 1% β-mercaptoethanol in 10 mM sodium phosphate, pH 7

Buffer II: 0.1% SDS and 0.1% β-mercaptoethanol in 10 mM sodium phosphate, pH 7.

Buffer III: 0.1% SDS, 0.1% β-mercaptoethanol, and 10% glycerol in 10 mM sodium phosphate, pH 7

Procedure

Adjust the concentration of the chromatin to 0.5 mg DNA/ml $A_{260} \sim 10$) with 10 mM Tris, pH 8. Maintain the chromatin and all solutions at 4° throughout the acid extraction procedure. While the chromatin is being stirred vigorously, add ¼ volume of 2 N H_2SO_4 dropwise. The final acid concentration is 0.4 N. Allow the precipitate to flocculate for 30 minutes on ice. Collect the precipitate by centrifugation at 17,000 g for 20 minutes. Decant and save the supernatant, which contains the histones. Wash the DNA–NHC protein pellet with cold 0.4 N H_2SO_4. Collect the material by centrifugation at 17,000 g for 15 minutes. Decant and discard the supernatant. Wash the pellet very quickly with cold 10 mM Tris, pH 8. (This is accomplished by adding about 0.5 ml of Tris to the centrifuge tube, allowing it to run over the pellet without disturbing the pellet, and removing the solution with a Pasteur pipette.) The pellet should now be allowed to warm up to room temperature. It is resuspended at a concentration of 1.5 mg DNA/ml in 1% SDS–50 mM Tris, pH 8 by homogenization in a glass–Teflon homogenizer. This and all subsequent operations are carried out at room temperature or warmer. The DNA–NHC protein solution is stirred overnight at 37°; the resulting solution should be clear. The preparation is then dialyzed at 37° to a final solvent of 0.1% SDS–10 mM Tris, pH 8. The DNA is removed by centrifugation at 50,000 rpm for 24 hours in a Spinco SW 50 rotor at 25° (ca. 200,000 g). The DNA forms a gelatinous pellet at the bottom of the tube; the supernatant is the NHC protein preparation. The NHC protein preparation can be dialyzed against buffer III and analyzed by SDS disc gel electrophoresis.

Practical Considerations

This procedure results in a clean separation of the histones and NHC proteins; over 95% of the acid-soluble protein has been removed.[17] Total recovery of protein is excellent, on the order of 95%. It should be noted that dodecyl sulfate will form insoluble precipitates with cations other than sodium or lithium; further, SDS at these concentrations will precipitate at 0–4°. Extraction with H_2SO_4 to quantitatively remove histones has given excellent results with many plant and animal tissues; furthermore, the histone sulfates can readily be recovered by precipitation with 4 volumes of ethanol at −20°.[1] However, in the case of *Drosophila* it has been found necessary to extract the chromatin with 0.67 volume of 5 M NaCl followed by 0.24 volume of 1 M HCl (final concentrations 1.6 M NaCl, 0.2 M HCl)[18] in order to avoid cross-contamination of the two protein fractions. Such histone chlorides must be recovered by dialysis to 10 mM acetic acid and lyophilization. The cold Tris wash of the DNA–NHC protein pellet is essential to ensure that all acid is removed from the pellet. The pellet will not go into solution in Tris–SDS if the pH is acidic. It should be noted that a significant fraction of the NHC proteins are of small molecular weight (10,000–16,000), and dialysis membranes must be such that these proteins are not lost. Note also that many of the more recent synthetic filters will not withstand this concentration of SDS. We use dialysis tubing from Union Carbide Corp., Food Products Division, Chicago, Illinois, which has been treated as follows. The tubing is soaked overnight in acetic anhydride. It is then washed extensively (overnight) with running water, washed ten times with distilled water, and boiled for 30 minutes in 0.1 M sodium bicarbonate–0.1 M EDTA, disodium salt. The tubing is stored in this solution and washed extensively in distilled water immediately before use. Such tubing is suitable for dialyzing chromatin and all fractions thereof.

Controls and Reproducibility

Advisable and easy controls for this procedure are to examine the histones and the total chromosomal proteins by disc gel electrophoresis. As noted above, the histones may be recovered quantitatively from acid solution by either alcohol precipitation or dialysis and lyophilization. Al-

[17] D. M. Fambrough and J. Bonner, *Biochemistry* **5**, 2563 (1966).
[18] C. W. Dingman and M. B. Sporn, *J. Biol. Chem.* **239**, 3483 (1964).

ternatively, aliquots of the acid extract may be dialyzed directly to the sample buffer of the given gel system. Histone samples should be analyzed both on SDS–phosphate gels (see below) and on a pI-dependent gel system, such as that of Bonner et al.[1] or Panyim and Chalkley.[19] In both cases one should observe the classical pattern of histones with little or no cross-contamination by NHC proteins. The histone analysis also allows one to check the quality of the chromatin; there should be no additional very basic proteins, such as basic ribosomal proteins, contaminating the preparation. If the observed level of histone I appears abnormally low, proteolytic degradation should be suspected.[20] It is advisable to prepare chromosomal proteins *immediately* after chromatin isolation to minimize such difficulties. Protease activity can be minimized by maintaining the chromatin at pH 6–7, adding 5 mM sodium bisulfite,[21] or lowering the salt concentration to 0.3 mM[22]; however, we have no experience concerning the extraction properties of chromatin under any of these conditions. Proteolytic degradation of NHC proteins in chromatin has not been observed.

Total chromosomal proteins may be examined by SDS-phosphate gel electrophoresis. To an aliquot of chromatin, 20% SDS and β-mercaptoethanol are added directly to obtain the desired final concentrations, 1% each. The sample is then dialyzed for 12 hours against buffer I at room temperature, for 12 hours against fresh buffer I at 37°, for 12 hours against buffer II at room temperature, and for 12 hours against buffer III at room temperature.[16] Virtually any protein-containing sample can be analyzed on gels after this treatment. Probably the procedure can be shortened appreciably for most preparations. If a sample is of low salt concentration, as chromatin generally is (10 mM Tris), it is usually sufficient to add SDS and β-mercaptoethanol to the sample, and dialyze it directly to buffer III at room temperature. It is advisable to heat the sample to 70–100° for a few minutes to ensure complete dissociation and reduction of the proteins. The total chromosomal proteins should give a band pattern on SDS gels equivalent to the sum of the histone and NHC protein band patterns. In our experience, this is always observed. The DNA causes some streaking of the gel pattern, but does not interfere with this conclusion. It should be noted that this control eliminates the possibility that any of the NHC protein bands are artifacts resulting from the exposure of chromatin to acid.

[19] S. Panyim and R. Chalkley, *Arch. Biochem. Biophys.* **130**, 337 (1969).
[20] J. I. Garrels, S. C. R. Elgin, and J. Bonner, *Biochem. Biophys. Res. Commun.* **46**, 545 (1972).
[21] S. Panyim, R. H. Jensen, and R. Chalkley, *Biochim. Biophys. Acta* **160**, 252 (1968).
[22] J. Bartley and R. Chalkley, *J. Biol. Chem.* **245**, 4286 (1970).

The SDS method for preparation of NHC proteins coupled with analysis by SDS polyacrylamide disc gel electrophoresis (see below) gives very reproducible results in terms of both the quantitative and qualitative features of the gel band pattern (see Fig. 1). In this instance, a given preparation of rat liver chromatin was split into two portions, and the NHC proteins were prepared from each by the above method. The two protein preparations were analyzed by SDS–phosphate gel electrophoresis with the same total amount of protein applied to each gel. Densitometric scans of the two gels have been superimposed. The results are shown in Fig. 1. It should be kept in mind that the standard deviation in determinations of the protein:DNA ratio of chromatin of a given tissue is 10%.[16] Figure 2 indicates the reproducibility of the method as a whole, including the preparation of chromatin. In this instance, rat liver chromatin was prepared on two different days by the same investigator using the same technique, the NHC proteins prepared by the SDS method, and the proteins analyzed by SDS–Tris–glycine gel electrophoresis (see below). The densitometric scans of the gels have been aligned for maximum homology. It can be seen that the method is qualitatively reproducible (bands with the same relative mobilities are always present) and roughly quantitatively reproducible (the relative peak heights remain approximately the same). This is independent of the total protein load on the gel. (One could not anticipate that the scans would show complete quantitative reproducibility independent of protein load since the dye binding coefficient for Coomassie Brilliant Blue R 250 is only linear over a limited range of protein concentrations, ca. 1–20 μg. Beyond this range the curve flattens out to give an overall S shape.[23]) One may conclude, then, that the reproducibility of the SDS method for the isolation of NHC proteins is excellent and the method is well suited for comparative studies of NHC proteins. Some quantitative variation in the gel band pattern has been noted in comparing results obtained by different investigators using very similar methods of chromatin preparation. The reasons for this have not yet been explored in any detail.

Preparation of NHC Proteins Using the SE Method[24]

Solutions

Formic acid–urea. Dissolve 220 g of Schwarz/Mann UltraPure urea in 250 ml of 98–100% formic acid. Adjust volume to 1000 ml with

---

[23] S. C. R. Elgin, Ph.D. Thesis, California Institute of Technology, 1971.
[24] S. C. R. Elgin and J. Bonner, *Biochemistry* **11**, 772 (1972).

10 M urea which has been deionized by passage through a mixed-bed ion exchange column (Barnstead D0803) immediately prior to use.

Formic acid–urea–salt solutions. NaCl is added directly to the formic acid–urea solution in the amount calculated to give the designated molarity, assuming no significant volume change of the solution.

Procedure

Adjust the concentration of the chromatin to 1.0 mg of DNA per milliliter or less. To dissociate the chromatin add 1.5 volumes of 98–100% formic acid at 0–4° with rapid stirring. Allow the preparation to stir in ice slowly for 0.5 hour. Remove the preparation from ice. Add 5 M NaCl to a final concentration of 0.2 M and solid urea (Schwarz/Mann Ultra-Pure) to a final concentration of 8 M, assuming a final volume such that the final concentration of formic acid is 25%. Thus if one wishes to dissociate 50 ml of chromatin, one would add 75 ml of formic acid, stir, and then add 12 ml of 5 N NaCl and 144 g of urea. The urea is added slowly to allow it to dissolve without supercooling the preparation. The final volume adjustment is made by adding water. The chromatin is now completely dissociated, and the DNA can be removed by centrifugation in a Spinco Ti 50 rotor for 18 hours at 200,000 g at 2–4°. The top 90% of the supernatant in each tube is removed by pipette and pooled for application to the resin. It should be noted that the formic acid–urea is a highly dissociating agent. It can safely be used with glass, polyethylene, and similar plastics, or Teflon. Thus in the above centrifugation Teflon O rings must be used to seal the rotor tubes, and in the following column chromatography Teflon contacts and pressure seals and polyethylene tubing must be used. While working with this solvent it is advisable to wear safety glasses and to wash one's hands frequently with dilute sodium bicarbonate.

Sephadex resin SE C-25 is prepared by soaking overnight in formic acid–urea solution. The resin is then combined with the above supernatant. Typically 150 mg of protein in 480 ml of solution is combined with 18 g (dry weight) of resin. The mixture is diluted with a 10-fold excess of formic acid–urea solution with rapid stirring (room temperature). The mixture is stirred at room temperature for 2 hours; the resin is collected by centrifugation at 1000 g and poured into a 2.5 by 30 cm column. All the supernatant protein is bound to the resin by this technique of salt dilution and can be quantitatively recovered by elution with increasing concentrations of sodium chloride in formic acid–urea. The following scheme was found to be most efficient in a study of rat liver NHC proteins. The column is first eluted with 50 ml of 0.1 N NaCl–formic acid–urea, then 75 ml of 0.2 M NaCl–formic acid–urea, then with a 300-ml gradient from 0.2 M to 0.8 M NaCl–formic acid–urea, and finally with 100 ml of formic acid. The flow rate is maintained at 15 ml per hour. The column is assayed by absorbance of fractions at 280 nm. Major peaks may be pooled, dialyzed to other solvents, and concentrated by pressure filtration or lyophilization for further fractionation.

Careful analysis by polyacrylamide gel electrophoresis and by chemical techniques has shown that this technique successfully separates NHC proteins from histones and also fractionates the NHC proteins into four groups. The fractionation appears to be on the basis of mole percent lysine. The protein eluted by 0.1 M NaCl appears to be a single, very acidic, major NHC protein. The protein fraction eluted by the 0.2 M NaCl and at the beginning of the gradient appears to include at least 4 major NHC proteins. These proteins have molecular weights of 50,000–100,000 and isoelectric points in the range of 5.0 to 6.5. The subsequent peak of NHC proteins, eluting at ca. 0.25–0.30 M NaCl, includes NHC proteins of lower molecular weight, 16,000–20,000, and higher isoelectric point, 6.5–8.0. It is also contaminated by histones, making further purification desirable. The bulk of the histones elute at ca. 0.35–0.4 M NaCl with a trailing shoulder of histone I. Eluted with the histones is one NHC protein, which can be separated from them by phosphocellulose chromatography. A residual 10% of the protein sticks to the column and is eluted only with 100% formic acid. By analysis on polyacrylamide gels, this fraction appears to contain mostly NHC protein similar to that which eluted at 0.2 N NaCl. It also includes some very high molecular weight proteins not observed in any other column fractions. It is not understood why this protein fails to elute with the earlier fractions.

Recently it has been observed that it is unnecessary to remove the DNA by centrifugation before binding the proteins to the resin. Under the conditions used, the chromatin is completely dissociated and the DNA does not bind to the resin after dilution of the salt, as do the proteins. The DNA can therefore be removed by washing the resin in 1 liter of 20 mM NaCl–formic acid–urea five times, collecting the resin each time by low speed centrifugation. When this procedure is used, it is generally advisable to elute the column first with 75 ml of 20 mM NaCl–formic acid–urea, and then proceed with the protein elution as described above.

Practical Considerations

This procedure is a particularly efficient one for preparing NHC proteins for chemical analysis. Large amounts of material may be processed; not only is the chromatin fractionated into its major components of DNA, histone, and NHC protein, but the NHC proteins are also fractionated into four classes. These protein classes show improved solubility characteristics and can be further fractionated by exclusion chromatography,[24] phosphocellulose chromatography,[24,25] or other types of ion exchange chromatography. Solvents developed to maintain ribosomal proteins in solution and chromatography methods developed to fractionate them appear very promising for the fractionation of NHC proteins.[25-27] These solvents frequently include urea, organic acids (formic, acetic), and small inorganic ions (lithium). Purification by exclusion chromatography has been successfully accomplished for the NHC protein fractions using Sephadex G-75 or G-100 resins and a solvent of 6 M urea–10 mM H_3PO_4 neutralized to pH 8 with methylamine–3 mM β-mercaptoethanol–0.1 M LiCl.[24] The NHC protein fractions are also soluble in all SDS solutions and in high concentrations of urea.

Cation exchange resins have also been used under milder conditions to separate the histones and NHC proteins; however, these techniques have not been reported to achieve a significant fractionation among the NHC proteins.[28-30] The SE method can also effect a significant saving in time over other methods by virtue of the ease of DNA removal (by washing from the resin) as discussed above. It is interesting to note that the best separations of the NHC proteins from each other observed to date have been accomplished at low pH, 2.5–3.2. The NHC proteins are fractionated on the SE cationic resin much more than on an anion exchange resin at neutral or slightly basic pH; the NHC proteins are well separated in the disc gel electrophoresis system at pH 3.2 originally developed to analyze histones.[19,28] Good fractionations in polyacrylamide gels at pH 8.6 have also been obtained, however.[30]

Controls and Reproducibility

In general the efficiency of binding of rat liver NHC protein to the SE column and recovery from the column are excellent, on the order of 95–100%. However, the state of the 10% of the protein eluted with 100% formic acid is unknown; no attempt has been made to characterize or further fractionate this material, so for practical purposes the recovery

---

[25] M. Ozaki, S. Mizushima, and M. Nomura, *Nature (London)* **222**, 333 (1970).
[26] H. Delius and R. R. Traut, in "Macromolecules, Biosynthesis and Function." *FEBS Symp.* **21**, 7 (1970).
[27] I. Hindennach, G. Stoffler, and H. G. Wittman, *Eur. J. Biochem.* **23**, 7 (1971).
[28] S. L. Graziano and R. C. C. Huang, *Biochemistry* **10**, 4770 (1971).
[29] E. A. Arnold and K. E. Young, *Biochim. Biophys. Acta* **257**, 482 (1972).
[30] R. J. Hill, D. C. Poccia, and P. Doty, *J. Mol. Biol.* **61**, 445 (1972).

of protein from the column becomes 85–90%. The pooled eluted fractions have been analyzed by disc gel electrophoresis; the major molecular weight classes of NHC proteins and the histones are well represented in the appropriate fractions, indicating little or no break down of protein by the solvent. However, a sample of lysozyme maintained in this solvent for equivalent times showed some liberation of glycine and serine as well as a strong lysine N-terminal, suggesting that a small amount of acid hydrolysis at susceptible peptide bonds may have occurred. The fractionation of rat liver chromosomal proteins by this method as monitored by disc gel electrophoresis is very reproducible; however, less satisfactory results have been obtained with *Drosophila* chromatin, the only other case which has been tested.

Analysis of NHC Proteins by Disc Gel Electrophoresis

NHC proteins isolated by the SDS method can probably only be analyzed in disc gels in the presence of SDS. While it is true that SDS can be successfully removed from proteins,[31] to attempt to do so with a preparation of total NHC proteins would probably result in aggregation, if not precipitation, of some of the proteins. Two methods of disc gel electrophoresis in SDS are given below; the first is based on the technique of Shapiro *et al.*[32] and the second on the technique of Laemmli.[33] Both techniques fractionate the proteins on the basis of molecular weight. The SDS–phosphate technique results in a distribution of protein such that the log of the molecular weight is linearly related to the relative migration distance; there is greater deviation from this principle using the SDS–Tris–glycine method. The latter method, however, results in a much better resolution of the NHC protein bands of greater than 18,000 molecular weight. Compare Fig. 1 (NHC proteins of rat liver run on SDS-phosphate gels) and Fig. 2 (NHC proteins of rat liver run on SDS–Tris–glycine gels). The subscripting of Greek letter names of the NHC protein bands given in Fig. 2 indicates the amount of increased resolution. Protein that appeared as single bands on the SDS–phosphate gels is frequently resolved into 3–4 bands on the SDS–Tris–glycine gels. NHC proteins prepared by the SE method or other methods cited can, in addition, be analyzed in urea–polyacrylamide gel systems in which the fractionation also reflects the pI of the protein.[19,30] These proteins can also be analyzed by isoelectric focusing in polyacryl-

[31] K. Weber and D. J. Kuter, *J. Biol. Chem.* **246**, 4504 (1971).
[32] A. L. Shapiro, E. Viñuela, and J. V. Maizel, *Biochem. Biophys. Res. Commun.* **28**, 815 (1967).
[33] U. F. Laemmli, *Nature (London)* **227**, 680 (1970).

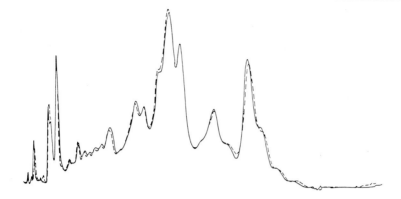

Fig. 1. Reproducibility of nonhistone chromosomal (NHC) protein preparation and analysis. Superimposed scans of sodium dodecyl phosphate gels of two preparations of rat liver NHC protein made from the same preparation of chromatin. Gels run left to right; stained with Coomassie Brilliant Blue, and scanned at 600 nm using a Gilford spectrophotometer with linear transport.

amide gels by the method given below. It should be pointed out that while disc gel methods are frequently carried out using gel running tubes of 5–6 mm i.d. and protein loads of 25–50 µg, the amount of protein required can be reduced to 10% of that amount by using gel running tubes of 2 mm i.d. Only a few modifications of procedure are required to use the 2-mm gels. Best results are obtained if the gels are run at one-half the power load used for the larger gels. The gels must be removed from the tubes either by extrusion with a smooth, well-fitting metal rod (for 5–10% acrylamide gels) or by breaking the tube in a vise (for 15–20%

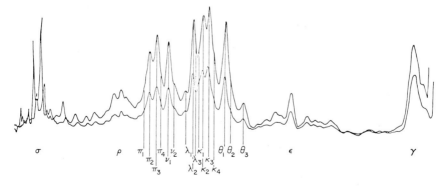

Fig. 2. Reproducibility of the total procedure. Aligned scans of sodium dodecyl sulfate–Tris–glycine gels of nonhistone chromosomal proteins from two independent preparations of rat liver chromatin.

acrylamide gels). Staining and fixing times are much shorter; it is generally advisable to elute unbound dye by diffusion only, rather than by horizontal electrophoresis, etc.

SDS–Phosphate Gels

Stock Solutions

Solution A: 0.4 ml of N,N,N',N'-tetramethylethylenediamine (TEMED) plus 99.6 ml of water. Store in a dark bottle in the refrigerator.

Solution B: 20 g of acrylamide plus 0.52 g of N,N-bismethylene acrylamide (bisacrylamide) plus 40 ml of 1 M sodium phosphate buffer, pH 7, plus 2 ml of 20% SDS, made up to 100 ml with water. Filter through Whatman No. 1 and store in refrigerator; it will be necessary before using the solution to warm it to redissolve the acrylamide.

Solution C: 0.12 g of ammonium persulfate made up to 100 ml with water. Make fresh just before using.

Tray buffer: 0.1% SDS–0.1 M sodium phosphate, pH 7

Sample buffer: Buffer III given above

Procedure. Mix 7.5 ml of C with 1.5 ml of A and 3 ml of B. The final gels should be 5% acrylamide, 0.13% bisacrylamide, 0.05% Temed, 0.1% SDS, 0.075% ammonium persulfate, and 0.1 M sodium phosphate, pH 7. Layer the gels with water and allow them to polymerize for 2 hours. Run the gels at constant voltage, 47 V, for 75 minutes (5 mm by 6 cm gels). Stain for 4 hours in 0.25% Coomassie Brilliant Blue R-250 (C. I. 42660) (Mann Research Laboratories) in 5:5:1 water:methanol:acetic acid. Destain the gels sideways electrophoretically in 17:1:2 water:methanol:acetic acid.

SDS–Tris–Glycine Gels

Stock Solutions

Solution A: same as above.

Solution B: 30.0 g of acrylamide plus 0.8 g of bisacrylamide made up to 100 ml with water.

Solution C: 0.3 g ammonium persulfate made up to 100 ml with water. Make fresh just before using.

Solution D: 18.17 g of Tris plus 50 ml of water; adjust pH to 8.8 with about 24 ml of 1 M HCl; add 0.4 g SDS and adjust volume to 100 ml with water.

Solution E: 3.02 g of Tris plus 20 ml of water; adjust pH to 6.8 with 1 M HCl and adjust volume to 50 ml with water.

Tray buffer: 6.0 g of Tris plus 28.8 g of glycine dissolved in water; add 5 ml of 20% SDS and adjust volume to 1000 ml with water. The pH should be 8.3.

Sample buffer: 6.2 ml of water plus 1.0 ml of 20% SDS plus 1.0 ml of glycerol plus 0.5 ml of β-mercaptoethanol plus 1.3 ml of solution E.

Procedure. Mix 1 ml of C with 0.67 ml of A, 3.33 ml of B, and 5.0 ml of D. The final gels should be 10% acrylamide, 0.27% bisacrylamide, 0.037% TEMED, 0.2% SDS, 0.03% ammonium persulfate, and 0.75 M Tris. Layer the gels with water and allow them to polymerize for 2 hours minimum. Run the gels at constant voltage, 50 V, for 2.5 hours (5 mm by 6 cm gels). Bromophenol blue, if included in the sample buffer, will run at the sample front. The gels may be stained as above; better results are obtained if the gels are fixed in 50% TCA, washed three times in 7% acetic acid (1 hour soak for each wash), and then stained in the Coomassie Brilliant Blue for 2 hours. Gels stained by this technique may be partially destained electrophoretically, but it is best to allow final destaining by diffusion.[34]

Isoelectric Focusing Gels[35,36]

Stock Solutions

Solution A: 7.0 g of acrylamide plus 0.4 g of bisacrylamide plus 48.0 g of urea dissolved and made up to 100 ml with water. Store in cold; stable for 2 weeks.

Solution B: 2.0 mg of riboflavin plus 60.0 mg of potassium persulfate plus 48 g of urea dissolved and made up to 100 ml with water. Store in a dark bottle in the cold; solution is stable for 1 week.

Anolyte: 10 mM phosphoric acid

Catholyte: 20 mM sodium hydroxide

Procedure. Seal tubes with a piece of dialysis tubing held in place by a ring of rubber tubing, and place them in the running apparatus. Mix 2.0 volumes of A, 0.2 volume of 8% ampholine, 0.8 volume of 10 M urea, and 1.0 volume of B; add 100 μl of TEMED per 100 ml of solution. Fill tubes to a height of 6 cm and then lower them to that depth in the lower reservoir filled with anolyte. Layer the gels with water and photopolymerize for 30 minutes with fluorescent light. After polymeriza-

---

[34] J. King and U. K. Laemmli, *J. Mol. Biol.* **62**, 465 (1972).
[35] G. R. Finlayson and A. Chrambach, *Anal. Biochem.* **40**, 292 (1971).
[36] P. Righetti and J. W. Drysdale, *Biochim. Biophys. Acta* **236**, 17 (1971).

tion remove the water and fill the upper reservoir with catholyte. NHC protein samples may be dissolved in 10 M urea and applied directly to the top of the gels without prior electrophoresis. Initiate focusing at a constant current of 1 mA per gel; as soon as a voltage of 200 V is reached, switch to constant voltage and continue the focusing at that setting for 6 hours total time. The gels may be stained by any conventional technique after extensive washing in 12.5% TCA to fix the proteins and remove the ampholyte. Alternatively, the proteins may be stained by methods that are not sensitive to the ampholyte. The gels may be stained for several days in a solution made of 200 ml of 5% TCA–5% sulfosalicylic acid plus 2 ml of 1% Coomassie Brilliant Blue in water plus 40 ml of methanol; after staining, the gels are stored in 5% TCA–5% sulfosalicylic acid.[37] In the second method the gels are stained for 1–2 hours in 10:45:45 acetic acid–ethanol–water with 0.2% fast green (C.I. 42053) and destained in 10:25:65 acetic acid–ethanol–water.[38] All three methods yield essentially the same results, although the dye binding coefficients will differ for a given protein with the different methods. The latter two methods are quicker but not as sensitive.

Analysis of NHC Proteins by Chemical Techniques

The isolated NHC proteins can be analyzed by all the conventional techniques of protein chemistry. We wish here only to point out that amino acid analysis, N-terminal analysis, and even sequencing of the N-terminal region of a protein can now be accomplished using samples isolated from SDS polyacrylamide gels. The direct amino acid analysis of stained bands from gels has been reported[39]; however, despite the use of purified reagents, we frequently observe spurious results on such analysis. Thus as a routine procedure it is preferable to recover the protein from gels by electrophoretic elution. In outline, the technique is as follows: The NHC proteins are fractionated on SDS polyacrylamide gels using the techniques described above. The protein bands are visualized under ultraviolet light using fluorescent markers; this avoids fixing the proteins in the gels. The desired band is cut from the gel using a razor blade; usually the slices of a given band from 6–12 identical gels are pooled. The gel slices from 5-mm gels are stacked in a 6-mm tube, and a new SDS gel poured around them. Care should be taken to ensure that fresh gel covers the bottom of the tube. The gel is allowed to polymerize

---

[37] E. M. Spencer and T. P. King, *J. Biol. Chem.* **246**, 201 (1971).
[38] R. F. Riley and M. K. Coleman, *J. Lab. Clin. Med.* **72**, 714 (1968).
[39] L. L. Housten, *Anal. Biochem.* **44**, 81 (1971).

for several hours. A small dialysis bag is fitted over the end of the tube, filled with tray buffer and secured with a rubber band; this is immersed in the bottom gel-running tray. Care should be taken to ensure that no bubbles are trapped in the dialysis bag on the bottom of the gel tube. The protein is then eluted from the gel into the bag by electrophoresis at the normal constant voltage settings. The required time can be estimated from the known molecular weight of the protein and the elution time of colored proteins such as myoglobin. After protein elution, the dialysis bag is removed and the protein solution is recovered. Yield is on the order of 50%. Excellent results have been obtained for the amino acid analysis of lysozyme recovered by this procedure.[40] Good results have also been reported using a technique of crushing and eluting the SDS gel, providing the gel has been dialyzed extensively after casting and prior to use.[41] Proteins dissolved in SDS can be readily used for the determination of the N-terminal amino acid[42] and N-terminal sequence.[41]

The SDS proteins on gels may be stained without being fixed by use of 0.003% 1-anilino-8-naphthalene sulfonate, magnesium salt (ANS) in 0.1 N sodium phosphate buffer, pH 6.8. (ANS, Mg^{2+} salt, is available from Eastman Kodak, No. 10990). Heavy protein bands (ca. 100 μg) are readily observed using a portable UV lamp.[43] However, in the case of unfractionated NHC proteins, the gel banding pattern is sufficiently complex to require a more sensitive technique. One satisfactory method is to dansylate the proteins. Note, however, that this results in the derivatization of the lysine residues in addition to the N-terminal residue, and thus corrections must be made if one wishes to use such proteins for amino acid analysis. The following method of dansylation has been found satisfactory.[44] A solution of NHC proteins at 1.5 mg/ml, prepared by the SDS method given above, is made 3% in SDS and dialyzed to 3% SDS–0.05 M sodium phosphate buffer, pH 7.4. Dansyl chloride, 0.1 ml in acetone, is added to 5 ml of the protein solution with vigorous mixing; the preparation is put in a boiling water bath for 3–4 minutes. Then 50 μl of β-mercaptoethanol are added, and the preparation is boiled for 1 more minute. After cooling, the protein is separated from the free dansyl chloride by exclusion chromatography on Sephadex G-25 in SDS–phosphate buffer. The protein fractions can be readily detected with a UV light; they will fluoresce bright yellow, compared to the green of the free dye. The labeled protein preparation can be concentrated by

[40] D. W. Miller and S. C. R. Elgin, unpublished results, 1972.
[41] A. W. Weiner, T. Platt, and K. Weber, *J. Biol. Chem.* **247**, 3242 (1972).
[42] W. R. Gray, this series, Vol. 25, p. 121.
[43] B. K. Hartman and S. Udenfriend, *Anal. Biochem.* **30**, 391 (1969).
[44] D. N. Talbot and D. A. Yphantis, *Anal. Biochem.* **44**, 246 (1971).

evaporation with a stream of nitrogen and subsequently dialyzed to buffer III or whatever SDS system is desired for the fractionation by gel electrophoresis. The individual NHC protein bands can then be used to obtain samples for further chemical characterization as discussed above.

Acknowledgments

The author is a Fellow of the Jane Coffin Childs Memorial Fund for Medical Research. Work from this laboratory and writing were supported by the Jane Coffin Childs Fund and National Science Foundation Grant GB 34160 to Dr. Leroy E. Hood.

[12] Methods for Isolation and Characterization of Chromosomal Nonhistone Proteins

Fractionation of Chromatin on Hydroxyapatite and Characterization of the Nonhistone Proteins by Ion Exchange Chromatography and Polyacrylamide Gel Electrophoresis

By A. J. MacGillivray, D. Rickwood, A. Cameron, D. Carroll, C. J. Ingles, R. J. Krauze, and J. Paul

There are at present numerous methods available for the separation of the nonhistone proteins from other constituents of chromatin. As we have described elsewhere,[1] many of these procedures are multistep, suffer from incomplete extraction of the nonhistone proteins, and expose the proteins to extreme conditions of pH and solvent environment. Since there is a requirement for methodology which can avoid these defects, we have been interested in utilizing the combination of salt and urea (e.g., 2–3 M NaCl and 5–7 M urea) which has been used as a solvent both for the dissociation and solubilization of chromatin and its specific reconstruction from previously isolated components.[2,3] The following methodology is based on the ability of hydroxyapatite to act as an ion exchange medium for macromolecules in the presence of high ionic strength and urea[4,5] and consists of a single column procedure, yielding high recoveries of chromatin constituents.

[1] A. J. MacGillivray *in* "Subnuclear Components" (G. D. Birnie, ed.). Butterworths, London, in press.
[2] I. Bekhor, G. M. Kung, and J. Bonner, *J. Mol. Biol.* **39**, 351 (1969).
[3] R. S. Gilmour and J. Paul, *J. Mol. Biol.* **40**, 137 (1969).
[4] G. Bernardi and T. Kawasaki, *Biochim. Biophys. Acta* **160**, 301 (1968).
[5] G. Bernardi, M. G. Giro, and C. Gaillard, *Biochim. Biophys. Acta* **278**, 409 (1972).

Methods

Chemicals

As far as possible these were of Analar grade and were largely obtained from British Drug Houses Ltd., Poole, Dorset. "Analar" grade urea was routinely prepared as an 8 M stock solution, filtered through Whatman No. 1 filter paper, and deionized immediately before use by passage through a column of AG 501-X8 (D) mixed bed resin (Bio-Rad Laboratories Ltd., St. Albans, Herts.). Calcium chloride and disodium hydrogen phosphate for preparing hydroxyapatite were purchased from Merck Chemicals Ltd., Darmstadt, Germany. QAE–Sephadex was purchased from Pharmacia (G.B.) Ltd., London. Acrylamide, N,N'-methylenebisacrylamide, and N,N,N',N'-tetramethylethylenediamine were purchased from Kodak Ltd., London. The acrylamide and N,N'-methylenebisacrylamide were recrystallized before use. Ampholines (pH range 3–10) were purchased from LKB Instruments Ltd., South Croydon. Standard marker proteins, bovine serum albumin, ovalbumin, chymotrypsinogen, trypsin, lysozyme, and ribonuclease were purchased from the Sigma Chemical Co. Ltd., London. Coomassie Brilliant Blue R and Carbowax (MW 15,000–20,000) were purchased from G. T. Gurr, High Wycombe, Bucks.

Preparation of Nuclei

Except where stated all procedures are carried out at 0–4°. Male mice (12 weeks old) were used in all experiments; mouse organs are excised, placed in ice-cold 0.25 M sucrose, and used immediately for the preparation of nuclei.

Nuclei are prepared by a modification of the method of Widnell and Tata.[6] The tissue is homogenized in three volumes of 0.25 M sucrose–3 mM CaCl$_2$ in a glass–Teflon Potter-Elvehjem type homogenizer using eight up and down strokes at 2000 rpm. After filtration through four layers of gauze, the homogenate is centrifuged at 1000 g for 10 minutes. The pellet is then homogenized in 2.4 M sucrose–3 mM CaCl$_2$ and centrifuged at 40,000 g for 1 hour. The supernatant is discarded, and the pelleted nuclei are washed in 0.25 M sucrose–3 mM CaCl$_2$ and collected by centrifugation at 1000 g for 10 minutes. The nuclei are suspended in the same medium containing 1% (v/v) Triton X-100, which removes about 90% of the inner and outer nuclear membranes,[7] as judged by elec-

---

[6] C. C. Widnell and J. R. Tata, *Biochim. Biophys. Acta* **123**, 478 (1966).
[7] A. D. Barton, W. E. Kisieleski, F. Wasserman, and F. Mackevicius, *Z. Zellforsch. Mikrosk. Anat.* **115**, 299 (1971).

tron microscopy. The Triton X-100 is removed by washing the nuclei twice in 0.25 M sucrose–3 mM CaCl$_2$. The nuclei are then stored as a pellet at $-20°$C.

Preparation of Chromatin

Chromatin is prepared from the purified nuclei by homogenizing the nuclei in a tight-fitting glass–Teflon homogenizer in 50 volumes of 0.14 M NaCl–50 mM Tris · HCl, pH 7.4 with eight up-and-down strokes at 2000 rpm. The suspended chromatin is stirred for 20 minutes and pelleted by centrifugation at 15,000 g for 10 minutes. The chromatin is extracted twice more with the same volume of saline–Tris buffer and pelleted as before.

Separation of Chromatin Components

The chromatin is homogenized in a glass–Teflon homogenizer in 30 volumes of 1 mM sodium phosphate, pH 6.8–2 M NaCl–5 M urea. The solubilized chromatin is centrifuged at 15,000 g for 10 minutes and the chromatin is reextracted with half the original amount of the same solution. The pooled chromatin extracts are then sonicated for 15 seconds using an MSE Ultrasonic Power Unit at a setting of 1.5 A and centrifuged at 10,000 g to remove any small residual material.

Hydroxyapatite, prepared by the method of Bernardi,[8] is fined several times in 1 mM sodium phosphate, pH 6.8–2 M NaCl–5 M urea by suspension and decantation. Columns of hydroxyapatite (25 cm $\times$ 1.6 cm) are packed and equilibrated with the above solution at room temperature and run at 10–12 ml per hour. The hydroxyapatite column is loaded with chromatin (up to 20 mg of DNA) dissociated in 1 mM sodium phosphate, pH 6.8–2 M NaCl–5 M urea; larger amounts of chromatin require proportionally larger columns. The chromatin fractionation used is outlined in Fig. 1. Some of the chromatin proteins are not retained by hydroxyapatite; this fraction (H1) contains no nucleic acid (Fig. 1), and analysis of these proteins by sodium dodecyl sulfate gels (Fig. 2) and amino acid analysis[9] shows that it consists of essentially pure total cell histone. After elution of the unretained material with equilibrium buffer, the bulk of the nonhistone proteins (fraction H2), together with a small amount of RNA (approximately 4% of the protein), is eluted with 50 mM sodium phosphate, pH 6.8 in salt–urea. The remainder of the nonhistone proteins (fraction H3) is eluted, together with an equal amount of RNA, by 200

[8] See this series, Vol. 21 [3].
[9] A. J. MacGillivray, A. Cameron, R. J. Krauze, D. Rickwood, and J. Paul, *Biochim. Biophys. Acta* **277**, 384 (1972).

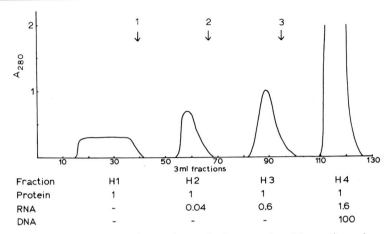

Fig. 1. Fractionation of chromatin on hydroxyapatite. Mouse liver chromatin is prepared from purified nuclei and dissociated in 1 mM sodium phosphate, pH 6.8–2 M NaCl–5 M urea as described in the text. The arrows indicate changes in the phosphate concentrations of the buffer: viz. (1) 50 mM, (2) 200 mM, (3) 500 mM. The column eluate is pooled as indicated to give fractions H1–H4. The analysis of each fraction is given in relation to the protein content.

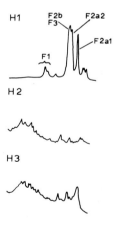

Fig. 2. Sodium dodecyl sulfate–polyacrylamide gel electrophoresis of mouse liver chromatin hydroxyapatite fractions. Fractions H1–H3 are obtained as described in Fig. 1 and dialyzed against urea–sodium dodecyl sulfate–2-mercaptoethanol–phosphate buffers as described in the text. Samples (80–150 μg of protein) are run on 15% polyacrylamide gels containing 0.1% (w/v) sodium dodecyl sulfate. The direction of migration is from left to right.

mM sodium phosphate, pH 6.8 in salt–urea. The proteins which are eluted in the H3 fraction appear to be generally more acidic, and contain a higher proportion of phosphoproteins than the proteins in the H2 fraction.[10] The remaining nucleic acid material can be eluted from the column by 0.5 M sodium phosphate, pH 6.8, in salt–urea (fraction H4). This last fraction contains a negligible amount of protein. The overall yield of total chromatin proteins using this fractionation procedure was 75–80%.

Note Added in Proof. Recent work shows that elution of the hydroxyapatite columns with 200 mM sodium phosphate, pH 6.8–2 M guanidine hydrochloride–5 M urea after recovery of fraction H3 removes essentially all the remaining nonhistone protein bound to the column. This fraction contains an amount of protein similar to that obtained in the H3 fraction and has an RNA to protein ratio of approximately 3. The major components are similar to some of those found by two-dimensional electrophoresis of the H3 fraction.

Likewise recoveries of nonhistone proteins in excess of 95% can be obtained by washing QAE–Sephadex columns with 4 M guanidine hydrochloride–5 M urea after the completion of the salt gradient.

For studying the *in vitro* phosphorylation of the nonhistone proteins, the procedure is slightly modified as previously described.[11] Triton-treated nuclei (approximately 300 mg wet weight) are phosphorylated *in vitro* as previously described in a 10-ml incubation mixture containing 0.25 M sucrose–0.1 M Tris · HCl, pH 8–25 mM NaCl–10 mM MgCl$_2$ and 10 μM [γ-^{32}P]ATP. After incubation at 37° for 5 minutes the nuclei were pelleted by centrifugation at 1000 g for 10 minutes at 4° and then washed twice with 20 ml of cold incubation medium containing 2.5 mM unlabeled ATP in place of the [γ-^{32}P]ATP. Chromatin was prepared from the nuclei by extraction with 0.14 M NaCl–50 mM Tris · HCl, pH 7.4, as previously described. Both free orthophosphate and ATP are adsorbed onto the hydroxyapatite,[11] enabling the phosphorylated histones to be isolated free of nonspecific ^{32}P contamination. No nonspecific binding, or exchange labeling of the nonhistone fractions occurred under the conditions used.

QAE–Sephadex Fractionation of Nonhistone Proteins

The hydroxyapatite nonhistone fractions H2 and H3 can be further fractionated on QAE-Sephadex essentially as described by Richter and

---

[10] A. J. MacGillivray and D. Rickwood, *Eur. J. Biochem.* **41**, 181 (1974).

[11] D. Rickwood, P. G. Riches, and A. J. MacGillivray, *Biochim. Biophys. Acta* **299**, 162 (1973).

Sekeris.[12] The H2 and H3 fractions were extensively dialyzed at 4° into 5 M urea–0.5 mM MgCl$_2$–1 mM EDTA–1 mM dithioerythritol and 10 mM Tris · HCl, pH 8.3, until the salt concentration, as determined by conductivity measurements, was less than 10 mM NaCl.

QAE–Sephadex is first washed with 0.1 M NaOH and then with distilled water until the washings are at neutral pH. Columns of QAE–Sephadex (44 cm × 1.6 cm) are equilibrated with 5 M urea–0.5 mM MgCl$_2$–1 mM EDTA–1 mM dithioerythritol and 10 mM Tris · HCl, pH 8.3 at 4°. The sample is loaded onto the column with a flow rate of 15–20 ml per hour. The H2 and to a lesser extent the H3 fraction contain proteins which are unretained by the QAE–Sephadex (see Fig. 3). Amino acid analysis showed that the unretained proteins were not histonelike and indeed they were rich in serine, glycine, and glutamic acid.[13] However, presumably amidation of the free carboxyl groups gives this group of proteins an alkaline isoelectric point. After washing the column with the equilibration buffer to elute the unretained proteins, a linear gradient (200 ml total volume) from 10 mM to 0.6 M NaCl in 5 M urea–0.5 mM MgCl$_2$–1 mM EDTA–1 mM dithioerythritol–10 mM Tris · HCl, pH 8.3, is passed through the column. The proteins are eluted from the column clearly separated from the nucleic acid material (see Fig. 3). Washing the column with 2 M NaCl in 5 M urea does not elute further significant amounts of either protein or nucleic acid material. The overall recovery of protein was routinely 75–80%.

Analysis of Chromatin Nonhistone Proteins

   a. *One-Dimensional Sodium Dodecyl Sulfate Polyacrylamide Gel Electrophoresis.* The system used is similar to that described by Laemmli[14] which consists of introducing sodium dodecyl sulfate to the double-gel discontinuous electrophoresis procedure of Davis.[15]

Aliquots of fractions obtained from the hydroxyapatite columns in phosphate–salt–urea buffers are dialyzed against three changes of 0.1% (w/v) sodium dodecyl sulfate for 48 hours at room temperature. After freeze-drying, each sample is dissolved in 8 M urea to give a final protein concentration of 2–4 mg/ml and dialyzed overnight against 8 M urea–1% (w/v) sodium dodecyl sulfate–1% (v/v) 2-mercaptoethanol–10 mM sodium phosphate, pH 7. Prior to electrophoresis, the samples are incubated at 37° for 3 hours and dialyzed for at least 2 hours against

---

[12] K. H. Richter and C. E. Sekeris, *Arch. Biochem Biophys.* **148**, 44 (1972).
[13] A. J. MacGillivray and D. Rickwood, *Biochem. Soc. Trans.* **1**, 686 (1973).
[14] U. K. Laemmli, *Nature (London)* **227**, 680 (1970).
[15] B. Davis, *Ann. N. Y. Acad. Sci.* **121**, 404 (1964).

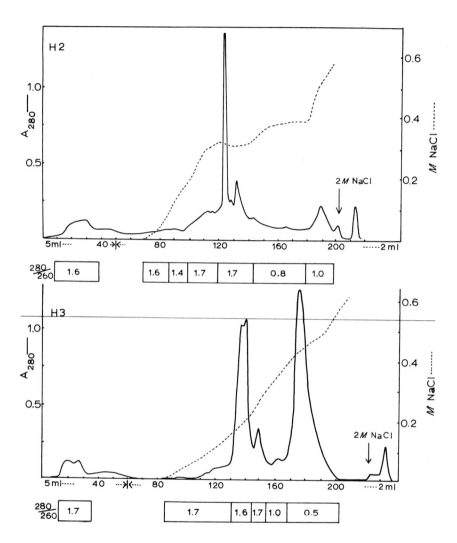

Fig. 3. Chromatography of mouse liver nonhistone proteins, fractions H2 and H3, on QAE–Sephadex. Each nonhistone fraction (up to 120 ml), is desalted by dialysis against 5 M urea–0.5 mM MgCl$_2$–1 mM EDTA–1 mM dithioerythritol–10 mM Tris · HCl, pH 8.3 and applied to a 44 cm × 1.6 cm column of QAE–Sephadex equilibrated with the same buffer. The unretained material is recovered and the material retained on the column is eluted by a salt gradient (200 ml total volume) from 10 mM NaCl to 0.6 M NaCl, both in 5 M urea–0.5 mM MgCl$_2$–1 mM EDTA–1 mM dithioerythritol–10 mM Tris · HCl, pH 8.3. The eluate is pooled into fractions as indicated; the ratio of A_{280}/A_{260} for each fraction is also shown.

the above buffer solution except that it contained 0.1% (w/v) sodium dodecyl sulfate. Up to 200 µl (80–150 µg protein) of each sample are then mixed with 3 µl of bromophenol blue marker dye and 5 µl of 2-mercaptoethanol and kept at room temperature for 1 hour. The gels are run in 0.1% (w/v) sodium dodecyl sulfate–4 M urea.

LOWER SMALL PORE GEL SOLUTIONS. Solution A: 48 ml of 1 M HCl, 36.6 g of Tris, 0.23 ml N,N,N',N'-tetramethylethylenediamine; water to 100 ml (final pH 8.9). Solution B: 40 g acrylamide, 0.6 g N,N'-methylenebisacrylamide; water to 100 ml. Solution C: 0.14% (w/v) ammonium persulfate–0.2% (w/v) sodium dodecyl sulfate–2 mM dithioerythritol in 8 M urea.

UPPER LARGE PORE GEL SOLUTIONS. Solution D: 5.98 g of Tris, N,N,N',N'-tetramethylethylenediamine, 0.46 ml; titrated to pH 6.7 with 1 M HCl and made up to 100 ml with water. Solution E: 10 g of acrylamide, 2.5 g of N,N'-methylenebisacrylamide; water to 100 ml. Solution F: 4 mg of riboflavin; water to 100 ml. Solution G: 0.2% (w/v) of sodium dodecyl sulfate–2 mM dithioerythritol in 8 M urea.

Small-pore gels are prepared by mixing 1 volume of solution A with 3 volumes of solution B and 4 volumes of solution C; 1.5-ml aliquots were pipetted into 9 cm $\times$ 0.6-cm tubes and overlaid with water. After 2 hours the water layer is removed and replaced by 0.2 ml of large-pore gel solution (1 volume solution D and 2 volumes solution E + 1 volume solution F + 4 volumes solution G). After overlaying with water, the gel is photopolymerized under a fluorescent lamp for 1 hour. The total length of the polymerized gels is 6.5 cm.

The water overlays are removed from the large-pore gels and replaced by the protein sample solutions. Electrophoresis is carried out at room temperature at 2 mA per gel using a buffer system which consists of 3 g of Tris, 14.4 g of glycine and 1 g of sodium dodecyl sulfate made to 1 liter with water. When the marker dye reaches the bottom of the small-pore gels (in approximately 2.5 hours) the gels are removed by cracking the glass tubes in a vice, the ion front is marked with a piece of wire and the gels are fixed overnight in methanol–20% acetic acid (1:1, v/v). They are then stained for 2 hours in 1% (w/v) naphthalene black in 7% (v/v) acetic acid and destained by elution in methanol–acetic acid–water (300:70:630, v/v/v) in the presence of charcoal (petroleum grade) as stain absorbant.

In order to calibrate the sodium dodecyl sulfate gels, proteins of known molecular weight were dissolved in 8 M urea–0.1% (w/v) sodium dodecyl sulfate–1% (v/v) 2-mercaptoethanol–0.01 M phosphate, pH 7, and applied to gels as described above. The standard proteins used were: bovine, albumin, ovalbumin, chymotrypsinogen, trypsin, lysozyme, and ribonuclease of molecular weights 68,000, 43,0000, 25,700, 23,300, 14,300, and 13,700, respectively.

b. *Two-Dimensional Electrophoresis.* This is carried out by using a combination of isoelectric focusing and electrophoresis in sodium dodecyl sulfate, i.e., utilizing differences in isoelectric points in the first dimension and differences in molecular size in the second.[10] In order to obtain reproducible separations in the isoelectric focusing system, it was found necessary to fully reduce the proteins. Initial experiments showed that reduction and maintenance in dithioerythritol gave results essentially identical to those obtained with proteins which had been reduced and blocked by, for example, carboxymethylation.

[32]P-labeled nonhistone proteins obtained as described above are concentrated to a protein concentration of 0.5–1 mg/ml by dialysis against Carbowax (MW 15,000–20,000) and are then dialyzed against several changes of 8 M urea–0.3 M Tris · HCl, pH 8.3 at room temperature. Dithioerythritol, 1 M in 8 M urea, is added to a final concentration of 50 mM; after mixing, reduction is allowed to take place for 2 hours at 37°. The concentration of reducing agent is lowered to 2.5 mM by dialysis against several changes of 8 M urea–2.5 mM dithioerythritol at room temperature. Isoelectric focusing of the reduced proteins is carried out as follows, after the method of Gronow and Griffiths.[16] The protein sample is mixed with acrylamide solution (4 g acrylamide and 0.2 g N, N'-methylenebisacrylamide in 40 ml of water), 40% ampholine solution (range pH 3–10) and ammonium persulfate solution (12 mg in 10 ml of 8 M urea) in the following proportions—1 : 1 : 0.12 : 0.37 (v/v), and 2-ml samples are placed in 10 cm $\times$ 0.5 cm siliconized glass tubes. After overlayering with water, the acrylamide is allowed to polymerize for 1 hour. The tubes are then placed in a disc electrophoresis apparatus containing 5% (v/v) 1,2-diaminoethane in the upper (cathode) chamber and 5% (v/v) orthophosphoric acid in the lower (anode) compartment. Focusing is carried out at 4° for 17 hours using a constant voltage of 100 V. During this period, the current drops 3 to 0.5 mA/gel.

The gels are removed by cracking the gel tubes and are immediately incubated at 40° with shaking for 30 minutes in each of a series of urea/phosphate/sodium dodecyl sulfate buffer solutions.[17] These are in turn 8 M urea–0.1 M sodium phosphate, pH 7–1% (w/v) sodium dodecyl sulfate–1 mM dithioerythritol; 8 M urea–10 mM sodium phosphate, pH 7–1% (w/v) sodium dodecyl sulfate–1 mM dithioerythritol and 8 M urea–10 mM sodium phosphate–0.01% (w/v) sodium dodecyl sulfate–1

---

[16] M. Gronow and G. Griffiths, *FEBS Lett.* **15**, 340 (1971).
[17] O. H. W. Martini and H. J. Gould, *J. Mol. Biol.* **62**, 403 (1971).

mM dithioerythritol. Electrophoresis in sodium dodecyl sulfate is then carried out by a procedure similar to that outlined above using a flat-bed apparatus already described by Hultin and Sjöquist[18] with the following modifications to the combined gel holder–cathode electrode compartment. The gel holder was constructed of Perspex with internal dimensions 10 cm × 10 cm × 0.35 cm and was sealed at the bottom by a greased rubber gasket and a retaining Perspex strip screwed to the gel holder. Between the gel holder and the cathode chamber (internal dimensions 13 cm × 6 cm × 5.5 cm) was fixed a small Perspex chamber of internal dimensions 10 cm × 1 cm × 0.4 cm which acts as a holder for the isoelectric focusing gel.

The small-pore gel solution containing 1 mM dithioerythritol (32 ml) is pipetted into the gel holder which is held in a vertical position (total height of gel approximately 9.5 cm). After overlayering with water the gel is allowed to set for 1 hour. The water layer is then removed and the top of the gel is rinsed with 1 ml of large-pore acrylamide solution containing 1 mM dithioerythritol. The sodium dodecyl sulfate-treated isoelectric focusing gel is drained of buffer and wetted with a few drops of large pore gel solution before being placed in a horizontal position in its holder on top of the small-pore holder. The acidic end of the isoelectric focusing gel is placed against one end of the holder so as to allow a sample slot former (1 cm × 0.3 cm × 0.3 cm) to be placed 1 cm from the basic end of the gel. The space between the small-pore gel and the top edge of the isoelectric focusing gel is then filled with large-pore gel solution to a final depth of approximately 2 cm. After overlayering with water, the large-pore gel is photopolymerized by exposure to a fluorescent light tube for 1 hour. The water overlay and sample slot former are carefully removed and after the rubber gasket and Perspex retaining strip have been detached from the bottom of the gel, the combined cathode chamber–gel tray is inserted into an anode chamber designed to hold two such trays according to Hultin and Sjöqvist.[18] This is filled with electrode buffer (composition 7.2 g of glycine, 1.5 g of Tris, 1 g of sodium dodecyl sulfate, made to 1 liter with water) which is maintained at 10° by circulation of the buffer through a cooling bath. After the upper cathode chamber is filled with the same buffer, 200 µl of the original protein sample treated with 0.1% sodium dodecyl sulfate as described above for one-dimensional gels is applied to the sample slot in the large pore gel. Electrophoresis is carried out at a constant current of 38 mA until the bromophenol blue marker dye reaches the bottom of the small-pore gel (in approximately 6 hours using two gels). The gel is rimmed with water from

[18] T. Hultin and A. Sjöqvist, *Anal. Biochem.* **46**, 342 (1972).

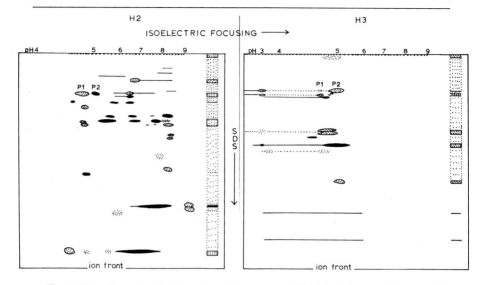

Fig. 4. Two-dimensional electrophoresis patterns of ^{32}P-labeled mouse liver nonhistone protein fractions H2 and H3. The phosphorylation and preparation of the nonhistone fractions is carried out as described in the text. Samples (400–800 μg of protein) of each fraction are electrofocused in a pH 3–10 gradient and after incubating the gel in urea–sodium dodecyl sulfate–dithioerythritol–phosphate buffers, the gel is polymerized in the large-pore spacer gel of a flat-bed 15% polyacrylamide gel containing 0.1% (w/v) sodium dodecyl sulfate, prior to electrophoresis in the second dimension. The shaded areas and those defined by solid lines indicate proteins detected by staining with Coomassie Blue. The dotted areas indicate the presence of phosphoproteins as detected by autoradiography.

a syringe needle and is then extruded from the tray using a sheet of Perspex of cross section 9.8 cm × 0.3 cm applied to the bottom of the gel. The position of the marker dye is fixed by insertion of a piece of thin wire, and the gel is then stained for 20 minutes at 65° in the Coomassie Blue dye solution of Vesterberg[19] (75 ml of methanol, 186 ml of water, 30 g of trichloroacetic acid, 9 g of sulfosalicylic acid, and Coomassie Brilliant Blue R to a final concentration of 0.1%, w/v). Destaining is carried out by elution in ethanol/water/acetic acid (100:260:32, v/v/v). After destaining each gel is sealed in a plastic bag. ^{32}P-labeled proteins are detected by autoradiography using Kodirex X-ray film. Figure 4 shows the distribution of stained and ^{32}P-labeled proteins after two-dimensional electrophoresis of mouse liver nonhistone proteins.

The complexity of the nonhistone proteins is demonstrated by the presence of polypeptides of differing isoelectric points in sodium dodecyl

[19] O. Vesterberg, *Biochim. Biophys. Acta* **243**, 345 (1971).

sulfate–protein complexes which appear as individual components in the one-dimensional sodium dodecyl sulfate system. In addition the two-dimensional separation procedure indicates that there is a considerable range of rapidly phosphorylated proteins in chromatin. The proteins comprising fractions H2 and H3 are markedly different since the latter possess more acidic isoelectric points. However, some overlap probably does occur between these fractions since the proteins designated as P1 and P2 (Fig. 4) appear common to both, the more phosphorylated forms in liver being present in fraction H3.

[13] Methods for Isolation and Characterization of Acidic Chromatin Proteins

By ELIZABETH M. WILSON and THOMAS C. SPELSBERG

The increasing awareness of the role of the acidic chromatin proteins as specific gene regulators has been obscured by difficulties inherent in their isolation and characterization. This method describes a rapid and quantitative isolation of chromatin acidic proteins by three simple steps: acid dehistonization of chromatin, hydrolysis of DNA by a short incubation with bovine pancreatic DNase I, and removal of soluble DNA fragments by acid which precipitates the protein. Greater than 95% recovery of the acidic protein of various chick tissue chromatin is possible with complete elimination of DNA. SDS-polyacrylamide gel electrophoresis of the isolated proteins reveals reproducible heterogeneous banding patterns which appear to be tissue specific.

Reagents

 Acidic Chromatin Protein Isolation
  Sulfuric acid, 4.0 N and 0.4 N
  Tris · HCl buffer, 0.1 M (pH 7.5), containing 2 mM MgCl$_2$ and 2 mM CaCl$_2$
  Bovine pancreatic DNase I, obtained from Worthington Biochem. Corp., Freehold, New Jersey (electrophoretically purified, code DPFF); 1 mg is dissolved in 1 ml of Tris buffer, and stored at $-10°$.
  Perchloric acid, 0.4 N
  Sodium dodecyl sulfate (SDS from Sigma Chem. Co.), 2% and 10 mM sodium phosphate buffer (pH 7.5)

SDS, 3%, containing 0.14 M 2-mercaptoethanol and 10 mM sodium phosphate (pH 7.5)

SDS Polyacrylamide Gel Electrophoresis
Acrylamide—Bio-Rad, electrophoresis grade
N,N'-Methylenebisacrylamide—Eastman Org. Chem.
N,N,N',N'-Tetramethylethylenediamine (TEMED)—Bio-Rad, electrophoresis grade
Separation gels (5 and 8.7% acrylamide):
  Tris · HCl, 3.0 M, containing 0.23% TEMED
  Acrylamide, 20%, containing 1.2% bisacrylamide
  Ammonium persulfate, $(NH_4)_2S_2O_8$, 0.15%, containing 0.2% SDS and 8 M urea (prepared before use)
Stacking gel (3% acrylamide):
  Tris-HCl, 0.49 M (pH 6.7), containing 0.46% TEMED
  Acrylamide, 12%, containing 1.2% bisacrylamide
  Riboflavin, 0.004%
  Urea, 8 M, containing 0.2% SDS
Dialysis buffer: 10 mM sodium phosphate (pH 7.4), containing 1% SDS, 10% glycerol (v/v), and 0.14 M 2-mercaptoethanol
Bromophenol blue, 0.05%
Reservoir buffer: 25 mM Tris · HCl (pH 8.3), containing 0.192 M glycine and 0.1% SDS
Methanol, 50%, containing 10% acetic acid
Coomassie blue, 0.25%, dissolved in 50% methanol (v/v) and 10% acetic acid (v/v)
Acetic acid, 7%
Methanol, 20%, containing 7% acetic acid

Procedure

Chromatin is isolated as previously described[1,2] from nuclei of various tissues. Chromatin (1 mg DNA) in 1 ml of 0.1 mM EDTA and 2 mM Tris · HCl (pH 7.5) is thawed, rehomogenized with a Teflon pestle–glass homogenizer and dehistonized by adding H_2SO_4 to give 12 ml of a 0.4 N H_2SO_4 solution. After standing for 15 minutes at 4°, the solution is centrifuged at 2000 g for 10 minutes. The supernatant containing the histones is decanted, and the pellet is resuspended in 10 ml of cold 0.4 N H_2SO_4 and centrifuged again. The tubes are thoroughly drained, and each pellet is homogenized in 2 ml of Tris buffer (pH 7.5) containing

---

[1] T. C. Spelsberg, A. W. Steggles, and B. W. O'Malley, *J. Biol. Chem.* **246**, 4186 (1971).

[2] T. C. Spelsberg and L. S. Hnilica, *Biochim. Biophys. Acta* **228**, 202 (1971).

2 mM Ca^{2+} and Mg^{2+}. The pH is adjusted to 7.5. To the dehistonized chromatin is added 25 μg bovine pancreatic DNase I in 25 μl of Tris buffer (pH 7.5). The suspension is routinely incubated for 30 minutes at 30°, with occasional shaking. The proteins are pelleted by adding 10 ml of 0.4 N perchloric acid, incubating 10 minutes at 4°, and centrifuging at 2000 g for 10 minutes. The pellet is resuspended in 0.4 N perchloric acid and recentrifuged. The final pellet is carefully drained and resuspended in 1–2 ml of the 2% SDS solution using a glass-Teflon homogenizer. Aliquots are taken for analysis of protein according to Lowry[3] and DNA by the diphenylamine reaction of Burton.[4] The solution is stored at −20°. The proteins may also be resuspended directly into the 3% SDS solution; however, a chemical analysis of protein concentration is complicated by the presence of mercaptoethanol.

For SDS polyacrylamide gel electrophoresis of the isolated nuclear acidic proteins, a gel system similar to that of Laemmli[5] is used, except that 4 M urea is added to the upper and lower separating gels, as described by MacGillivray et al.[6] The 6.5-cm by 6-mm gels consist of a 0.5 cm stacking gel of 3% acrylamide, and 2 separating gel phases of 5 and 8.7%, each 3 cm long. The lower, small-pore gel of 8.7% acrylamide is made from the Tris, acrylamide, and persulfate solutions (cf. Reagents) in a volumetric ratio of 2:7:7, respectively. The 5% acrylamide gel is layered on top of the 8.7% phase and is made from the same solutions of Tris, acrylamide, persulfate, and water using a volumetric ratio of 1:2:4:1, respectively. After gelation of the 5% phase, the large-pore (3% acrylamide) stacking gel is layered on top. The 3% phase is made from the Tris, acrylamide, riboflavin, and urea solutions (cf. reagents) in a volumetric ratio of 1:2:1:4, respectively. UV light is required for polymerization of the stacking gel. All solutions except those containing SDS are stored at 5°.

An aliquot of the isolated acidic proteins in the 2% SDS solution is dialyzed overnight at room temperature against the 1% SDS dialysis buffer. Directly before electrophoresis, the samples are heated to 60° for 20 minutes to enhance solubilization and one-tenth volume of 0.05% bromophenol is added to each as a tracking dye; 50 to 150 μg protein is layered on the gels. Standard protein samples are suspended directly into the 1% SDS dialysis buffer. Electrophoresis is carried out for 4–5

---

[3] O. H. Lowry, N. J. Rosebrough, A. L. Farr, and R. Randall, *J. Biol. Chem.* **193**, 265 (1951).

[4] K. Burton, *Biochem. J.* **62**, 315 (1956).

[5] V. K. Laemmli, *Nature (London)* **227**, 680 (1970).

[6] A. J. MacGillivray, A. Cameron, R. J. Krauze, D. Richwood, and J. Paul, *Biochim. Biophys. Acta* **277**, 384 (1972).

hours at 2 mA per gel until the marker band (bromophenol blue) has migrated to the bottom of the gel. The gels are removed and soaked overnight in the 50% methanol and 10% acetic acid solution to remove excess SDS. The gels are subsequently stained for 1 hour with Coomassie blue and destained electrophoretically and stored in 7% acetic acid. Residual stain is removed using 20% methanol and 7% acetic acid.

Remarks

The DNase procedure described here for isolating the acidic proteins of chromatin is advantageous in that it is quick, quantitative, and reproducible. No long centrifugation or repeated dialysis steps are required as in many procedures of the past.[1,6-10] Also, no proteolytic activity could be detected either in the DNase preparation[11] or in the acidic proteins during isolation. The latter was confirmed by adding [^{3}H]ovalbumin to dehistonized chromatin, carrying out the isolation and electrophoresis, and subsequently counting gel slices.[12]

The necessity of removing DNA from chromatin preparations for effective resolution of the acidic proteins on SDS polyacrylamide gels is demonstrated in Fig. 1. Electrophoresis of whole chromatin results in streaking and faint bands (gel 1), in contrast to the distinct bands of a preparation of dehistonized chromatin which, prior to gel electrophoresis, was treated with DNase (gel 2). Also shown are the gel patterns of the histones (gel 3) and bovine pancreatic DNase used in the isolation procedure (gel 4). The tissue-specific banding patterns of the acidic proteins are displayed in Fig. 2 for chick liver, spleen, erythrocyte, heart, and oviduct.

Of the various parameters checked for optimal removal of DNA and recovery of protein, it was found that Ca^{2+} ions enhance DNA hydrolysis relative to that in the presence of only Mg^{2+} ions.[12] These results are in agreement with previous studies which report that Ca^{2+} protects the active conformation of the DNase enzyme from proteolytic breakdown by trypsin[13] and that Ca^{2+} promotes the simultaneous cleavage of both

---

[7] K. Shelton and V. G. Allfrey, *Nature (London)* **228**, 132 (1970).
[8] S. Levy, R. T. Simpson, and H. A. Sober, *Biochemistry* **11**, 1547 (1972).
[9] H. W. J. van den Broek, L. D. Nooden, J. S. Sevall, and J. Bonner, *Biochemistry* **12**, 229 (1973).
[10] S. C. Elgin and J. Bonner, *Biochemistry* **11**, 772 (1972).
[11] T. C. Spelsberg, unpublished observations, 1972.
[12] E. M. Wilson and T. C. Spelsberg, *Biochim. Biophys. Acta* **322**, 145 (1973).
[13] P. A. Price, T.-Y. Liu, W. H. Stein, and S. J. Moore, *J. Biol. Chem.* **244**, 917 (1969).

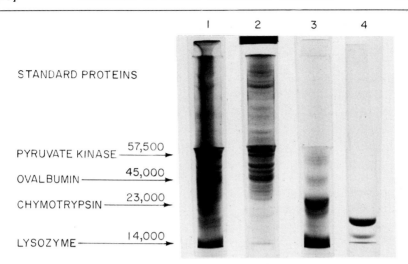

Fig. 1. Sodium dodecyl sulfate (SDS)-polyacrylamide gel electrophoretic patterns of whole chromatin and the acidic proteins and histones separated from DNA. The following gels are represented: (1) Whole chick oviduct chromatin (100 μg of protein); (2) total acidic proteins (100 μg) of oviduct chromatin isolated by the DNase procedure; (3) histones (50 μg) extracted from oviduct chromatin with acid; and (4) bovine pancreatic DNase I (50 μg). Oviduct chromatin was obtained from 7-day-old chicks (Rhode Island Reds) which had received daily injections (subcutaneous) of 5 mg of diethylstilbestrol for 15 days. Whole chromatin was resuspended directly in 3% SDS solution described below. Histones were isolated from oviduct chromatin using 0.4 N H_2SO_4 as described. The histone solution was dialyzed against deionized water and lyophilized. Each protein preparation and the whole chromatin were resuspended in 3% SDS, 0.14 M 2-mercaptoethanol, 10 mM sodium phosphate (pH 7.5) to make a solution of 1 mg of protein per milliliter, and subsequently dialyzed overnight against 100 volumes of the 1% SDS dialysis buffer. The molecular weights of some standard proteins are indicated at their characteristic position of migration. From E. Wilson and T. Spelsberg, *Biochim. Biophys. Acta* **322**, 145 (1973).

strands of DNA.[14] Although the resulting DNA fragments are not readily dialyzable, they are soluble when dilute perchloric acid is used. When compared to dilute acetic and sulfuric acid, perchloric acid is optimal in precipitating the acidic protein after DNA hydrolysis.

It has been found that a DNase I concentration of at least 12 μg per 0.9 mg of dehistonized chromatin DNA is sufficient for complete DNA hydrolysis.[12] The minimal time required is 10 minutes when 25 μg DNase is incubated with 0.9 mg dehistonized chromatin. Hydrolysis of whole chromatin DNA, where the histones have not been dissociated with acid, requires longer time periods or higher concentrations of DNase.

[14] T. L. Poulos and P. A. Price, *J. Biol. Chem.* **247**, 2900 (1972).

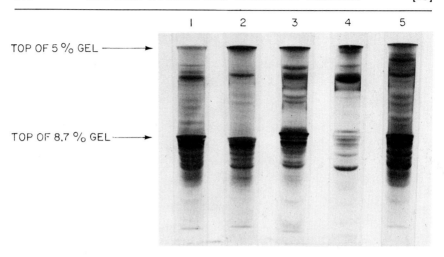

FIG. 2. Sodium dodecyl sulfate (SDS)-polyacrylamide gel patterns of various chick tissues. Liver, spleen, oviduct, and heart tissue were removed from chicks which were treated as described in Fig. 1. Erythrocytes were obtained from 2-year-old hens. Total acidic proteins were isolated by the DNase procedure. The following chick tissues are represented: (1) liver, (2) spleen, (3) erythrocyte, (4) heart, and (5) oviduct. About 100 μg of protein was applied to each gel. From E. Wilson and T. Spelsberg, *Biochim. Biophys. Acta* **322**, 145 (1973).

Preliminary application of the DNase procedure to other animal tissues has proved highly successful. As with chick chromatin preparations, high acidic protein yields with no DNA or histone contaminations were obtained with rat testis, liver, and brain. However, DNase treatment of calf thymus dehistonized chromatin consistently gave rather poor protein recovery ($\simeq 60\%$).

The main limitation of the procedure described here is that acid treatment of chromatin has been shown previously to alter secondary structure.[15] Although this is of little consequence when studying the spectrum of proteins by SDS polyacrylamide gel electrophoresis, it may limit biological applications to the isolated proteins.

Acknowledgments

This work is supported by Public Health Service Grant CA-13065-01 and CA-14920-01 from the National Cancer Institute. E. M. Wilson is an NDEA predoctoral fellow, and T. C. Spelsberg is a fellow of the National Genetics Foundation.

---

[15] T. C. Spelsberg, W. M. Mitchell, and F. Chytil, *Molec. Cell. Biochem.* **1**, 242 (1973).

[14] Methods for Analysis of Phosphorylated Acidic Chromatin Protein Interactions with DNA[1]

By LEWIS J. KLEINSMITH and VALERIE M. KISH

It is now known that phosphorylated proteins are a major component of the nonhistone (acidic) chromatin protein fraction of the cell nucleus, and many of their properties suggest that they play a role in the specific regulation of gene activity. For example, changes in their phosphorylation correlate with changes in gene activity,[2-5] they are heterogeneous and tissue specific,[6,7] they increase the rate and type of RNA synthesis in cell-free systems,[7-10] and finally they can be shown to bind specifically to DNA.[7,11] This chapter will be concerned with methods for the isolation of these phosphorylated nonhistone proteins and techniques for studying their interaction with DNA.

Isolation of Phosphorylated Acidic Chromatin Proteins

Salt Extraction

The method most commonly employed in our laboratory for purification of these proteins involves salt extraction and selective purification on calcium phosphate gel.[8,12] The procedure will be described for rat liver, but it can be adapted to a wide variety of tissues and cell types. We normally employ nuclei which have been isolated by a dense sucrose pro-

---

[1] Studies on this subject in our laboratory have been supported by grants from the National Science Foundation (GB-8123 and GB-23921). V.M.K. held a predoctoral fellowship from U.S. Public Health Service Training Grant 5-T01-GM-72-15.
[2] L. J. Kleinsmith, V. G. Allfrey, and A. E. Mirsky, *Science* **154**, 780 (1966).
[3] E. L. Gershey and L. J. Kleinsmith, *Biochim. Biophys. Acta* **194**, 519 (1969).
[4] R. W. Turkington and M. Riddle, *J. Biol. Chem.* **244**, 6040 (1969).
[5] K. Ahmed and H. Ishida, *Mol. Pharmacol.* **7**, 323 (1971).
[6] R. D. Platz, V. M. Kish, and L. J. Kleinsmith, *FEBS Lett.* **12**, 38 (1970).
[7] C. S. Teng, C. T. Teng, and V. G. Allfrey, *J. Biol. Chem.* **246**, 3597 (1971).
[8] T. A. Langan, in "Regulation of Nucleic Acid and Protein Biosynthesis" (V. V. Koningsberger and L. Bosch, eds.), p. 233. Elsevier, Amsterdam, 1967.
[9] M. Kamiyama, B. Dastugue, and J. Kruh, *Biochem. Biophys. Res. Commun.* **44**, 1345 (1971).
[10] N. C. Kostraba and T. Y. Wang, *Biochim. Biophys. Acta* **262**, 169 (1972).
[11] L. J. Kleinsmith, J. Heidema, and A. Carroll, *Nature (London)* **226**, 1025 (1970).
[12] E. L. Gershey and L. J. Kleinsmith, *Biochim. Biophys. Acta* **194**, 331 (1969).

cedure[13] as described below, but other types of nuclear preparations can often be substituted.

Male Sprague-Dawley rats, weighing 150–250 g, are injected with [^{32}P]orthophosphate in 0.9% NaCl at a dose of 3 mCi per 100 g body weight. After 2 hours, the animals are sacrificed and their livers are removed and quickly chilled. All subsequent operations are performed at 4°. Thirty grams of liver are finely minced with scissors and added to 300 ml of 0.32 M sucrose–3 mM MgCl$_2$. The tissue is homogenized for 2 minutes at 6000 rpm in a Sorvall Omni-Mixer with a small-bladed chamber. The resulting homogenate is filtered through double-napped flannelette, and centrifuged for 7 minutes at 1000 g. The resulting pellet is resuspended in 225 ml of 2.4 M sucrose–1 mM MgCl$_2$ by homogenizing for 2 minutes at 2000 rpm in the Sorvall Omni-Mixer using the large-bladed chamber. The resulting suspension is centrifuged for 60 minutes at 70,000 g. The nuclear pellets are collected and washed twice by resuspending in 10 mM Tris · HCl, pH 7.5–0.25 M sucrose–4 mM MgCl$_2$ and centrifuging for 7 minutes at 1000 g.

These purified nuclei serve as the starting material for subsequent purification of phosphorylated acidic chromatin proteins. The soluble proteins of the nuclear sap are removed by suspending the nuclei in 10 mM Tris · HCl, pH 7.5 for 20 minutes. The nuclei are collected by centrifugation at 10,000 g for 10 minutes, and are then suspended in 1.0 M NaCl–20 mM Tris · HCl, pH 7.5 to a final protein concentration of 2 mg/ml. The suspension is dispersed using a Polytron homogenizer (Brinkmann) for 20 seconds at setting 3. The resulting viscous solution is mixed with 1.5 volumes of 20 mM Tris · HCl, pH 7.5, and the precipitated nucleohistone is removed by centrifugation at 200,000 g for 2 hours. Bio-Rex 70 (Na$^+$), which has previously been equilibrated with 0.4 M NaCl-20 mM Tris · HCl, pH 7.5, is then added to the supernatant at a ratio of 20 mg of Bio-Rex per milligram of protein.[14] After stirring slowly for 10 minutes, the suspension is centrifuged for 10 minutes at 6000 g and the supernatant withdrawn. The resin is washed by suspending it in 10–15 ml of 0.4 M NaCl–20 mM Tris · HCl, pH 7.5 and centrifuging again for 10 minutes at 6000 g. The two supernatants are then combined and calcium phosphate gel[15] added at a ratio of 0.46 mg of gel per milli-

[13] A. O. Pogo, V. G. Allfrey, and A. E. Mirsky, *Proc. Nat. Acad. Sci. U.S.* **56**, 550 (1966).

[14] In several steps during this procedure it is necessary to know the concentration of protein in solution. Since extreme accuracy is not required, it is advisable to use one of the direct spectrophotometric methods (see this series, Vol. 3 [73]) which have considerable advantages in terms of ease and speed. We routinely employ the formula: protein conc. (mg/ml) = $1.55 A_{280} - 0.76 A_{260}$.

[15] D. Keilin and E. F. Hartree, *Proc. Roy. Soc., Ser. B* **124**, 397 (1938).

gram of protein. After slowly stirring for 20 minutes the suspension is centrifuged for 5 minutes at 6000 g, and the supernatant is discarded. The gel is washed by resuspension in 10–15 ml of 1.0 M $(NH_4)_2SO_4$–0.05 M Tris · HCl, pH 7.5 using a motor-driven stirrer at 1000 rpm, followed by centrifugation at 6000 g for 5 minutes. The supernatant is again discarded, and the gel is then dissolved by gentle homogenization with a Teflon Potter-Elvehjem tissue grinder in 0.3 M EDTA, pH 7.5–0.33 M $(NH_4)_2SO_4$ in a ratio of 0.2 ml of solution per milligram of gel. The suspension is allowed to stand for 1 hour in the cold with occasional rehomogenization, and the insoluble residue is then removed by centrifugation for 15 minutes at 33,000 g. The supernatant is then dialyzed overnight against 50 mM Tris · HCl, pH 7.5. The final protein product contains about 1.3% phosphorus by weight, and is generally referred to as the phosphorylated nonhistone (acidic) protein fraction.

Phenol Extraction

The other method commonly utilized in the isolation of phosphorylated acidic nuclear proteins is based on their solubility in phenol. The procedure has been developed primarily in Allfrey's laboratory,[7,16] and is a modification of a technique designed originally for the solubilization of phage proteins.[17] The technique will be described starting at the stage of purified nuclei, which can be prepared as described above or by other suitable procedures.

All procedures are carried out at 4° unless otherwise noted. After measurement of packed nuclear volume, the nuclei are suspended by hand homogenization in 25 volumes of 0.14 M NaCl using 10 strokes of a Teflon Potter-Elvehjem tissue grinder. The mixture is transferred to a beaker and stirred for 30 minutes prior to centrifugation at 3000 g for 5 minutes. The pellet is reextracted as described above, the supernatants are combined, and aliquots are taken for measurement of the 0.14 M NaCl-soluble nuclear proteins. The packed volume of the nuclear residue is measured at this step. The nuclear residue is suspended in 25 volumes of 0.25 N HCl by the method described for the saline-soluble proteins and is then transferred to a beaker and stirred for 2 hours. After centrifugation at 3000 g for 5 minutes the extraction is repeated and the supernatant phases are combined and analyzed for total acid-extractable protein. The volume of nuclear residue is again determined prior to removal of lipids via suspension in 10 volumes of 1:1 (v/v) chloroform–methanol

---

[16] K. Shelton and V. G. Allfrey, *Nature (London)* **228**, 132 (1970).
[17] E. Viñuela, I. D. Algranati, and S. Ochoa, *Eur. J. Biochem.* **1**, 3 (1967).

containing 0.2 N HCl. In this step the residue is suspended to a paste using a glass rod, and then the organic solvent mixture is added dropwise with continual mixing until the total amount has been added and the residue is evenly suspended. This suspension is stirred for 10 minutes and then centrifuged at 5000 g for 5 minutes. The supernatant is discarded, and the pellet is extracted with 10 volumes of 2:1 chloroform–methanol containing 0.2 N HCl and centrifuged as before. The resulting pellet is washed with 10 volumes of ether, centrifuged, the ether removed, and the volume of the pellet then measured.

The residue remaining after extraction of the 0.14 M NaCl-soluble proteins, acid-soluble proteins, and lipids is suspended in 5 volumes of 0.1 M Tris · HCl, pH 8.4 containing 10 mM EDTA and 0.14 M 2-mercaptoethanol (buffer A) by adding this solution dropwise to the residue while stirring with a glass rod. The whole process requires approximately 30 minutes per 5 ml of buffer A. This suspension is stirred while adding dropwise an equal volume of cold phenol saturated with buffer A, and the mixture is stored at 2° overnight. The resulting solution contains a copious white precipitate and is homogenized in a Teflon-glass Potter-Elvehjem tissue grinder rotating at 500 rpm for 4 strokes prior to centrifugation at 12,000 g for 10 minutes. The resulting aqueous phase is collected and re-extracted with an equal volume of phenol-saturated buffer A and centrifuged as above. The phenol phase from the first centrifugation is carefully collected using a needle and syringe, avoiding the thick layer of precipitated protein above it. It is then added to the phenol layer from the second centrifugation and the combined extracts are dialyzed for 3 hours against 100 volumes of 0.10 M acetic acid containing 0.14 M 2-mercaptoethanol. The dialysis buffer is changed, and dialysis is continued until the phenol phase is reduced to one-fifth the original volume. The dialysis tubing is then opened, and the aqueous layer above the phenol phase is removed using a needle and syringe. The remaining phenol is dialyzed at room temperature for 24 hours against 50 mM acetic acid containing 9.0 M urea and 0.14 M 2-mercaptoethanol. Dialysis is continued for 2 hours in 0.1 M Tris · HCl, pH 8.4 containing 8.6 M urea, 10 mM EDTA, and 0.14 M 2-mercaptoethanol. This dialysis procedure restores the phenol-soluble proteins to the aqueous phase. Aliquots of the solution are analyzed for total protein content according to a modification of the method of Lowry.[18] The resulting protein fraction has an average

---

[18] Since both urea and phenol interfere with development of color in the Lowry assay, protein samples are first precipitated with 5% trichloroacetic acid containing 2 mM silicotungstic acid. The resulting pellets are dissolved in 0.1 ml of 1.0 N NaOH and analyzed via the Lowry procedure using bovine serum albumin as standard (see this series, Vol. 3 [73]).

phosphorus content of 0.94% by weight and an amino acid content which confirms its acidic nature.

Comparison of Procedures

Although both techniques for preparation of the phosphorylated protein fractions yield products with similar phosphorus contents, the procedures should not be considered equivalent. First of all, electrophoretic analysis in polyacrylamide gels indicates that although both preparations are highly heterogeneous, there are significant differences between them (Fig. 1). In addition, the phenol treatment involves proteins which have been extensively denatured due to their exposure to both HCl and phenol, and thus would probably not be the method of choice if one were interested in functional or enzymatic activities. Unfortunately, the high ionic strength utilized in the salt extraction technique may also be deleterious to some protein properties, so that in the end it will be useful to have data from both types of protein preparations for comparative purposes.

The extended dialysis associated with preparation of the phenol extractable proteins results in a total preparation time of approximately 50 hours, while the salt extraction technique requires approximately 30 hours. This large discrepancy in time may in itself be an important variable to consider when choosing the preferred method of nonhistone chromatin phosphoprotein preparation.

Yields of protein vary with the two procedures; the salt extraction procedure generally yields approximately 5 mg of nonhistone chromatin phosphoprotein per 100 g of rat liver (or 4–5 ml of packed nuclear volume). Teng et al.[7] report a yield of 6.7 mg of phenol-soluble proteins from an estimated 14–28 g of fresh rat liver. We have used this procedure to isolate the acidic protein fraction from both fresh and frozen bovine liver, and in our hands the yield obtained is considerably less, although this difference may be a function of the different tissue source. We have also noted that if frozen bovine liver is used, the yield is approximately 25% of that obtained using fresh bovine liver. This is in contrast to the results obtained via the salt extraction procedure using fresh and frozen tissue; in this case the yields of material are somewhat greater if the tissue is frozen first.

Binding of Phosphorylated Proteins to DNA

DNA-Cellulose Chromatography

An extremely sensitive method for studying the binding of these nuclear protein fractions to DNA involves the technique of DNA-cellu-

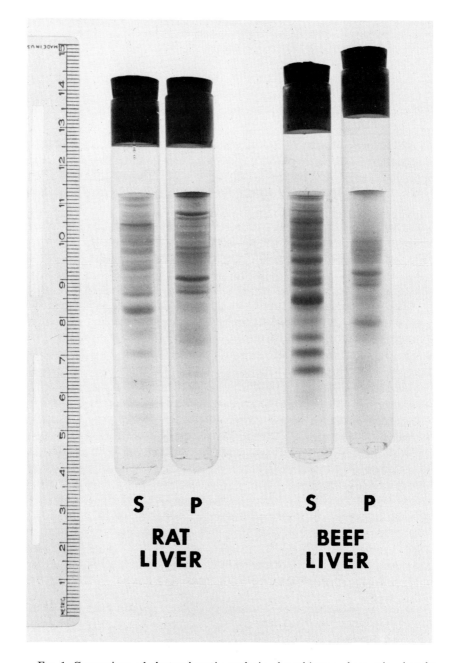

FIG. 1. Comparison of electrophoretic analysis of nonhistone chromatin phosphoproteins isolated from rat and bovine liver by either the salt extraction (S) or phenol isolation (P) technique. Electrophoresis was performed in 10% acrylamide gel in the presence of the detergent sodium dodecyl sulfate as described by K.

lose chromatography developed by Alberts (this series, Vol. 21 [11]). DNA is purified using the phenol procedure of Kirby (this series, Vol. 12B [98]), which gives a DNA product free from protein. DNA-cellulose is prepared by mixing 6 ml of DNA solution (1 mg/ml in 10 mM Tris · HCl, pH 7.2–1 mM Na$_3$EDTA) with 2 g of Munktell 410 cellulose (Bio-Rad) which has previously been washed with ethanol. The material is allowed to dry overnight in an open petri dish, and is then put in a vacuum desiccator for 1 week. It is then suspended in cold Tris-EDTA (concentrations as above) for one day, washed twice by centrifugation in the same buffer, and stored frozen in Tris-EDTA containing 0.14 M NaCl. The final product contains about 1 mg of DNA bound per gram of cellulose.

Binding of phosphorylated nonhistone proteins to DNA-cellulose is routinely studied via the technique of gradient dialysis.[19,20] The ^{32}P-labeled protein preparation is first incubated for 15 minutes at 25° with deoxyribonuclease (1 μg/ml) in the presence of 1 mM Mg^{2+}, and is then dialyzed against several changes of cold 0.6 M NaCl, 1 mM Na$_3$ EDTA, 1 mM 2-mercaptoethanol, 10% glycerol, 20 mM Tris · HCl, pH 7.5 ("dialysis buffer"). An aliquot of the protein solution containing 50–400 μg of protein in 0.5–1.0 ml is mixed with 250 mg of DNA-cellulose which has been previously washed with the same dialysis buffer. The mixture is then dialyzed for 2 hours against dialysis buffer containing 5 M urea, and then against successive 2-hour changes of dialysis-buffer–5 M urea in which the NaCl concentration is sequentially lowered to 0.4, 0.2, and 0.14 M. The mixture is finally dialyzed overnight against 0.14 M NaCl dialysis buffer without urea. All dialyses are performed at 4°. The DNA-cellulose is then placed in a 0.5-cm diameter glass column, and washed successively with dialysis buffer made up with 0.14, 0.6, and 2.0 M NaCl. Fractions (0.5 ml) are collected and counted in Bray's solution[21] in a liquid scintillation counter.

The original phosphorylated protein fraction contains a mixture of proteins with varying affinities for DNA. In order to focus attention on those proteins which bind in a specific fashion, we routinely prerun our protein fraction with DNA-cellulose made from salmon sperm DNA. The

---

[19] Phosphorylated proteins can also be applied directly to the DNA-cellulose column, but in our hands the results tend to be less reproducible than when gradient dialysis is employed.

[20] I. Bekhor, G. M. Kung, and J. Bonner, *J. Mol. Biol.* **39**, 351 (1969).

[21] G. A. Bray, *Anal. Biochem.* **1**, 279 (1960).

Weber and M. Osborn [*J. Biol. Chem.* **244**, 4406 (1969)]. Note that the protein binding patterns are considerably different in the protein fractions prepared by the two different techniques.

runoff fraction is collected and is our starting material for all subsequent studies, thereby leaving behind the proteins that bind to salmon sperm DNA and are thereby relatively nonspecific in their interactions with DNA. Figure 2 shows that when such pretreated proteins are chromatographed on rat DNA-cellulose, about 1% of the radioactivity binds to the column and is eluted at 0.6 M NaCl. Much lower levels of binding are seen if a bacterial DNA is substituted for the rat DNA, demonstrating the specific nature of the protein–DNA interaction. Control columns of cellulose containing no DNA exhibit negligible binding.

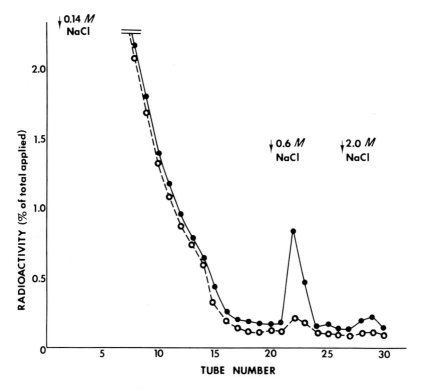

FIG. 2. Binding of ^{32}P-labeled phosphorylated nonhistone proteins to DNA-cellulose. Phosphorylated proteins from rat liver prepared by the salt extraction technique were mixed with DNA-cellulose and run through gradient dialysis as described in the text. After the mixture was poured into a glass column, the cellulose was eluted with increasing ionic strength. Although most of the protein does not stick to the column made from rat DNA (●——●), a small reproducible quantity of material does bind and elutes as a distinct peak of radioactivity. Much less binding is observed when columns containing *Escherichia coli* DNA-cellulose are employed (○---○), thus demonstrating the specificity of the protein–DNA interaction.

Sucrose Gradient Analysis

Binding of phosphorylated proteins to DNA can also be monitored by centrifugation in sucrose density gradients. ^{32}P-Labeled phosphoproteins are dialyzed overnight at 4° against 100 volumes of 10 mM Tris · HCl, pH 8.0 containing 5.0 M urea and 1.0, 0.8, and 0.6 M NaCl, by the Marmur procedure[22] involving deproteinization with chloroform followed by treatment with ribonuclease and pronase. Approximately 100 μg of ^{32}P-labeled phosphoprotein and 200 μg of DNA are mixed in a final volume of 0.5 ml in dialysis buffer, and the salt concentration of the mixture is progressively lowered by stepwise dialysis against 10 mM Tris · HCl, pH 8.0 containing 5.0 M urea and 1.0, 0.8, and 0.6 M NaCl, allowing 2 hours' dialysis at each salt concentration. The mixture is dialyzed overnight against the same dialysis buffer containing 0.4 M NaCl, followed by a 2-hour dialysis against 10 mM Tris · HCl, pH 8.0 containing 10 mM NaCl. After dialysis, 0.5 ml of the sample is carefully layered on top of a 4-ml 5 to 25% sucrose gradient prepared in 10 mM Tris · HCl, pH 8.0 containing 10 mM NaCl, and after centrifugation at 358,000 g for 2.5 hours at 20°, 0.2-ml fractions are collected. Each fraction is diluted to 0.4 ml, and the DNA content is measured by ultraviolet absorption at 260 nm. Radioactivity is monitored by adding 10 ml of Bray's scintillation cocktail to each fraction followed by counting in a liquid scintillation counter.

Evidence for the specificity of binding using this technique is illustrated in Fig. 3, as reported by Teng et al.[7] The sucrose density profiles demonstrate that rat liver phenol-soluble ^{32}P-labeled proteins bind to rat liver DNA (as evidenced by the movement of ^{32}P activity downward into the denser regions of the gradient), but not to calf thymus DNA. The binding of homologous phosphoprotein and DNA under these conditions is dependent on the ionic strength during centrifugation, since in the presence of buffer containing 50 or 15 mM NaCl no binding occurs.

Membrane Filter Analysis

A third technique for monitoring binding of phosphorylated proteins to DNA involves trapping the complex on nitrocellulose membrane filters in a manner similar to the assay developed for measuring the binding of *lac* repressor protein to *lac* operon DNA.[23] Since uncomplexed native DNA passes through the filter, this permits the monitoring of the extent of DNA–protein interaction with relative ease. Unfortunately, the non-

---

[22] J. Marmur, *J. Mol. Biol.* **3**, 208 (1961).
[23] A. D. Riggs, H. Suzuki, and S. Bourgeois, *J. Mol. Biol.* **48**, 67 (1970).

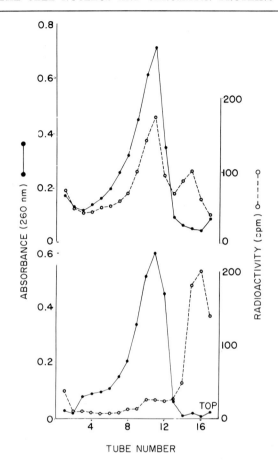

Fig. 3. Binding of ^{32}P-labeled phosphorylated nonhistone proteins to DNA as monitored by sucrose density gradient analysis. Phosphorylated proteins from rat liver prepared by the phenol isolation technique were mixed with various DNA's, subjected to gradient dialysis, and then applied to sucrose density gradients as described in the text. Note that the rat liver phosphoprotein (○---○) binds to rat DNA (top, ●——●) but not to calf thymus DNA (bottom, ●——●); indicating the specificity of the protein–DNA interaction. Data were taken from C. S. Teng, C. T. Teng, and V. G. Allfrey, *J. Biol. Chem.* **246**, 2597 (1971).

histone chromatin phosphoprotein fraction alone is retained by the membrane filter, so that in order to follow DNA-protein binding the DNA must be labeled prior to use in the binding assay.

Radioactive DNA of high specific activity is easily prepared from a wide variety of sources by *in vitro* methylation with labeled dimethyl sulfate. The method routinely employed in our laboratory is a modification of that reported by Tan and Miyagi[24] for the methylation of rat DNA. We have found this procedure works well using commercial calf thymus and *E. coli* DNA's as well as bovine liver DNA prepared by the phenol method of Kirby (this series, Vol. 12B [98]). The reaction is carried out at room temperature in a screw-cap vial containing 4 ml of DNA (250 µg/ml) in 0.5 M NaCl to which 0.26 ml of tri-N-butylamine and 250–5000 µCi of [^{3}H]dimethyl sulfate (235 mCi/mmole) dissolved in 0.2 ml of ether are added. The mixture is shaken vigorously for 4 hours using an Eberbach automatic shaker and then extracted 10 times at 4° with an equal volume of ether. After overnight dialysis, the specific activity of the DNA is monitored by a modification of the method reported by Sibatani.[25] Aliquots of the [^{3}H]DNA are pipetted onto Whatman filter paper discs, and the discs are washed twice for 10 minutes at room temperature in 50 mM EDTA–0.16% cetyltrimethylammonium bromide (CTAB), followed by two 5-minute washes in ether. This method is 2–3-fold more sensitive than a similar procedure employing cold 10% TCA washes followed by ethanol:ether (3:1 ratio) and ether. The specific activity of the DNA methylated by the above method ranges from 100 to 1000 cpm/µg, depending on the source of the DNA as well as on the amount of [^{3}H]dimethyl sulfate added to the reaction mixture.

Another method for methylating DNA has been reported by Akiyoshi and Yamamoto.[26] This procedure, however, requires 24–48 hours of incubation, depending on the sources of the DNA, and results in DNA of lower specific activity.

Binding of phosphorylated proteins to radioactive DNA is measured employing the following procedure. All DNA and protein samples are dialyzed overnight at 4° against 100 volumes of 10 mM Tris · HCl, pH 7.4 containing 10 mM NaCl–1 mM EDTA–5 mM 2-mercaptoethanol in 5% dimethyl sulfoxide (DMSO). DMSO is added to the buffer to reduce background counts on the filters. Initial experiments included 10 mM magnesium acetate in the buffers; however, it was later found that this divalent cation is not a prerequisite for binding, so it is left out of the procedure described here.

The binding curves are run using a constant amount of DNA and varying levels of phosphoprotein. The reaction mixture contains dialysis buffer to 1.0 ml, 3–4 µg of [^{3}H]DNA, competing DNA's, and varying levels of phosphoprotein, added in the order stated. After incubation at 22–24° for 15 minutes, two 0.4-ml aliquots are removed from each sample and are filtered through 25-mm nitrocellulose membrane filters

---

[24] C. H. Tan and M. Miyagi, *J. Mol. Biol.* **50**, 641 (1970).
[25] A. Sibatani, *Anal. Biochem.* **33**, 279 (1970).
[26] H. Akiyoshi and N. Yamamoto, *Biochem. Biophys. Res. Commun.* **38**, 915 (1970).

(Schleicher and Schuell, B-6, 0.45 μm) which are presoaked in dialysis buffer at room temperature for at least 30 minutes. The filtration rate should be adjusted so that the filters do not dry out, since drying at this stage leads to erratic results. After each filter has been washed twice with 3 volumes of the dialysis buffer, filters are dried using low heat and counted in Bray's scintillation cocktail.[21] The nitrocellulose dissolves in the presence of this fluid, resulting in a slow increase in counts in each vial over a period of time. We have found that a plateau is reached after approximately 14 hours, so we routinely allow the vials to sit overnight prior to counting.

Results of a typical binding assay are indicated by the uppermost curve in Fig. 4. With the amount of labeled bovine liver DNA held constant, increasing the concentration of bovine liver phosphoprotein results in an initial linear increase in the amount of [³H]DNA retained on the

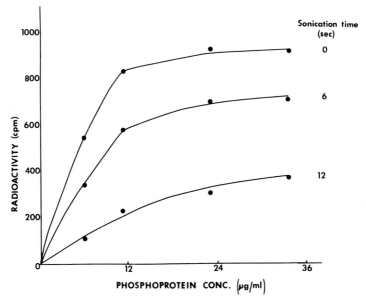

Fig. 4. Binding of phosphorylated nonhistone proteins to DNA as measured by nitrocellulose membrane filtration. Bovine liver phosphoprotein prepared by salt extraction was incubated with 10 μg of [³H]DNA and filtered as described in the text. In some cases, [³H]DNA was pretreated by sonication in the cold for 6 or 12 seconds using a Bronwill Biosonik (5 mm probe) at minimum power. [³H]DNA which binds to protein during the incubation procedure is retained on the membrane filters. The amount of radioactivity on the filters is plotted as a function of phosphoprotein concentration in the incubation mixture. Note that the binding of DNA to the protein is reduced as the size of the DNA molecules is decreased by progressive sonication.

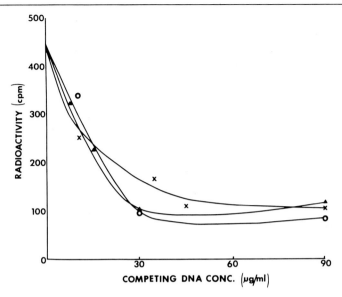

Fig. 5. Competition by unlabeled DNA's from various sources against binding of bovine liver [^{3}H]DNA to bovine liver nonhistone phosphoprotein. [^{3}H]DNA and salt-extracted phosphoprotein from bovine liver were held constant at 4 μg and 28 μg, respectively, in the incubation medium. Prior to incubation, all DNA's were phenol extracted, treated with RNase and pronase, and sonicated for 6 seconds as described in the text and in the legend to Fig. 4. Varying amounts of unlabeled DNA's from various sources were mixed with the [^{3}H]DNA from bovine liver prior to incubation with phosphoprotein. Note the decrease in counts retained on the nitrocellulose filter as increasing concentrations of competing DNA are added. However, similar competition curves are obtained by adding unlabeled DNA from bovine liver (○——○), salmon sperm (×——×), and *E. coli* (▲——▲), thus demonstrating the relative lack of specificity of binding observed in this assay system under these conditions.

filter, followed by a plateau at higher levels of phosphoprotein. Sonication of the [^{3}H]DNA results in a significant reduction in the binding of phosphoprotein, indicating the importance of the size of the DNA in the binding reaction. Thus, in experiments where one wishes to compare binding of protein to different types of DNA, it is extremely important to control for the size of the different DNA's.

Competition experiments can be easily run employing this assay system, where increasing concentrations of unlabeled DNA are used to compete with the radioactive DNA for binding sites on the phosphorylated proteins. When DNA's from different species are used to compete against binding of radioactive bovine DNA to bovine phosphoprotein, similar competition curves result (Fig. 5). These data emphasize that under the conditions employed, the membrane binding assay does not detect the highly specific type of DNA-binding observed with both DNA-cellulose chromatography and sucrose gradient analysis.

Comparison of Procedures

Each of the three procedures which have been used to investigate the binding of the nonhistone chromatin phosphoprotein fraction to DNA incorporates a number of advantages and disadvantages which should be considered prior to choosing the most appropriate method of analysis for a given situation. DNA-cellulose column chromatography appears to be the most sensitive of the three methods in terms of its ability to detect extremely low levels of specific binding. In some situations it has been used to detect specific binding of proteins which represent less than 0.1% of the original sample applied.[11] However, its use as a preparative-scale technique is quite limited, and it is difficult to run a large number of samples simultaneously under the same conditions. Sucrose gradient analysis permits larger amounts of material to be handled, but again it is difficult to run large numbers of samples simultaneously. The potential sensitivity of sucrose gradient analysis is also considerably less than that of DNA-cellulose chromatography.

One problem encountered in both these procedures is that, owing to the long time involved in sample preparation and analysis, meaningful kinetic data are difficult to obtain. This obstacle is largely overcome in the membrane filter technique, which has the added advantages of requiring relatively small amounts of material and allowing a large number of samples to be run simultaneously. Unfortunately, the sensitivity of the membrane technique in terms of detecting specific binding seems to be the lowest of the three procedures.

It should be pointed out that the assay conditions for binding in the three procedures are quite different in terms of buffer, salt, and divalent cation requirements, and that these variables may be responsible for some of the different characteristics of binding observed in the three systems. For example, the fact that much higher levels of binding have been observed by Teng *et al.*[7] using sucrose gradient analysis than by Kleinsmith *et al.*[11] using DNA-cellulose chromatography may be at least partially explained by the higher ionic strength (0.14 M NaCl) employed in the latter studies. However, this higher ionic strength has the advantage of approximating the physiological situation more closely, so that the binding observed under these conditions may be more significant. Thus, the choice of assay conditions, the purity and physical state of the DNA, and the technique chosen to monitor protein–DNA binding are all important variables that must be considered when attempting to study the interaction of phosphorylated nonhistone proteins with DNA.

[15] Immunochemical Characteristics of Chromosomal Proteins

By F. CHYTIL

This chapter describes procedures for production of antibodies against histones and nonhistone chromosomal proteins, as well as against the nonfractionated chromatin. The conditions for testing the antigenicity of the chromosomal proteins by the method of quantitative complement fixation will be given also. This method appears to be a method of choice as it requires relatively small amounts of the chromosomal material which become very often a limiting factor in the determination of the properties of these proteins.

The reader should consult first the introduction in general immunochemical techniques published in this series[1] as well as the detailed description of the method of quantitative complement fixation published earlier.[2]

Immunogens

Histones

To facilitate the production of antibodies against these basic proteins it is advisable either to bind them covalently to serum albumin[3] or to complex them with phosphorylated serum albumin[3] or to form a complex with RNA or DNA,[4] though free histones apparently could be used also.[5]

Coupling the Histones with Human Serum Albumin (HSA).[3] HSA, 100 mg, and 40 mg of whole histone are dissolved in 3 ml of H_2O. Three milliliters of 1-cyclohexyl-3-(2-morpholinoethyl) carbodiimide metho-*p*-

---
[1] See this series, Vol. 11 [91].
[2] See this series, Vol. 11 [92].
[3] A. L. Sandberg, M. Liss, and B. D. Stollar. *J. Immunol.* **98**, 1182 (1967).
[4] B. D. Stollar and M. Ward, *J. Biol. Chem.* **245**, 1261 (1970).
[5] P. Rumke and M. Sluyser, *Biochem. J.* **101**, 1c (1966).

toluenesulfonate[6] in H_2O (75 mg/ml) are added, and the solution is stirred for 30 minutes at room temperature and then dialyzed for 24 hours against distilled H_2O at 4° (25 mg of HSA, 10 mg of histone fraction, and 75 mg of the above carbodiimide can be used). Conjugates of HSA and the carbodiimide without histones can be also prepared.

Coupling of Histones with Phosphorylated Bovine Serum Albumin (BSA)

PHOSPHORYLATED BSA.[7,8] BSA, 1.5 g is dissolved in 55 ml of 5% Na_2HPO_4 containing one drop of a phenolphthalein solution and placed in an ice-salt bath. Three milliliters of $POCl_3$ dissolved in 25 ml of carbon tetrachloride are added dropwise to the stirring BSA solution. Simultaneously, 1 M NaOH is added dropwise to maintain a pH of approximately 9 during the 4-hour reaction period. The reaction mixture is then poured into a dialysis bag and dialyzed overnight in the cold room against three changes of H_2O. By twisting the dialysis bag at a level well above the carbon tetrachloride–water interphase, the aqueous phase could be separated. Alternatively, the carbon tetrachloride layer could be removed by centrifugation. Aliquots of the reaction mixture (at pH 8 to 9) are delivered into small tubes and stored frozen. The phosphorylated BSA can be isolated by acidification of the aqueous layer in the cold with 1 N hydrochloric acid to achieve maximum precipitation. The phosphorylated BSA is separated by centrifugation.

COUPLING[7] Phosphorylated BSA, 250 μg in 50 μl of water, is added to 2 ml of 0.15 sodium chloride solution containing 250 μg of histone. The resulting solution is used for immunization.

Histone Complexes with Nucleic Acids.[4,9] Equal amounts of yeast RNA and histones in 0.14 NaCl, 10 mM sodium phosphate buffer, pH 6.8, are mixed. In another procedure 600 μg of each histone fraction is mixed with 200 μg of RNA, or 2 mg of whole histone in 2 ml is added to 0.6 mg of DNA in 0.75 ml.

Nonhistone Proteins

It is the author's experience that acidic proteins still complexed with DNA or free are good immunogens.[10,11]

---

[6] T. L. Goodfriend, L. Levine, and G. Fasman, *Science* **144**, 1344 (1964).
[7] H. Van Vunakis, J. Kaplan, H. Lehrer, and L. Levine, *Immunochemistry* **3**, 393 (1966).
[8] M. Heidelberger, B. Davis, and H. P. Treffers, *J. Amer. Chem. Soc.* **63**, 498 (1941).
[9] M. Bustin and B. D. Stollar, *J. Biol. Chem.* **247**, 5716 (1972).
[10] F. Chytil and T. C. Spelsberg, *Nature (London) New Biol.* **233**, 215 (1971).
[11] F. Chytil, unpublished observations.

Chromatin

The preparation can be used directly for immunization.[11]

Immunization Procedure

Rabbits: New Zealand White rabbits weighing between 3 and 4 kg are usually used. Complete Freund adjuvant is mixed with the solution of immunogen (1:1 or 1:1.5 v/v) in a syringe or homogenized in a small Teflon pestle glass homogenizer. Usually the immunization schedule used in this laboratory involves a toe pad injection of as much of the immunogen as possible on the first and eighth days, the rest being administered intramuscularly in multiple sites. About 10 days after administration of the complete Freund adjuvant, inflammation of the toe pads is observed. Then the injection in the toes has to be avoided, and the immunogen is administered intramuscularly only.

Immunization Doses. The table shows the immunization doses for different chromosomal components.

Intravenous Booster. Often an intravenous injection into the marginal vein is given (Freund adjuvant has to be omitted) a week before bleeding.

Bleeding. The animals are usually bled a week after the intravenous injection from the ear vein, and the sera are prepared as described earlier.[1]

Antisera. Usually the sera can be used for testing without purification. However, sometimes, especially when a low dilution of the serum has to be employed, a high fixation of the complement in the absence of an antigen is observed in the complement fixation assay. In order to lower this property ("anticomplementarity") the globulin fraction can be purified either by ammonium sulfate precipitation[12] or by ion exchange chromatography (DEAE-cellulose).[12] In the author's laboratory the following procedure was found to remove substantially the "anticomplementarity." The sera are diluted with 9 volumes of 0.14 M sodium chloride and centrifuged at 105,000 g at 4° for 60 minutes. The serum is then carefully siphoned off, sterilized by Millipore filtration, and stored at −20°.

Antigens. Varying amounts of antigens are used by diluting with the diluent used for the complement fixation assay. The antigens are stored frozen at −20°. The histones and chromatins do not exert any "anticomplementarity." On the other hand, the nonhistone protein–DNA complexes may show limited binding of complement in the absence of anti-

[12] H. H. Fudenberg, *in* "Methods in Immunology and Immunochemistry" (C. A. Williams and M. W. Chase, eds.), Vol. 1, p. 306. Academic Press, New York, 1967.

PRODUCTION OF ANTI-CHROMOSOMAL PROTEIN ANTIBODY IN RABBITS

Immunogen	Form	Injected[a]	Dose per injection (mg)	Interval (days)	Number of injections	Booster (µg)	References
Histones	Free	i.m.	1–10	7	3	No	b
	Covalently bound to serum albumin	s.c.	5	7–14	6–10	No	c
	Complexed with phosphorylated serum albumin	s.c.	5	7–14	6–10	No	c
	Complexed with nucleic acids	i.d.	0.45–0.60	7	3	Yes, 450–600	d,e
Nonhistone proteins	Free	s.c. and i.m.	0.10	7	3–5	Yes, 25	f
	Complexed with DNA	s.c. and i.m.	0.10	7	3–6	Yes, 25	g
Chromatin	Nonfractionated	s.c. and i.m.	0.10	7	4	Yes, 25	f

[a] i.m., intramuscularly; s.c., subcutaneously; i.d., intradermally.
[b] R. Rumke and M. Sluyser, *Biochem. J.* **101**, 1c (1970).
[c] A. L. Sandberg, M. Liss, and B. D. Stollar, *J. Immunol.* **98**, 1182 (1967).
[d] B. D. Stollar and M. Ward, *J. Biol. Chem.* **245**, 1261 (1970).
[e] M. Bustin and B. D. Stollar, *J. Biol. Chem.* **247**, 5716 (1972).
[f] F. Chytil, unpublished observations.
[g] F. Chytil and T. C. Spelsberg, *Nature (London) New Biol.* **233**, 215 (1971).

body, which can be subtracted from that observed in the presence of antibody–antigen complexes. Repeated thawing and freezing of the antigens might induce a rise in "anticomplementarity." It is therefore advisable to store the antigens in small aliquots. Low "anticomplementarity" of nonhistone protein–DNA complexes is observed when the histones are removed by high urea, high salt treatment.[10] On the other hand, when histones are extracted with acid the "anticomplementarity" of these preparations is frequently very high.

Testing the Antigenicity

The method of quantitative microcomplement fixation described earlier[2] is employed with small modification which consists in using one-fifth of the incubation volume. Briefly, to disposable glass tubes 13 × 100 mm the following components are added: 0.6 ml of the diluent, 0.2 ml of diluted antibody, 0.2 ml of the antigen, and 0.2 ml of diluted complement to make the total volume 1.2 ml. After an overnight incubation at 4°, 0.2 ml of the sensitized sheep red blood cells are added and the hemolysis is stopped usually after 45–60 minutes. The dilution of the antisera was found to range between 1/200 and 1/3200. The maximum complement fixation by chromosomal protein–antibody complexes varies from 0.5 to 2.0 μg per assay tube. This amount is substantially higher than that found for another antigen.[2] The titer of the complement which has to be determined for each shipment varies in this laboratory between 1/125 and 1/175.

Anti-histone Antibody. Figure 1 shows typical curves obtained when complement fixation assay was employed to test the antigenicity of antihistone antibody in the presence of different preparations of F1 histone. These results show the extent of reproducibility of this method. This laboratory has similar experience when different batches of nonhistone protein–DNA complexes were tested against homologous antibody.

Anti-nonhistone Protein Antibody. Nonhistone proteins are apparently good immunogens. Complement-fixing antibody can be obtained when the rabbits are immunized with these proteins still attached to DNA[10] or with the free form.[11] Histones, double- or single-stranded DNA, do not interact with these antibodies. The antibodies react also, but to a smaller extent, with the chromatin preparation from the same organ[10] and show tissue specificity, which is demonstrated in Fig. 2. They react also with free nonhistone proteins.[13]

Anti-chromatin Antibody. Antibody against the whole chromatin can be induced rather easily.[11] The results from this laboratory show in Fig. 3

[13] F. Chytil and T. C. Spelsberg, unpublished observations.

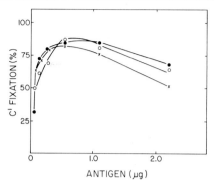

FIG. 1. C' fixation of antiserum to whole calf thymus F1 histone with varying preparations of whole calf thymus F1 histone. ○, An old chromatographically purified preparation frozen and thawed several times; ●, a fresh chromatographically purified preparation frozen and thawed once; ×, a preparation before purification by column chromatography. Serum dilution was 1:1000. From M. Bustin and B. D. Stollar, *J. Biol. Chem.* **247,** 5716 (1972).

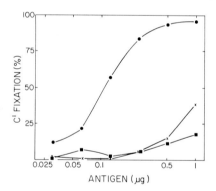

FIG. 2. C' fixation of antiserum to nonhistone protein–DNA complexes from chick oviduct by varying quantities of nonhistone protein–DNA complexes from chick oviduct, ●; heart, ×; and liver, ■. Serum dilution was 1:400. From F. Chytil and T. C. Spelsberg, *Nature (London) New Biol.* **233,** 215 (1971).

that the immunization with chromatin leads to antibody reacting also with free histones. Moreover, these antibodies fix complement in the presence of nonhistone protein–DNA complexes and do not react with DNA.

Testing the Nuclear Origin of the Immunogen

In order to obtain additional evidence as to whether the antichromosomal protein antibody is directed against the nuclear material, and not

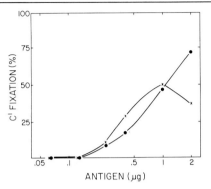

FIG. 3. C' fixation of antiserum to chromatin from chick oviduct by varying quantities of chromatin, ●, and whole histone, ×. Serum dilution was 1:400. From unpublished data of F. Chytil and T. C. Spelsberg.

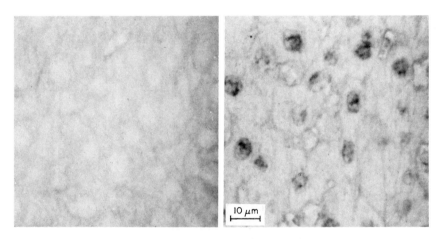

FIG. 4. Intracellular localization of antiliver nonhistone protein–DNA antibodies in liver slices by peroxidase bridge analysis described in this series (Vol. 37 [9]). *Left:* Adult rat liver slices incubated with control serum. *Right:* With anti-liver nonhistone protein antibody. From unpublished experiments of F. Chytil and P. K. Nakane.

against some cytoplasmic components that could have been adsorbed to the nuclei or chromatin during the isolation, it is advisable to localize the antibody in the cell histochemically. Figure 4 shows the intracellular localization of anti-nonhistone protein antibody by the peroxidase bridge technique.[14] The immunogen was a nonhistone protein–DNA complex isolated from adult rat liver by the method described earlier.[10] It is evident

[14] See this series Vol. 37 [9].

that the antibody reacts with the nuclear material. Alternatively, fluorescent-labeled antisera can be used for this purpose.[15]

[15] L. S. Desai, L. Pothier, G. E. Foley, and R. A. Adams, *Exp. Cell Res.* **70**, 468 (1971).

[16] Chromatin Protein Kinases[1]

By VALERIE M. KISH and LEWIS J. KLEINSMITH

The phosphorylation of nonhistone chromatin proteins has been suggested to play a key role in the regulation of gene activity in higher organisms[2-12] (also see Kleinsmith and Kish, this volume [14]). Currently, however, there is little information concerning the enzyme(s) involved in the phosphorylation of these proteins. The nonhistone chromatin phosphoprotein fraction is known to contain an endogenous protein kinase activity which catalyzes the phosphorylation of these proteins in the absence of added exogenous substrate.[3,13] This chapter is concerned with methods for fractionation and assay of these chromatin-associated protein kinases, and will show that chromatin contains a broad spectrum of different types of protein kinases with differing substrate specificities and sensitivity to control by cyclic AMP. The general experimental approach employed involves purification of phosphorylated nonhistone

[1] Studies on this subject in our laboratory have been supported by grants from the National Science Foundation (GB-8123 and GB-23921). V.M.K. held a predoctoral fellowship from U.S. Public Health Service Training Grant 5-T01-GM-72-15.
[2] L. J. Kleinsmith, V. G. Allfrey, and A. E. Mirsky, *Proc. Nat. Acad. Sci. U.S.* **55**, 1182 (1966).
[3] T. A. Langan, in "Regulation of Nucleic Acid and Protein Biosynthesis" (V. V. Koningsberger and L. Bosch, eds.), p. 233. Elsevier, Amsterdam, 1967.
[4] E. L. Gershey and L. J. Kleinsmith, *Biochim. Biophys. Acta* **194**, 519 (1969).
[5] R. W. Turkington and M. Riddle, *J. Biol. Chem.* **244**, 6040 (1969).
[6] K. Ahmed and H. Ishida, *Mol. Pharmacol.* **7**, 323 (1971).
[7] R. D. Platz, V. M. Kish, and L. J. Kleinsmith, *FEBS Lett.* **12**, 38 (1970).
[8] C. S. Teng, C. T. Teng, and V. G. Allfrey, *J. Biol. Chem.* **246**, 3597 (1971).
[9] M. Kamiyama, B. Dastugue, and J. Kruh, *Biochem. Biophys. Res. Commun.* **44**, 1345 (1971).
[10] N. C. Kostraba and T. Y. Wang, *Biochim. Biophys. Acta* **262**, 169 (1972).
[11] P. B. Kaplowitz, R. D. Platz, and L. J. Kleinsmith, *Biochim. Biophys. Acta* **229**, 739 (1971).
[12] L. J. Kleinsmith and V. G. Allfrey, *Biochim. Biophys. Acta* **175**, 136 (1969).
[13] L. J. Kleinsmith and V. G. Allfrey, *Biochim. Biophys. Acta* **175**, 123 (1969).

chromatin proteins via salt extraction, chromatography on phosphocellulose columns involving a salt/pH step gradient, and assay for protein kinase activity in the absence of exogenous substrate.

Fractionation Procedure

Preparation of Phosphocellulose Columns

Ten grams of phosphocellulose (Whatman P-11) are suspended in 300 ml of 0.5 N NaOH and stirred for 30 minutes at 22°. The fines are removed by suction, and the slurry is then transferred to a small Büchner funnel fitted with a circular disk of Whatman No. 1 filter paper. The phosphocellulose is washed on the funnel with distilled water until the pH of the filtrate is 8 as tested by pH paper. The cake is then transferred to another beaker and stirred with 300 ml of 0.5 N HCl for 30 minutes at 22°. The slurry is then washed on the Büchner funnel as described previously until the filtrate approximates a pH of 4. The phosphocellulose is again suspended in 0.5 N NaOH and stirred, and is then washed to pH 8 as described above. The resulting cake is suspended in 50 mM Tris · HCl, pH 7.5, containing 0.3 M NaCl and the pH of the solution is then carefully adjusted to 7.5 using concentrated HCl. The phosphocellulose is stirred at 22° for several hours, and the pH is checked at intervals. Phosphocellulose washed in this manner is then immediately used to pack the column. The excess cellulose is stored at 4° in a tightly closed screw-capped bottle and can be used for packing subsequent columns within the next week. The day before use the phosphocellulose is allowed to come to room temperature and the pH is checked prior to packing.

Freshly washed phosphocellulose gives the cleanest separations of proteins. Storage of washed phosphocellulose at 4° for periods of more than 7–10 days results in protein profiles, which are not as sharply resolved as those resulting from chromatography on freshly washed cellulose. Washed phosphocellulose is slurried in 50 mM Tris·HCl, pH 7.5 containing 0.3 M NaCl in a ratio of approximately 1:2. A 0.9 × 15 cm column is filled approximately halfway with 50 mM Tris · HCl, pH 7.5 containing 0.3 M NaCl. After removing bubbles from the bottom section, the bottom of the column which is fitted with a 2-inch section of polyethylene tubing is clamped off and phosphocellulose is pipetted into the column and allowed to settle for 5 minutes. The phosphocellulose should be finely dispersed, with no visible chunks of material present. The bottom is then slowly opened over a period of 10 minutes, and additional slurry is added as the meniscus recedes from the top of the column. When the level of packed phosphocellulose has reached a height approximately

1 cm from the top of the column, the bottom is closed off and the column is replumbed in the cold room. An elution flask, fitted with a capillary tube and containing starting buffer at room temperature is then attached to the column at a differential of 65–70 cm and the column bottom is opened. After approximately 2 hours the differential is lowered to approximately 4 cm and the column is allowed to temperature equilibrate at 4° overnight. The column is washed with 75–100 ml of the starting buffer and the optical density and pH of the eluate are checked prior to protein application. Changes in the permeability of the upper surface of the phosphocellulose usually occur during the overnight wash, as evidenced by the obvious clumping of the cellulose in the top 2–5 mm. To avoid irregular bands caused by uneven permeability of the surface, the top few centimeters of adsorbent are stirred up and removed, leaving a smooth, flat surface. The height of the bed is routinely adjusted to 12–12.2 cm. It should be pointed out that column flow should not be interrupted at any time from the beginning of the buffer wash to the end of fraction collecting.

Column Chromatography of Phosphorylated Nonhistone Proteins

The nonhistone chromatin phosphoprotein fraction, with its associated protein kinase activities, is prepared according to the salt extraction method described elsewhere in this volume (see Kleinsmith and Kish [14]). After concentration at 50 psi of nitrogen in an Amicon ultrafilter equipped with a UM-10 membrane, the proteins are either frozen at $-25°$,[14] or are immediately dialyzed for 12–15 hours at 4° against 100 volumes of 50 mM Tris · HCl, pH 7.5 containing 0.3 M NaCl prior to fractionation. The fractionation procedure is a modification of that reported by Takeda *et al.*[15] for fractionation of rat liver nuclear protein kinases. All procedures are carried out at 4° unless otherwise noted.

Approximately 1.5–3 mg of the phosphoprotein fraction (670–980 µg/ml) is applied by pipette directly to the top of the bed after the last of the starting buffer has entered the adsorbent. The entire surface is covered with sample quickly to permit uniform penetration of the sample into the bed. When the last of the sample has entered the cellulose, the surface of the bed and the column wall above it are washed with three 0.5-ml portions of starting buffer, each wash being permitted to sink into the cellulose before the next is applied. After the last buffer wash has entered the cellulose, the column is filled and the elution flask containing

[14] Freezing at this step does not appear to impair protein kinase activity.
[15] M. Takeda, H. Yamamura, and Y. Ohga, *Biochem. Biophys. Res. Commun.* **42**, 103 (1971).

starting buffer is attached. Fractions of 1.0–1.2 ml are collected at 15–20 ml per hour at a constant differential of 24–30 cm until 25–28 ml of starting buffer has passed through the column. The buffer above the bed is then changed to 50 mM Tris, pH 8.1 containing 0.6 M NaCl and elution is continued for another 34–36 ml. At a total volume of 64–66 ml the last step in the gradient is initiated using 50 mM Tris·HCl, pH 8.1 containing 1.0 M NaCl and an additional 30–35 ml is collected. The protein content of each fraction is measured by ultraviolet absorbance at 280 nm prior to enzyme assay. Since the protein kinases are extremely labile after fractionation (storage at $-70°$ or overnight at $2°$ results in a significant loss of activity) the enzyme assay is carried out immediately.

Assay of Protein Kinase Activity

Protein kinase activity in each fraction is measured in the presence of endogenous substrate immediately after the protein profile is read. The reaction mixture of 0.3 ml contains 13 μmoles of Tris·HCl, pH 7.5; 0.1–2 nmoles of [γ-^{32}P]ATP (770–3650 mCi/mmole); 7.5 μmoles of magnesium acetate, and 0.2 ml of each column fraction. When 3′,5′-cyclic adenosine monophosphate (cAMP) is added to the reaction (0.3 nmoles) it is first preincubated for 2 minutes at 30° in the presence of protein kinase and buffer before the remaining components of the reaction mixture are added. The reaction is initiated by the addition of magnesium and after 10 minutes in a 30° shaking water bath, 3.0 ml of cold 1 mM ATP are added followed by 3.3 ml of cold 10% trichloroacetic acid (TCA) containing 3% sodium pyrophosphate. The diluted reaction mixture is then filtered under vacuum through nitrocellulose membrane filters (Schleicher and Schuell, B-6, 0.45 μm, 25 mm diameter) which are presoaked in 1 mM ATP for at least 30 minutes at room temperature. Each filter is washed twice with 5 ml of 5% TCA containing 1.5% sodium pyrophosphate, oven dried, and counted in 5 ml of toluene scintillation fluid.

The final assay conditions were derived on the basis of results of several preliminary experiments. A comparison of nitrocellulose membrane filters with glass fiber filters (Reeve Angel 934-H and 934-AH) in the presence and in the absence of bovine serum albumin as carrier protein demonstrates lower background counts using the nitrocellulose filter; therefore we have routinely used this type of filter in our assay. In order to minimize still further the background radioactivity on nitrocellulose filters, a variety of conditions were employed, with the results summarized as follows: (1) Unlabeled ATP is the most effective agent in which to soak the filters; neither potassium phosphate buffer (20 mM)

nor sodium pyrophosphate (3%) results in any significant diminution of background counts, thus suggesting that the ring structure of the ATP may be important in this regard. (2) Terminating the reaction with 5% TCA in the presence or in the absence of 3% sodium pyrophosphate, followed by unlabeled ATP leads to a high background, whereas adding the ATP first, followed by TCA (our standard procedure) results in low background radioactivity. The fact that the order of addition is critical, coupled with the observation that TCA if used alone results in high background levels, suggests that there may be some interaction between the acid and the nitrocellulose filter causing the increased nonspecific retention of labeled ATP. (3) Use of 2 M NaCl to stop the reaction (rather than TCA) does not decrease the high background.

It should be pointed out that other methods which monitor the degree of incorporation of ^{32}P into proteins have been reported. Comparison of the nitrocellulose filter procedure with a modification of that described by Reimann et al.[16] involving filter paper washed in cold 10% TCA + 1% sodium pyrophosphate followed by propanol:ether substantiates the superiority of the former procedure. Another method which has been reported[17] involves the precipitation of acid-insoluble ^{32}P-labeled material with 10% TCA followed by dissolution of the centrifuged pellet in 0.1 ml of 1 N NaOH and reprecipitation by 5% TCA. This precipitate is then collected on glass fiber filters. However, since the phosphodiester bond is alkali labile, it is unclear how this method can be used to monitor ^{32}P incorporation, since much of the ^{32}P will remain in the supernatant after the second acid precipitation. In conjunction with these studies we discovered that the presence of bovine serum albumin (200 μg added after the TCA precipitation step), although not altering the background level of radioactivity, resulted in a significant decrease in the amount of radioactivity retained after filtration of the samples containing protein kinase. Since attempts to replace the albumin with a phosvitin carrier resulted in elevated backgrounds, we examined the retention of ^{32}P in the absence of carrier protein. Under these conditions the observed retention of counts in experimental samples is significantly increased, hence carrier protein is routinely left out of the procedure.

Acrylamide Gel Electrophoresis

The nonhistone chromatin phosphoprotein fractions are phosphorylated with [γ-^{32}P]ATP using the same assay conditions as described

[16] E. M. Reimann, D. A. Walsh, and E. G. Krebs, *J. Biol. Chem.* **246**, 1986 (1971).
[17] J. Erlichman, A. H. Hirsch, and O. M. Rosen, *Proc. Nat. Acad. Sci. U.S.* **68**, 731 (1971).

above. The reaction is stopped by adding solid urea to a final concentration of 4.0 M, followed by dialysis at 22° overnight against 200 volumes of 10 M sodium phosphate, pH 7.0 containing 0.1% 2-mercaptoethanol and 0.1% sodium dodecyl sulfate (SDS). Electrophoresis is performed in a 10% SDS-acrylamide gel as described by Weber and Osborn.[18] Gels are sliced at 1-mm intervals, dehydrated at 68° for 1 hour, and counted in 5 ml of toluene scintillation fluid.

In order to assay protein kinase activity in the gels, a modification of the procedure reported by Rubin et al.[19] is employed. Gels containing 7.5% acrylamide are prepared from a solution containing 22.2 g of acrylamide and 0.6 g of N,N'-methylenebisacrylamide, which are dissolved in water to 100 ml and then filtered. The gels are prepared by deaerating 10 ml of 0.26 M Tris·HCl, pH 8.1 followed by addition of 6.75 ml of the acrylamide solution and 2.25 ml of water. After additional deaeration, 0.025 ml of N,N,N',N'-tetramethylethylenediamine and 1.0 ml of a freshly made 1% solution of ammonium persulfate are added. The gels are polymerized in 6 mm (i.d.) $\times$ 75 mm glass tubes after layering 3–4 mm of water above the gel surface. Protein kinase samples are dialyzed overnight at 4° against 200 volumes of 0.13 M Tris·HCl, pH 8.1 prior to use. Before protein is added, the polymerized gels are prerun for 30–60 minutes at 2.5 mA/gel using 0.13 M Tris·HCl, pH 8.1 in both the upper and lower buffer chambers. Dialyzed samples of 50–100 μl are mixed with an equal volume of 40% sucrose containing bromophenol blue marker and are then layered on top of the gel column. Electrophoresis is carried out at 2.5 mA/gel until the marker reaches the bottom of the gels (approximately 8 hours). Gels are then sliced at 1-mm intervals, and each slice is eluted overnight at 4° in 250 μl of 50 mM Tris·HCl, pH 7.5 containing 4 mM 2-mercaptoethanol. Aliquots of the eluate are then withdrawn and assayed for protein kinase activity as described previously.

Typical Results[20]

When the purified nonhistone phosphoproteins are chromatographed on phosphocellulose columns and the fractions are monitored for protein kinase activity as described above, a complex profile of heterogeneity is seen which routinely consists of at least 12 distinct regions of enzyme activity (Fig. 1). This fractionation procedure thus emphasizes the multiplicity of the protein kinases associated with the nonhistone chromatin

---

[18] K. Weber and M. Osborn, *J. Biol. Chem.* **244**, 4406 (1969).
[19] C. S. Rubin, J. Erlichman, and O. M. Rosen, *J. Biol. Chem.* **247**, 36 (1972).
[20] For a detailed description of results see V. M. Kish and L. J. Kleinsmith, *J. Biol. Chem.* **249**, 750 (1974).

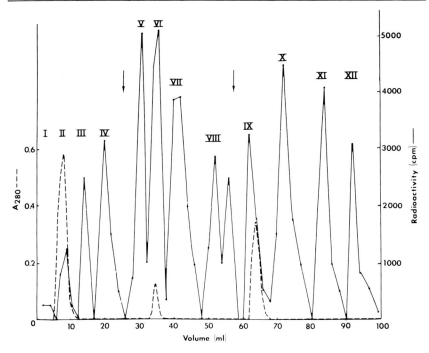

FIG. 1. Phosphocellulose column chromatography of protein kinases associated with the nonhistone chromatin phosphoprotein fraction of beef liver. Approximately 2.5 mg of the protein fraction is applied to the column as described in the text. Arrows mark the salt/pH steps in the gradient. Each column fraction is assayed for protein content by ultraviolet absorption at 280 nm (- - -) and for protein kinase activity by monitoring incorporation of radioactivity from [γ-^{32}P]ATP into protein (●——●) as described in the text. Twelve distinct regions of protein kinase activity are reproducibly distributed throughout the profile, demonstrating the heterogeneous nature of these nuclear enzymes.

phosphoprotein fraction, and also demonstrates the presence of endogenous substrate within each column fraction.

In order to investigate the characteristics of these protein kinases in terms of substrate specificities and response to cyclic nucleotides, it is most convenient to pool various regions eluting from the column. These pools are found to differ from each other in terms of their abilities to catalyze the phosphorylation of different protein substrates. For example, the table shows the results obtained when 4 of the different kinase pools are compared in terms of their abilities to catalyze the phosphorylation of casein, histone, or nonhistone proteins. These data are typical of the rest of the enzyme pools as well, in that each pool has a unique pattern

SUBSTRATE SPECIFICITIES AND EFFECTS OF CYCLIC AMP ON SELECTED NUCLEAR PROTEIN KINASE FRACTIONS[a]

	Protein kinase activity							
	Fraction I		Fraction VI		Fraction VIII		Fraction XI	
Treatment	cpm	% Control	cpm	% Control	cpm	% Control	cpm	% Control
Control	546	100	1029	100	647	100	2113	100
+ cAMP	2883	517	479	47	1910	295	545	26
+ Casein	5	1	1770	172	1039	160	20	1
+ Casein + cAMP	5	1	25	2	3039	469	1751	83
+ Histone	161	30	46	4	6	1	222	11
+ Histone + cAMP	206	38	1800	175	2911	450	530	25
+ Nonhistone protein (NHP)	525	96	1280	124	3950	610	566	27
+ NHP + cAMP	4010	735	1440	140	3882	600	580	27

[a] Fractions I, VI, VIII, and XI from the phosphocellulose column (see Fig. 1) were each pooled and immediately concentrated to a volume of approximately 1 ml by ultrafiltration using a PSAC Pellicon membrane (Millipore). Each pooled fraction was then desalted by passage through a 0.8 × 12 cm column of Bio-Gel P-4 equilibrated with 50 mM Tris · HCl, pH 7.5. Under these conditions the protein kinase activity is eluted in the exclusion volume. The above procedures were all carried out at 4°. Protein kinase activity was measured for each pooled fraction in the absence of added substrate as described in the text. This was considered the "control" system. The effects of adding 100 μg of bovine casein (Hammarsten), 180 μg of total calf thymus histone (Type IIA, Sigma), 9–24 μg of a purified nonhistone chromatin phosphoprotein (NHP) fraction (pool II, heated at 60° for 3 minutes to destroy endogenous protein kinase activity), and 1 μM 3′,5′-cyclic AMP (cAMP) were also tested in this system. Note that the different kinase fractions have differing substrate specificities and responses to cAMP.

of substrate specificity. These results suggest that each of the pools represents a distinctly different protein kinase activity.

This conclusion is further substantiated when one employs different kinase pools as sources of enzyme for the phosphorylation of nonhistone proteins. This nonhistone protein fraction is known to be highly heterogeneous, containing at least several dozen phosphorylated components which are separable by SDS-acrylamide gel electrophoresis.[7] Thus the obvious question to be asked is whether these protein kinases which have been separated are specific for different components within this nonhistone fraction. When different kinase pools are used to catalyze the phosphorylation of nonhistone proteins and the pattern of labeling is examined by SDS-acrylamide electrophoresis, it is seen that different labeling patterns are obtained (Fig. 2). These results lend further support to our conclusion that the different protein kinase pools actually represent different enzymes.

The effects of the addition of cyclic AMP on the activities of these protein kinases are also uniquely different for each kinase pool. As can be seen in the table, the effects of cyclic AMP in this system are quite complex. Some of the phosphorylation reactions are stimulated, while others are inhibited. The existence of an inhibitory or stimulatory effect is not dependent on the source of the kinase alone, since the same kinase pool may be either activated or inhibited by the cyclic nucleotide, depending on the substrate being phosphorylated. These results again point to the unique characteristics of the different protein kinase pools.

Although we have been routinely dealing with 12 kinase pools for the sake of experimental simplicity, this does not necessarily imply that each of these 12 fractions actually represents a single species of protein kinase. In fact, we have some evidence which suggests that these individual pools may contain more than one enzyme within them. If, for example, pool V is collected and electrophoresed in acrylamide gel under conditions where protein kinase activity is not destroyed, enzyme activity is found in at least two distinct regions of the gel (Fig. 3).

The overall picture which emerges, then, is an extraordinarily large degree of heterogeneity and complexity within the nuclear protein kinases. In order to observe this heterogeneity, however, it is necessary to closely follow the procedures outlined here, since other investigators using somewhat different techniques have not seen this level of complexity.[15,21] Aside from the obvious differences in tissue source and methods of enzyme assay, factors which in themselves may be very important considerations, the most fundamental difference between our work and

[21] R. W. Ruddon and S. L. Anderson, *Biochem. Biophys. Res. Commun.* **46**, 1499 (1972).

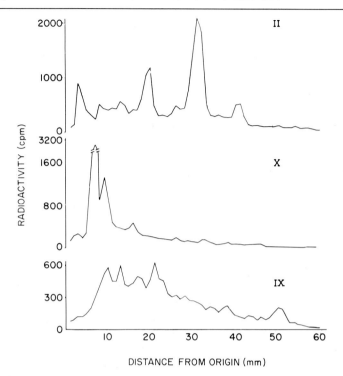

FIG. 2. Phosphorylation of nonhistone chromatin phosphoproteins by different protein kinase fractions. Protein kinase pools II, IX, and X were collected and concentrated as described in the table. The total nonhistone chromatin phosphoprotein fraction used as substrate was preheated at 60° for 3 minutes to destroy its endogenous protein kinase activity. The incubation conditions were as described in the text, except that $MgCl_2$ was used and the incubation was carried out at 37°. After labeling, the proteins were dialyzed and then applied to 10% SDS-acrylamide gels. Note that the distribution of radioactivity in the gels shows that the different kinase fractions are phosphorylating different components of the nonhistone phosphoprotein fraction.

that of others involves the purity of the starting material used for phosphocellulose chromatography. The highly purified nonhistone phosphoprotein fraction which we start with has a very high protein kinase specific activity, thus allowing us to resolve enzyme activities which are not seen when starting with cruder material.

In addition, we have included an additional step in our column elution which increases the complexity of the protein kinase profile by a substantial amount. These enzyme activities are presumably left on the column under conditions employed by others.[15,21] Another factor which may also be involved is the extreme lability of these kinases, which might allow

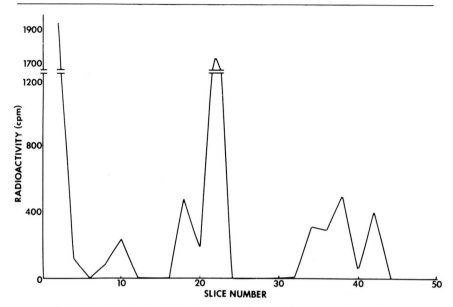

FIG. 3. Electrophoretic analysis demonstrating the heterogeneity of the protein kinase activity present in pooled fraction V. Pool V was collected and concentrated as described in the table, dialyzed against 0.13 M Tris · HCl, pH 8.1, and then electrophoresed in a 7.5% acrylamide gel as described in the text. The gel was sliced, each slice was eluted, and protein kinase assays were performed on the eluates in the presence of [γ-^{32}P]ATP as described in the text (no exogenous substrate was added). Although some activity is retained at the very top of the gel, the results show at least two major regions of protein kinase activity which migrate into the gel, thus demonstrating the multiplicity of protein kinases present even within a single pooled fraction (V).

them to become inactivated by other components in the relatively crude starting material usually used in such studies. Other investigators also routinely employ exogenous substrates, such as casein and histone, when assaying for protein kinase, and we have found that some of our enzyme activities are actually inhibited by these substrates. Thus, the present methodology has several distinct advantages over the techniques previously used, and although the exact extent of the multiplicity of nuclear protein kinases is not yet known, the present results indicate the existence of a large number of distinctly different enzymes involved in the phosphorylation of nonhistone chromatin proteins.

[17] Circular Dichroism Analysis of Nucleoprotein Complexes

By THOMAS E. WAGNER, VAUGHN VANDEGRIFT, and DEXTER S. MOORE

Although most spectroscopic methods (e.g., infrared, nuclear magnetic resonance, fluorescence) are particularly sensitive to the atomic composition of chemical groups and only grossly affected by the geometric arrangement of these groups or the geometry of their environment, optical activity spectroscopy is uniquely sensitive to these geometric aspects of molecular structure. For this reason, optical activity spectroscopy is the spectral method of choice for studies of molecular geometry.

One of the most biologically significant and interesting complexes is the eukaryote chromosome. In conjunction with genetic activity studies and chemical studies of this structure, much interest has been demonstrated recently in studies of the geometry or conformation of chromatin and simpler nucleoprotein complexes. The recent application of optical activity spectroscopy to the study of nucleoprotein conformation is now of sufficient interest to warrant a detailed descriptive discussion of this method and its specific application to the study of nucleoproteins.

The following sections are presented in order to acquaint researchers in the field of nucleoprotein chemistry and biology with some aspects of optical activity theory and instrumentation as well as a discussion of the analysis of nucleoprotein optical activity spectra.

Theory

Although optical activity spectroscopy began with optical rotatory dispersion measurements, this spectral method has been largely replaced by the more sensitive and reliable method of circular dichroism. An understanding of circular dichroism spectroscopy requires both a knowledge of the polarized energy used and the nature of its interaction with matter.

Plane-Polarized Light

Light energy is considered to consist of radiation of periodically varying electric and magnetic fields whose behavior is described by Maxwell's equations for an electromagnetic field. The electric and magnetic fields of this electromagnetic radiation oscillate at right angles to each other in a plane perpendicular to the direction of propagation of the light beam:

Light restricted to only one wavelength or frequency is termed monochromatic. Plane-polarized light is monochromatic light restricted so that its electric vector oscillates (i.e., goes through maxima and minima) in one and only one direction (azimuth) in the plane perpendicular to the direction of propagation of the beam. The magnetic vector of plane-polarized light will therefore oscillate in a plane mutually perpendicular to the

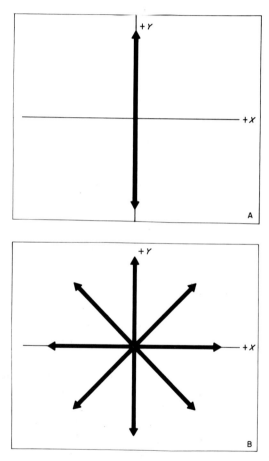

Fig. 1. (A) Diagrammatic representation of polarized electromagnetic radiation, showing the electric vectors polarized in the Y-Z plane. The direction of propagation, the $+Z$ axis, is perpendicular to the plane of the paper. The magnetic vector (not shown) would be polarized in the X-Z plane perpendicular to the plane of the electric vectors. (B) Diagrammatic representation of unpolarized oscillations of electromagnetic radiation in all directions perpendicular to the direction of propagation. The direction of propagation is perpendicular to the plane of the paper. The magnetic vectors are not shown.

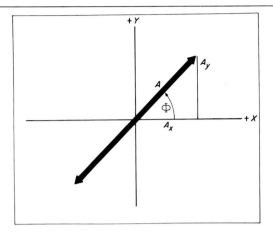

FIG. 2. The resolution of a plane-polarized wave motion into two mutually perpendicularly polarized wave components. The x component has an amplitude $Ax = A \cos \Phi$.

direction of propagation of the light beam and azimuth of the electric vector oscillation. Conventionally, the $+Z$ direction is taken as the direction of propagation of the light, the Y axis is taken as the azimuth of electric vector oscillation making the Y-Z plane the electric plane of polarization. The X axis is taken as the azimuth of magnetic vector oscillation making the X-Z plane the magnetic plane of polarization (see Fig. 1). Traditionally the plane of electric vector polarization (Y-Z plane) is taken as the polarization direction of the polarized light beam as a whole. In Fig. 1A is shown a representation of an electric vector (the $+Z$ axis is perpendicular to the plane of the page) oscillating in the Y azimuth (polarized in the Y direction). Unpolarized electric vectors are shown in Fig. 1B.

A plane-polarized light beam in any arbitrary azimuth may be thought of as the resultant of two, in-phase, wave motions which are plane polarized along any two mutually perpendicular reference axes,[1] X and Y (Fig. 2). If Φ is the angle between the X axis and the azimuth of polarization of a wave of amplitude A, then the amplitudes of the in-phase components of the wave, Ax and Ay are given by: $Ax = A \cos \Phi$ and $Ay = A \sin \Phi$ (Fig. 2). Thus it is important to think of the amplitudes of real plane-polarized vectors (electric or magnetic) as the resultants of two perpendicular, in-phase, wave motions. In Fig. 3 is shown a plane-polarized wave with an azimuth that is 45° to the X axis. In this example, the resultant plane-polarized wave is due to in-phase, X

[1] E. M. Slayter, "Optical Methods in Biology." Wiley (Interscience), New York, 1970.

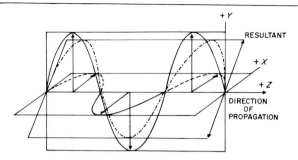

Fig. 3. A three-dimensional representation of the plane-polarized wave motion (resultant) resolved into mutually perpendicularly polarized wave components in the Y-Z and X-Z planes.

and Y components which are of equal magnitudes, i.e., $Ax = Ay$ (Fig. 2). Various other angles of inclination of the emergent plane-polarized beam are described as a result of variation in the amplitudes of the two in-phase, plane-polarized, wave motions (Fig. 3) according to the above relation. For azimuths of oscillation of the plane-polarized wave other than 45° [45° + n 90° (where n = 1, 2, 3, . . .)], the absolute amplitudes of the two in-phase components are unequal, $|Ax| \neq |Ay|$.

Circularly Polarized Light

Light rays cannot interfere unless they are polarized in the same plane. The observed intensity of two superimposed light beams which are polarized in different planes is simply the sum of the individual intensities. The term "circularly polarized light" is simply a convenient "pictorial" representation of the resultant wave modes of vibration which are produced when the intensities of the two superimposed wave components, which are polarized in different planes, are summed.[1] Circularly polarized light is produced by selectively retarding one of the two components, either the X or Y polarized component, of plane-polarized light one-quarter of a wavelength. In our previous example (Figs. 2 and 3) the components of the 45° plane polarized beam were polarized along the X-Z (X-component) and Y-Z (Y component) planes. Figure 4A shows the resultant "circularly polarized" light when the Y component is retarded a quarter-wave (90°) relative to the X component. When viewed down the direction of propagation (+Z axis), the resultant electric vectors rotate, as the beam impinges upon an observation point, in a circular, counterclockwise direction around the Z axis. This situation describes the production of an emergent left-circularly polarized beam. Figure 4B shows the emergent left-rotating electric vector E_L at two arbitrary times

t_1 and t_2. It should be noted that the position of E_L at t_1 and t_2 may also be described by the circular motion of the tip of vector E_L designated C_{L1} and C_{L2} (circular components of the left circularly polarized light). By selectively retarding the X component of the entering 45° plane-polarized light by a quarter wavelength relative to the Y component, right-circularly polarized light results. The resultant when viewed down the $+Z$ axis is an electric vector of constant amplitude which rotates in a circular, clockwise direction around the Z axis (Fig. 5A). Figure 5B

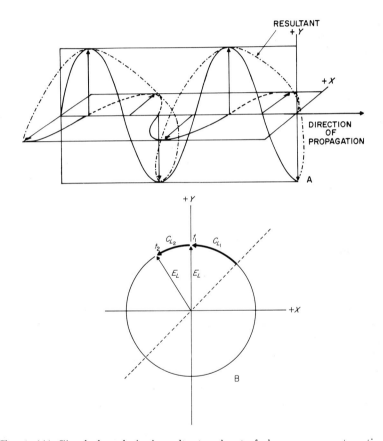

Fig. 4. (A) Circularly polarized resultant and out-of-phase component motions as seen in three dimensions. The component motion in the Y-Z plane has been retarded a quarter-wave relative to the component motion in the X-Z plane to produce a resultant left-circularly polarized vector. (B) The resultant left-circularly polarized electric vector as viewed down the direction of propagation (see A). The resultant electric vector (E_L) is shown at two arbitrary times (t_1, and t_2). The counterclockwise rotation of the tip of the vector (E_L) can be described by the circular component (C_{L1} and C_{L2}).

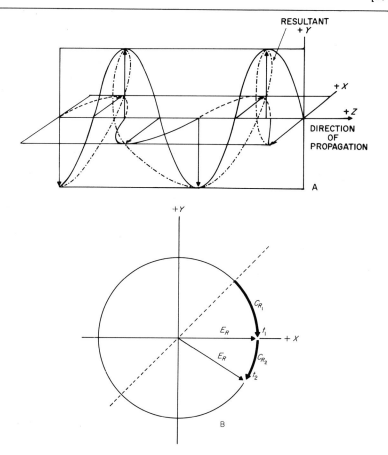

Fig. 5. (A) Circularly polarized resultant and out-of-phase component motions as seen in three dimensions. The component motion in the X-Z plane has been retarded a quarter-wave relative to the component motion in the Y-Z plane to produce a resultant right-circularly polarized vector. (B) The resultant right-circularly polarized electric vector as viewed down the direction of propagation (see A). The resultant electric vector (E_R) is shown at two arbitrary times (t_1 and t_2). The clockwise rotation of the tip of the vector (E_R) can be described by the circular component (C_{R1} and C_{R2}).

shows the emergent right-rotating electric vector E_R at two arbitrary times t_1 and t_2 along with the circular components C_{R1} and C_{R2}.

Circular Dichroism

When the *refractive index* of matter differs for left and right circularly polarized light, the emergent plane of polarization will be rotated with respect to the incident plane of polarization.[2] This phenomenon is *optical rotation*. Optical rotation is usually measured as a function of wavelength by the spectroscopic technique known as optical rotatory dispersion (ORD). *Circular dichroism* is the phenomenon due to the *differential absorption* by matter of left and right circularly polarized light. Matter which displays optical rotation and circular dichroism is said to be optically active.

In this discussion we have restricted our topic to circular dichroism measurements. A circular dichroism spectrum is a plot of the extinction coefficient for left circularly polarized light minus the extinction coefficient for right circularly polarized light ($\Delta\epsilon = \epsilon_l - \epsilon_r$) versus wavelength. When circularly polarized light energy equal to the energy of a molecular transition from ground state to excited state interacts with optically active matter to cause this transition, the extinction coefficient for the absorption of left circularly polarized light will not be equal to that for right circularly polarized light.[2,3]

The intensity of a circular dichroism band associated with an electronic transition is given by the quantum mechanical expression for the rotational strength[4]

$$R_{OA} = \text{Im} \ (\mu_{OA} M_{AO})$$

where O is the polymer ground state, A is the polymer excited state, μ_{OA} is the electric dipole transition moment vector, M_{AO} is the magnetic dipole transition moment vector and Im denotes the imaginary part of the dot product of the two vectors. The dipole operator for each transition moment is the sum of the corresponding monomer operators μi and mj, so that for the polymer[5]

$$R_{OA} = \text{Im} \sum_i \langle O|\mu i|A\rangle \sum_j \langle A|mj|O\rangle \tag{1}$$

where $|O\rangle$ and $|A\rangle$ are the molecular wave functions for states O and A. The magnetic dipole transition moment operator is defined as follows:

$$mj = (e/2mc) \ (Rj \ X \ Pj) + M'j$$

where Rj is the vector distance from an arbitrary polymer origin to the origin of monomer j; Pj is the linear momentum operator of the monomer; $M'j$ is the magnetic moment operator of monomer j relative to the origin

---

[2] B. Jirgenson, "Optical Rotatory Dispersion of Proteins and Other Macromolecules." Springer-Verlag, Berlin and New York, 1969.
[3] H. Eyring, H. C. Liu, and D. Caldwell, *Chem. Rev.* **68**, 525 (1969).
[4] L. Rosenfeld, *Z. Phys.* **52**, 161 (1928).
[5] W. C. Johnson, Jr., and I. Tinoco, Jr., *Biopolymers* **7**, 727 (1969).

of j (taken as the center of monomer j to minimize the contribution of $M'j$, since this is expected to be small). Since[5] $Pi_{OA} = (-2\pi i m/e)\, \nu_{OA}\mu_{iOA}$ and since the indices are arbitrary and may be exchanged, the relation for the rotational strength[5] becomes

$$R_{OA} = (-\pi\, \nu_{OA}/2c)_i \Sigma_j\, R\, i\, j\, \langle O|\mu i|A\rangle\, X\, \langle O|\mu j|A\rangle \tag{2}$$

where ν_{OA} is the frequency of the transition and $R\, i\, j$ is the vector distance between monomers i and j. Thus, it can be seen that the circular dichroism arises from the interaction between electric transitions located asymmetrically or dissymmetrically with respect to one another.

Classical mechanics describes the phenomenon of circular dichroism by analogy to a simplified model of an electron on a helix. A detailed description from this point of view is presented by Kauzman.[6]

Ellipticity

Even though circular dichroism instruments measure differential absorption ($\epsilon_l - \epsilon_r$) directly, they usually convert this measurement electronically to a quantity known as ellipticity. The phenomenological relationship between differential absorption and ellipticity is easily shown. When the circularly polarized components of the incident beam are differentially absorbed by a molecular transition the emergent left and right circularly polarized components are different in amplitude (Fig. 6A and 6B). The tip of the resultants of these vector components traces out an ellipse as the light beam travels along the $+Z$ axis (Fig. 6C). In this figure is shown the resultant electric vectors R_1 and R_2 on the elliptical path which results if the right circularly polarized component is absorbed to a greater extent than the left component (i.e., $\Delta E = E_L - E_R$ takes on a negative value). The resulting elliptically polarized light has an ellipticity (θ) defined as the arctangent of the ratio of the minor axis of the elliptical path (Fig. 6C) to its major axis [i.e., θ = arctangent (minor/major)]. Since the minor axis is simply $E_R - E_L$ and the major axis is $E_R + E_L$ then the equation defining ellipticity becomes

$$\theta = \arctan\,[(E_R - E_L)/(E_R + E_L)] \tag{3}$$

Mathematical and Dimensional Relationships between θ and $\Delta\epsilon$

CD data are commonly presented as a plot of the molar ellipticity, $[\theta]$, versus the wavelengths. While an equation such as

$$\Delta A = \Delta\epsilon\, bc$$

[6] W. Kauzman, "Quantum Chemistry." Academic Press, New York, 1957.

seems a straightforward extension from isotropic absorption spectroscopy, many researchers find the relationship between these familiar quantities and the molar ellipticity, $[\theta]$, a source of confusion. Even more puzzling to many is the origin of the units of $[\theta]$, namely, degrees centimeter^2 decimole^{-1}. Adding to the problem is that nowhere in the available literature is there an explicit account of the dimensional relationships between the quantities encountered in CD measurements. Therefore, the purpose of this section is to set out these relationships using dimensional analysis and to suggest useful working formulas for the calculation of experimental CD data. Figure 6C depicts the elliptically polarized radiation caused by differential absorption in a chiral medium of the two circularly polarized components of a linearly polarized wave.

The tangent of the angles of ellipticity, θ, is obtained from the ratio of the minor to the major axes of the ellipse [Eq. (3), Fig. 6C]. The major and minor axes are the sum and difference, respectively, of the amplitude of both circular components upon emerging from the chiral medium according to Eq. (3).

The amplitude of each component is related to the absorption index K through the Beer-Lambert law:

$$I/I_o = e^{-4\pi Kb/\lambda} \quad (4)$$

so that

$$E/E_o = e^{-4\pi Kb/\lambda} \quad (5a)$$

and

$$E_R \text{ or } E_L = E_o e^{-2\pi Kb/\lambda} \quad (5b)$$

where λ = vacuum wavelength of the radiation; b = pathlength (cm); E_o = incident amplitude; E = emergent amplitude.

Substitution of the appropriate form of Eq. (5b) into Eq. (3) gives

$$\tan \theta = \frac{E_o e^{-(2\pi K_r b/\lambda)} - E_o e^{-(2\pi K_l b/\lambda)}}{E_o e^{-(2\pi K_r b/\lambda)} + E_o e^{-(2\pi K_l b/\lambda)}}$$

$$= \frac{1 - e^{-(2\pi b/\lambda)(K_l - K_r)}}{1 + e^{-(2\pi b/\lambda)(K_l - K_r)}} \quad (6)$$

In the usual case, i.e., when the pathlength b is less than about 2 cm, the ratio of minor to major axes is small (the major axis is $\gg$ minor axis), therefore the ratio minor axis/major axis $\simeq \theta$ and since $\tan \theta$ = minor/major then $\tan \theta \simeq \theta$. The exponentials in the Taylor series expansion of Eq. (6) can be replaced by the first two terms of the series, so that Eq. (6) becomes

$$\theta = \frac{(2\pi b/\lambda)(K_l - K_r)}{[2 - (2\pi b/\lambda)](K_l - K_r)} \quad (7)$$

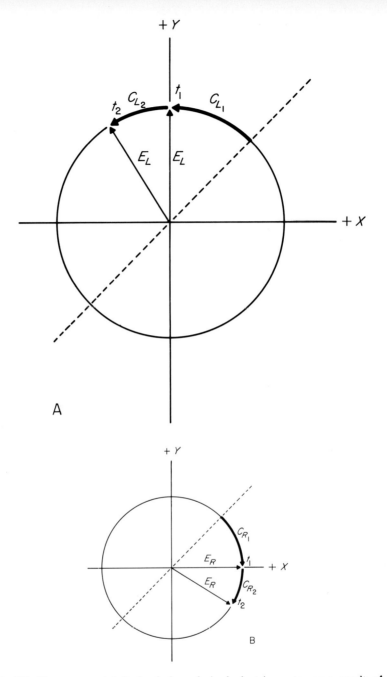

FIG. 6. (A) The emergent left-circularly polarized electric vector as a result of differential absorption of the left and right circularly polarized components. (B) The emergent right-circularly polarized electric vector as a result of differential absorption of the left- and right-circularly polarized components. This component is represented as smaller in magnitude than the left-circularly polarized component to indicate that the optically active medium absorbed the right component to a greater degree than the left component (A).

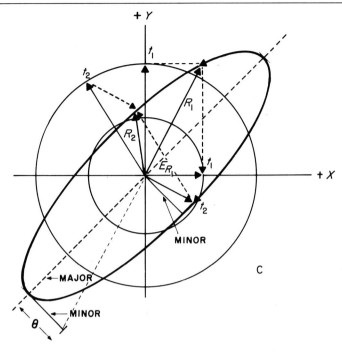

Fig. 6. (C) The resultant elliptically polarized radiation as a result of the differential absorption of the left circularly polarized component (Fig. 6A) and the right circularly polarized component (Fig. 6B). R_1 and R_2 are the resultant electric vectors at times t_1 and t_2. The ellipticity (θ) is the angle described by the arctangent of the ratio of the minor to major axes of the resultant elliptical path.

Since $K_l - K_r$ is small then $(2\pi b)/\lambda(K_l - K_r)$ may be neglected compared to 2 in the denominator of Eq. (7), so that

$$\theta = (\pi b/\lambda)(K_l - K_r) \tag{8}$$

In the cgs system of units, θ has the dimension of radians. It must be multiplied by $180/\pi$ to have the more common units of degrees. Therefore

$$\theta = (\pi b/\lambda)(K_l - K_r)(180/\pi)$$

and $\hfill (9)$

$$\theta = (180b/\lambda)(K_l - K_r)$$

where b = pathlength in cm; θ = degrees. Thus, θ is the ellipticity produced by a pathlength b cm long.

Ordinary absorption follows the Beer-Lambert law according to

$$I = I_o \times 10^{-A}$$

where $A = \epsilon cb$ and

$$\epsilon = \frac{A}{[c \text{ (mole/liter) } b \text{ (cm)}]} \qquad (10)$$

The absorption index K is defined by Eq. (4), and the molar absorptivity ϵ is defined by the relationship

$$I = I_o \times 10^{-\epsilon c b} \qquad (11)$$

where c = molar concentration.

Therefore,

$$e^{-4\pi Kb/\lambda} = 10^{-\epsilon c b}$$

and

$$[-(4\pi Kb)/\lambda] \ln e = -\epsilon cb \ln 10$$

Since $\ln e = 1$, and for equal pathlengths b,

$$(4\pi K)/\lambda = \epsilon c \, 2.303$$

and

$$K = [(2.303\lambda \, c)/4\pi] \, \epsilon \qquad (12)$$

or

$$K_l - K_r = [(2.303\lambda c)/4\pi] \, \Delta\epsilon \qquad (13)$$

Substitution of Eq. (13) into Eq. (9) gives

$$\theta = [(180b)/\lambda][(2.303\lambda c)/4\pi] \, \Delta\epsilon$$
$$= 414.54/4\pi \, bc \, \Delta\epsilon$$
$$\theta = 33bc \, \Delta\epsilon = 33\Delta A \qquad (14)$$

where θ = degrees, b = cm, c = moles/liter, and $\Delta\epsilon$ = liters/mole cm.

The molar (molecular) ellipticity, $[\theta]$, is defined by analogy with the molar rotation as

$$[\theta] = (MW/100)(\theta/C' \, P) \qquad (15)$$

where $[\theta]$ = (degree centimeter^2/decimole), θ = degrees, C' = g/cm^3, P = 10 cm (1 decimeter), MW/100 = g/centimole.
The units of Eq. (15) arise as follows[7]:

$$[\theta] = \frac{g}{\text{centimole}} \frac{cm^3}{g} \frac{\text{deg}}{10 \text{ cm}} = \frac{\text{deg cm}^2}{\text{decimole}}$$

The derivation of the units of molar ellipticity from Eq. (14) requires another course. Since the molar ellipticity is a function of the concentration and pathlength of the sample, Eq. (14) becomes

$$[\theta] = \theta/b \, c = 33 \text{ deg } \Delta\epsilon \, [\text{liter}/(\text{mole cm})] \qquad (16)$$

[7] L. Verbit, presented at "The Advanced Study Institute on Fundamental Aspects and Recent Developments in Optical Rotatory Dispersion and Circular Dichroism," NATO Scientific Affairs Division, Tirrenia, Italy, 1971.

The units of $\Delta\epsilon$ in Eq. (16) are converted to the desired units through the following relationship:

$$\text{liter/(mole cm)} = 1000 \text{ cm}^3/\text{(mole cm)} = 100 \text{ cm}^2/\text{decimole} \quad (17)$$

Thus, from Eqs. (16) and (17)

$$[\theta] = 3300 \Delta\epsilon \text{ (degree cm}^2 \text{ decimole}^{-1}) \quad (18)$$

Agreement is lacking as to whether it is preferable to use $[\theta]$ or $\Delta\epsilon$ in representing CD data. Although the molar ellipticity, $[\theta]$, is directly proportional to $\Delta\epsilon$ by Eq. (18), it is felt that $[\theta]$ is the preferable quantity to use. In part this preference arises from the advantage of not having to deal with the decimals of typical $\Delta\epsilon$ values, and in part because ellipticity values allow a more facile comparison of circular dichroism (CD) with optical rotatory dispersion (ORD) data by integral transforms of the Kronig-Kramers type; a comparison which will become more important in the future. One may convert experimental ΔA values to molar ellipticity by use of Eq. (19), which is a straightforward combination of $A = \Delta\epsilon\, cb$ with Eq. (18).

$$[\theta] = 3300\, [(\Delta A)/(cb)] \quad (19)$$

where $c = $ moles/liter, $b = $ cm.

However, most CD instruments are now capable of reading out ellipticity, θ, directly. Since almost all available instruments actually measure ΔA, this modification may be effected following Eq. (14). The following example is illustrative. Assume an instrument has a scale which is calibrated[8] to read $0.002A$ per 100 mm of chart paper. The scale is expanded electronically (usually by increasing the ac gain) by a factor of 3.3 so that it now reads $0.000606\, \Delta A/100$ mm of chart. By the use of Eq. (14), we have $\theta = 33\, \Delta A = 33(0.000606) = 0.020°/100$ mm of chart paper. We now have a scale which reads directly in terms of observed ellipticity, θ, in natural units of degrees. It is important to bear in mind that θ is still a function of path length and concentration. A dimensional manipulation of Eq. (16) gives a convenient form for obtaining molar ellipticity values from the observed ellipticities[7]

$$[\theta] = 10^2\theta/(cb) \quad (20)$$

where $c = $ moles/liter, $b = $ cm.

Instrumentation

In 1960 Grosjean and Legrand described the first instrument for the automatic recording of circular dichroism.[9] Their description was soon

[8] J. Y. Cassim and J. J. Yang, *Biochemistry* **8**, 1947 (1969).
[9] M. Grosjean and M. Legrand, *C. R. Acad. Sci.* **251**, 2150 (1960).

TABLE I
Instruments for Measuring Circular Dichroism

Instrument and manufacturer	Wavelength range	Sensitivity levels[a]	Features
Cary Model 1402, accessory for Cary Model 14, Cary Instruments	200–700 nm	2.5×10^{-4} units absorbance	UV-visible absorption measurements made with accessory attached
Cary Model 60, Cary Instruments	185–600 nm	RMS noise at 250 nm 0.0005° (time const 10 seconds)	Magnetic CD attachment available
Cary Model 61, Cary Instruments	185–600 nm	RMS noise at 250 nm 0.0004° (pen period 30 sec)	Differential CD capability with accessory
Durrum-Jasco Model J-20, Durrum Instrument Corporation	185–800 nm	RMS noise at 250 nm 0.0007° (pen period 5 sec)	Differential CD accessory, magnetic CD attachment, automatic slit program
Roussel-Jouan Model CD-185, Roussel-Jouan Corporation	185–600 nm	RMS noise at 250 nm 0.0005°	Magnetic CD temp. control attachments available

[a] 1 millidegree ellipticity equals 3×10^{-5} absorbance units. Note: rms values may vary according to experimental conditions, thereby making it difficult to compare these values directly.

followed by the appearance of several instruments, some of which were modified from ORD measurement to CD measurement. A listing of circular dichrographs designed primarily for CD measurement is found in Table I. This is by no means a comprehensive list, and the reader is directed to recent reviews that discuss in detail several instruments which measure CD as well as ORD.[10–12] Because of the availability of these extensive reviews, instruments discussed herein are those circular dichrographs that have achieved wide usage among researchers in the field.

Principles of Circular Dichroism Measurement

Although some instruments are known which measure ellipticity directly, most instruments presently in use measure differential absorption

---

[10] I. Tinoco, Jr. and C. R. Cantor, in "Methods of Biochemical Analysis" (D. Glick, ed.), p. 81. Wiley (Interscience), New York, 1970.

[11] C. F. Chignell and D. A. Chignell, in "Methods in Pharmacology" (C. F. Chignell, ed.), Vol. 2, p. 111. Appleton, New York, 1972.

[12] J. T. Yang, in "A Laboratory Manual of Analytical Methods of Protein Chemistry" (P. Alexander and H. P. Lundgren, eds.), p. 25. Pergamon, Oxford, 1969.

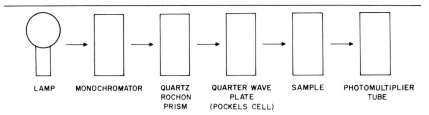

Fig. 7. Diagrammatic representation of the optical system found in most circular dichrographs.

of left- and right-circularly polarized light. Instruments that measure CD by differential absorption have an optical arrangement as shown in Fig. 7. Light from a deuterium or xenon arc lamp becomes monochromatic by passing through a double prism monochromator. The monochromatic light then passes through a polarizer, usually consisting of a quartz Rochon prism. The linearly polarized light is then circularly polarized by passage through a quarter-wave plate exhibiting the electrooptic or Pockels effect. Such a quarter-wave plate is composed of a crystal, normally uniaxial, whose optical properties are modified by application of an electric field. The electric field induces the crystal to become biaxial. Oscillation of the electric field alternately produces left- and right-circularly polarized light from the incident plane polarized light. A specific voltage for each wavelength is supplied by a variable transformer connected to the monochromator. The circularly polarized light produced at the quarter-wave plate then passes through the sample and undergoes a difference in absorption of the alternating left- and right-circularly polarized light if the sample cell contains an optically active sample. The light then impinges on the photomultiplier tube, causing an ac modulation. The amplitude of the modulation corresponds to the magnitude of the circular dichroism. The phase of the modulation corresponds to the sign of the CD. The signals are amplified and then transmitted to a recorder. Instruments of this type may be electronically calibrated to read directly in degrees of ellipticity as a function of either wavelength or time, according to relationships discussed above in the section on mathematical and dimensional relationships between θ and $\Delta\epsilon$. The readout does not alter the fact that these instruments measure differential absorption.

Several instruments measure CD in the manner described above. Included among these are the Durrum-Jasco J-20, the Cary 61, and the Roussel-Jouan Model CD-185. The optical diagram for the Durrum-Jasco J-20 (Fig. 8) and Cary 61 (Fig. 9) are included so that the reader may appreciate the simplification of our description.

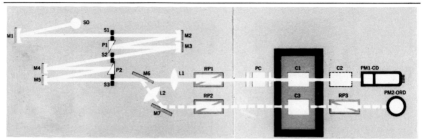

Fig. 8. Optical diagram of the Durrum-Jasco Recording Spectropolarimeter, Model J-20 (Courtesy of Durrum-Jasco Instrument Corporation). S0, Xenon source; M1, source mirror; S1, 2, 3, monochromator slits; M2, 3, 4, 5, collimating mirrors; P1, 2, monochromator prisms (fused silica); M6, retractable ORD mirror; M7, plane mirror; L1, CD lens; L2, ORD lens; RP1, CD polarizer (Rochon prism); RP2, ORD polarizer (Rochon prism); RP3, ORD analyzer (Rochon prism); PC, CD modulator and quarter-wave plate (Pockels cell); C1, CD cell position, normal; C2, CD cell position for light-scattering samples; C3, ORD cell position; PM1, CD photomultiplier (end-window); PM2, ORD photomultiplier (side-window).

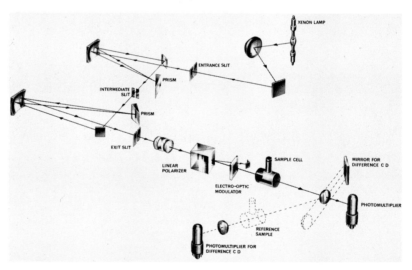

Fig. 9. Optical diagram of the Cary 61 Recording CD Spectropolarimeter (Courtesy of Cary Instruments Division of Varian Inc.).

Instrumental Modifications

In a number of studies including studies of effector molecules, such as hormones on nucleoprotein complexes, it is advantageous to instrumentally record the difference CD between two samples both of which have optical activity, even though they may be slightly different. Although difference absorption measurements are easily made by simply placing one sample in the sample beam of a dual beam spectrophotometer and the other sample in the reference beam, differential CD measurement is more difficult because all presently available instruments are single-beam systems.

There are at least two commercially available instruments with provision for difference CD measurement. The Durrum-Jasco J-20 difference CD accessory consists of a double Fresnel rhomb housed between two water-jacketed cell holders. The first cell contains the reference to be studied; the second cell contains the reference to be subtracted. In the Jasco system, light exiting from the quarter-wave plate passes through the sample, which may preferentially absorb different portions of the left- or right-handed component of the circularly polarized light. The beam then passes through a double Fresnel rhomb positioned to provide the effect of a half-wave retardation plate. Light passing out of the Fresnel rhomb is polarized 180° out of phase with respect to light entering the rhomb. Therefore, when left or right circularly polarized light is entering the rhomb, light of the opposite circular polarization is exiting. The exiting light then impinges on the reference sample, which interacts with the light in a manner similar to the first interaction except that the two are 180° out of phase. If the reference sample is a duplicate of the initial sample, then the light reaching the photomultiplier tube will be constant. Should the reference contain only a similar sample, however, the effect is the subtraction of only the contribution which is identical in each sample. In such a case, only the difference CD will be recorded by the instrument.

In the Cary 61 difference CD design a control device is introduced which moves a mirror into the optical path between the sample cell and the standard photomultiplier tube (see Fig. 9). The mirror reflects the light leaving the sample cell and directs it to the reference sample. Light leaving the reference sample then impinges on a photomultiplier tube. The net effect of this design is to record the difference in CD between a sample in the standard position and a sample in the reference position.

Solutions of nucleoprotein complexes occasionally display extensive light-scattering phenomena. Artifacts due to light scattering are many and complicated, but one sure effect is that light scattered outside the range of the phototube will appear instrumentally as absorbed light. Theory of the interaction of light and matter suggests that if a substance differentially absorbs left- and right-circularly polarized light it will also differentially scatter left- and right-circularly polarized light. This different scattering of left- and right-circularly polarized light may well introduce a serious artifact in CD measurements of turbid solutions of optically active material. This particular artifact of turbid solutions may

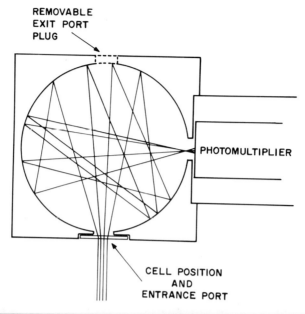

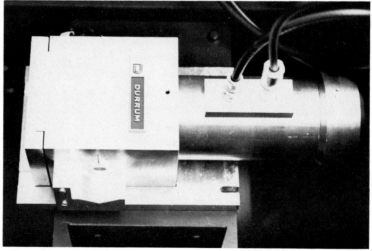

Fig. 10. Schematic diagram and photograph of the integrating sphere detection system for circular dichroism measurement.

be overcome by using an integrating sphere detection system. The sample is placed at the entrance of a reflecting sphere which contains a photomultiplier tube at right angles to the incident beam. All light, incident and scattered, which is not absorbed by the sample enters the sphere and

TABLE II
Calibration of Circular Dichrometer with d-Camphor Derivatives[a]

Sample[b]	Solvent	λ_{max} (nm)	Δe_{max}	$[\theta]_{max}$	Av. absolute % deviation
d-10-Camphorsulfonic acid	Water	290.5	+2.20	+7260	1.61
d-Camphor	Methanol	295.5	+1.54	+5080	1.56
d-Camphor	Dioxane	299.5	+1.69	+5570	1.66

[a] From J. Y. Cassim and J.-T. Yang, *Biochemistry* **8**, 1947 (1969).
[b] d-10-Camphorsulfonic acid: of reagent grade (not purified) prepared by Eastman Kodak. d-Camphor: of reagent grade (purified by repeated sublimation) prepared by Eastman Kodak.

is reflected into the photomultiplier window. In this manner the differential scattering artifacts of turbid solutions may be overcome. It is important to remember that other artifacts due to light scattering may still remain even when the integrating sphere detection system is used. Although no integrating sphere detection systems are presently commercially available we have built and studied such a system (see Fig. 10) in collaboration with the Durrum Instrument Corporation.

Calibration of CD Instruments

Camphor derivatives have recently become widely used for the calibration of circular dichrographs. Numerical values concerning camphor derivatives can be found in Table II. Care should be exercised in handling d-10-camphorsulfonic acid since it is apt to absorb moisture. Computer programs have been recently developed to aid in calibration. Details of this method may be found in a review by Yang.[12]

Experimental Parameters

Preparation of Nucleoprotein Complexes

Nucleoprotein complexes thus far studied using circular dichroism have included complexes of DNA with histones, with protamines, and with model histones, such as polylysine. Complexes may be formed by two principal methods. The first method involves the association of protein with DNA,[13,14] whereas the second method involves selective dissoci-

---

[13] R. C. C. Huang, J. Bonner, and K. Murray, *J. Mol. Biol.* **8**, 54 (1964).
[14] E. W. Johns and J. A. V. Butler, *Nature (London)* **204**, 853 (1964).

ation of protein (i.e., histones) from the DNA component of chromatin.[15]

The association method involves formation of nucleoprotein complexes either by direct addition of components or by gradient dialysis. In the direct addition method, protein in buffer is added directly to DNA in buffer. This procedure is normally accompanied by vigorous stirring of the protein-DNA solution during addition (particularly if the protein solution is concentrated) to prevent local precipitation of nucleoprotein complexes. The direct addition method is ill suited for the formation of complexes where, owing to buffer conditions, the protein or nucleoprotein exhibits aggregation behavior. For this reason, a gradient dialysis method has been used where protein and DNA are mixed in high ionic strength buffer and dialyzed in a series of steps until the desired ionic strength is reached. Modifications of this approach which have resulted in the formation of well defined complexes of DNA with purified histone fractions involve gradient dialysis in the presence of 5 M urea or GuCl.[16] The urea or GuCl is finally removed by extensive dialysis against the buffer system desired.

Nucleohistone complexes resulting from selective dissociation of histone from chromatin may be prepared for CD study. Gradually increased concentrations of dissociating agents, such as sodium chloride or sodium deoxycholate, are used to selectively remove histone fractions from chromatin or deoxynucleoprotein. Sodium chloride selectively removes the lysine-rich histone fractions (histone I) at lower concentrations and the intermediate and arginine-rich histones at higher concentrations.[17] Sodium deoxycholate removes the intermediate histone fractions first, followed by the arginine-rich and last the lysine-rich fractions.[18] The remaining deoxynucleoprotein may be separated from the dissociated histone by gel filtration or by differential sedimentation.

During preparation of nucleoprotein complexes, care should be taken to specify and control pH, ionic strength, and denaturant conditions. Blanks consisting of the proteins and DNA used in the formation of the nucleoprotein complexes should be run because of the greatly varying effects of pH and ionic strength on DNA and protein conformation. In order to produce complexes with identical dialysis histories, linear gradients such as the one recommended by Carroll may be used.[19]

[15] K. Murray and A. R. Peacocke, *Biochim. Biophys. Acta* **55**, 935 (1962).
[16] T. Y. Shih and J. Bonner, *J. Mol. Biol.* **48**, 469 (1970).
[17] H. H. Ohlenbusch, B. M. Olivera, P. Tuan, and N. Davidson, *J. Mol. Biol.* **25**, 299 (1967).
[18] J. E. Smart and J. Bonner, *J. Mol. Biol.* **58**, 675 (1971).
[19] D. Carroll, *Anal. Biochem.* **44**, 496 (1971).

Cells

Fused silica cells selected for low strain are best suited for circular dichroism work. High strain cells may exhibit optical activity of their own and may, therefore, cause confusion if adequate compensation is not made. Cemented cells should be avoided, particularly since occasional necessary soakings in a 1:1 mixture of sulfuric and nitric acids may cause deterioration of the cement. Cells come in a variety of shapes and sizes from 0.01 to 10 cm path lengths. However, for most work with nucleoproteins, a cell with a 1.0 mm or 1.0 cm path length should suffice. The most common type of cell in use is the 22 mm diameter cylindrical fused cell. Such a cell is reasonably convenient to fill and clean. For closely controlled temperature work, water-jacketed cells of 22 mm diameter are available in path lengths ranging from 1 to 10 mm. These cells are provided with ground-glass stoppers although a parafilm seal may serve the same purpose without introduction of strain.

Cells should always be placed in the same position in the cell compartment when spectra are being recorded. The position of the cell is a key point in the detection of weak CD bands. Care should be taken to keep the cell window free of dirt, dust, and solvent residue. Carefully wiping the faces of the cell with gauze soaked in alcohol should help. The cell should not have air trapped in it, and the meniscus should not be in the path of the incident light.

Optical Parameters

For nucleoprotein complexes, in which the amount of optical activity per unit absorbance is low, the limiting factor for consideration in sample preparation is the absorbance. High sample concentrations will result in greater measurement of the optical effect, but lesser signal since a larger proportion of the light is absorbed. Therefore, a balance must be struck to provide the optimum signal-to-noise ratio. For instruments similar in design to the Cary 61, it has been calculated that the optimum is reached at an absorbance of 0.868.[16] In practice, excellent results can be obtained if the absorbance is kept between 0.5 and 1.5 in the spectral region to be studied. Absorbances above 1.5 not only affect the signal-to-noise ratio adversely, but require much greater slit widths as well. We have found that excellent results may be obtained if the DNA component of the nucleoprotein complex is kept at a concentration on the order of 1×10^{-3} for a 1 mm path length cell. To check the concentration of the DNA component in nucleoprotamine or nucleohistone complexes, ultraviolet absorption may be used after the protein has undergone dissociation by treatment with 0.1% sodium dodecyl sulfate.[20] The extinction coefficient for DNA is 6600 per mole of phosphate at 260 nm.

A significant parameter in CD analysis of nucleoprotein complexes is light scattering. At the moment of this writing, there is considerable confusion regarding the best way in which compensation can be made for this phenomenon. Such confusion is to be expected, however, considering the complexity of the theoretical considerations involved. At the very least, the loss of light intensity due to scattering may cause artifacts due to a decrease in the signal-to-noise ratio. Since scattered light not reaching the photomultiplier tube will be treated by the instrument as absorbed light, the best practice would be to avoid the use of turbid solutions completely. This may be accomplished by appropriate filtrations or centrifugation of the sample. Circular dichroism spectra of samples known to be turbid (i.e., nucleoprotein complexes with appreciable absorbance at 320 nm) should be run at several concentrations and, at the very least, should be interpreted very carefully. Further work with integrating sphere detectors may help to alleviate this confusion.

As previously discussed, it is important to adjust the absorbance of the sample to keep within acceptable signal-to-noise ratios for the spectral region surveyed. Other parameters for consideration include scanning speed, time constants for pen response, and slit width.

Wavelength scanning speed and the time constants for pen response are closely related. For example, if the wavelength scanning speed is too high in relation to the time constant, then the spectrum recorded will be shifted toward shorter wavelength and peak heights will be slightly lowered. Most instrument manuals contain tables and suggestions which recommend the scan speeds and time constants which are best suited for optimum signal-to-noise ratios. Analytical efficiency is increased by using lower scanning speed for scanning spectral ranges where the photomultiplier voltage is high and by using higher scanning speed where the voltage is low. Normally, lower scan speeds will necessitate the use of longer time constants.

Slit widths on some instruments are automatically controlled in accordance with wavelength as specified by a previously set program. In cases where slit width is set manually, caution should be exercised to avoid setting the slit width wider than required. When the slit width selected is too wide, resolution is lowered. Excess slit width will also cause degradation of the photocathode area of the photomultiplier tube. The slit width should be adjusted manually so as to set the photomultiplier tube voltage no lower than values specified by the instruction manual.

[20] T. Y. Shih and G. D. Fasman, *Biochemistry* **11**, 398 (1972).

Dry nitrogen purging of the light source and of the sample compartment is necessary on most instruments. This practice prevents the formation of ozone and prolongs instrument life. Purging is also advisable because the oxygen molecule absorbs light very strongly below 195 nm. We have found that the use of a large dewar containing liquid nitrogen provides the necessary 2 cubic feet per hour for purging through heat leaks alone. For the 8–12 cubic feet per hour required for purging when the lamp is lit, a resistor immersed in the dewar and connected to a calibrated variable electrical supply may be used. Most instruments also require a supply of water to cool the lamp during operation.

Data

As previously described, most data are reported in the literature in terms of the molar ellipticity versus wavelength. The minimum and maximum values of circular dichroic bands are recorded along with the exact location of crossover points. Data should be collected by scanning approximately 50 nm before the wavelength where CD bands occur. Since CD instruments are very sophisticated electronically and in consideration of the nature of nucleoprotein complexes, there is much opportunity for artifacts during a single spectral scan. For this reason data are usually presented as the average of several spectral scans. Many laboratories collect a large number of spectral scans using a digital electronic tape recording system and use a computer to average these data for presentation.

Relationships of Structure and Circular Dichroism Spectra

Recording spectra of any type is a relatively straightforward procedure. The difficulty in spectroscopy comes in interpretation of spectra recorded. This is doubly true for circular dichroism spectroscopy of nucleoprotein complexes. Although the literature is filled with circular dichroism studies of nucleoprotein complexes where alterations in CD spectra are correlated with undefined changes in molecular geometry, it is the goal of CD spectroscopists studying nucleoprotein conformation to define the specific geometric parameters of these complexes using CD spectroscopy. Because a nucleoprotein complex is, by definition, at least a two-component system, the problem of interpreting CD spectra from these complexes is further complicated. We shall approach this problem first by analyzing DNA structure in nucleoproteins, then protein structure, and finally nucleoprotein structure.

Relationships between DNA CD Spectra and DNA Structure

DNA circular dichroism shows a strong positive CD band at approximately 275 nm and a crossover point at 257 nm when the DNA molecule is in a mild aqueous medium.[21] Since no major protein CD bands are present above 250 nm, the region of a nucleoprotein CD spectrum between 250 nm and 300 nm is indicative of DNA geometry alone.[22,23] As long as spectral analysis is carried out exclusively in this region of the spectrum, there is no reason to analyze the DNA component of a nucleoprotein complex any differently from isolated DNA.

DNA is an extremely conformationally plastic molecule which may assume a wide range of helical structures, all of which are optically active and display circular dichroism.[24] Three of these structures have been prepared as single fibers and analyzed by X-ray diffraction. The structures of A-, B-, and C-form DNA are known from these studies.[25-27] In order to describe the geometry of these and other forms of DNA, it has been necessary to develop certain geometric conventions. Five geometric parameters are sufficient to completely describe any helical structure of DNA. The positions of DNA base pairs in any particular level along the DNA chain are described by the number of degrees of twist around the twist axis (Fig. 11), the number of degrees of tilt around the tilt axis (Fig. 11), and the distance (Dx) (Fig. 11) from a fixed position on the base pair.[28] The geometric relationship between any two adjacent base pairs is given by the turn angle (Φ) about the helix axis and the base pair separation (Dz) along the helix axis. These five geometric parameters for A-, B-, and C-form DNA are shown in Table III. The B-form DNA fiber structure is observed at very high humidity conditions with NaDNA and is believed to be the structure of DNA in mild aqueous media. As shown in Table III this conformation of DNA has little base tilt or twist and has the centers of its base pairs directly on the helix axis ($Dx = 0$). The A and C forms of DNA differ from B-form DNA owing to base twist, tilt and displacement from the helix axis. These three

---

[21] J. Brahms and W. F. H. M. Mommaerts, *J. Mol. Biol.* **10**, 73 (1964).
[22] S. Beychok, *Science* **154**, 1288 (1966).
[23] G. D. Fasman, B. Schaffhausen, H. Goldsmith, and A. J. Alder, *Biochemistry* **9**, 2814 (1970).
[24] D. S. Studdert, M. Patroni, and R. C. Davis, *Biopolymers* **11**, 761 (1972).
[25] D. A. Marvin, M. Spencer, M. H. F. Wilkins, and L. D. Hamilton, *J. Mol. Biol.* **3**, 547 (1961).
[26] R. Landridge, D. A. Marvin, W. E. Seeds, M. R. Wilson, C. W. Hooper, W. H. F. Wilkins, and L. D. Hamilton, *J. Mol. Biol.* **2**, 38 (1960).
[27] W. Fuller, M. H. F. Wilkins, M. R. Wilson, and L. D. Hamilton, *J. Mol. Biol.* **12**, 60 (1965).
[28] D. S. Moore and T. E. Wagner, *Biopolymers* **12**, 201 (1973).

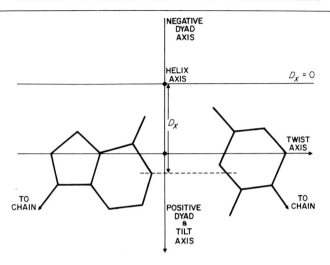

FIG. 11. The relationship of the paired purine-pyrimidine bases to the helix axis. Dx is measured along the dyad axis from the perpendicular to the dyad and helix axes ($Dx = 0$) to the N-1 nitrogen of the purine base. The twist axis is defined as passing through the C-8 and C-6 carbons of the purine-pyrimidine pair, respectively.

TABLE III
Geometric Parameters of DNA

Poly-nucleotide	Dx (distance from helix axis) (Å)	Tilt of bases	Twist of bases	Φ (turn angle between base pairs)	Dz (translation of bases along helix axis) (Å)
DNA-A	5.1	$-20°$[a]	8°	32.7°	2.56
DNA-B	0	2°	5°	36°	3.36
DNA-C	-1.5	6°	5°	38.7°	3.32

[a] The negative sign for the tilt indicates that the base on the right in Fig. 11 is tilted downward below the plane of the paper (toward the negative Z axis) and the base on the left is tilted upward (toward the positive Z axis). A positive value for the tilt indicates the opposite.

forms of DNA all differ from each other in Φ and Dz. In order to correlate DNA geometry with DNA circular dichroism, Tunis-Schneider and Maestre studied the circular dichroism of films of DNA known by X-ray diffraction studies to be A, B, or C conformations.[29] The B-form DNA film displays a CD spectrum (Fig. 12) which has a positive band

[29] M. J. B. Tunis-Schneider and M. F. Maestre, *J. Mol. Biol.* **52**, 521 (1970).

at 275 nm, a crossover point at 257 nm, and a negative band at 245 nm. Because the positive and negative bands in this spectrum are of approximately the same intensity, this spectrum is termed conservative. The A-form DNA film displays a CD spectrum (Fig. 12) which has a large positive band at approximately 265 nm, a crossover point at about 245 nm, and a small negative band at 240 nm. This spectrum is termed a positive nonconservative spectrum. The circular dichroism shown by the C-form DNA film describes a negative nonconservative spectrum (Fig. 12). Since A, B, and C forms of DNA differ in all five characteristic geometric parameters, it is of interest to determine which of these differences are responsible for the quite different CD spectra displayed by these three forms of DNA. This problem has recently been investigated by two of the authors.[28]

Since the base pairs of DNA are symmetrical and show no circular dichroism outside the helical DNA secondary structure, the CD of DNA is a result of the dissymmetric helical arrangement of these base pairs in relation to each other. More precisely, the circular dichroism is a result of the interaction of electronic transition moments and polarizabilities in one DNA base with bases dissymmetrically arranged above and below it in the helical structure of DNA. Consequently, the circular dichroism of any helical arrangement of a DNA polymer is due directly to the distance and angle between the transition moments and polarizability of any given base and the transition moments and polarizabilities of bases above and below it in the DNA structure in question. Using theory and

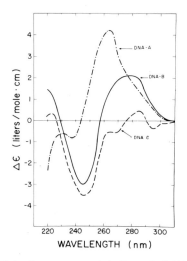

Fig. 12. Circular dichroism spectra of A, B, and C DNA films [M.-J. B. Tunis-Schneider and M. F. Maestre, *J. Mol. Biol.* **52**, 521 (1970)].

calculations developed by Johnson and Tinoco,[5] the circular dichroism of any DNA structure may be calculated. Calculations of A, B, and C form DNA CD spectra yield spectra remarkably similar to those measured by Tunis-Schneider and Maestre.[29] Using a computer simulation technique, geometrically and sterically feasible DNA structures intermediate between DNA B-form and A-form or C-form were produced.[28] Five structures were studied between DNA B-form and A-form each differing from B-form by only one characteristic geometric parameter. Five similar structures were produced intermediate between DNA B-form and C-form. By calculating the CD of these computer produced structures, it becomes possible to determine the magnitude of the effect of each of the five geometric parameters on the CD of DNA.[28] The results of these studies suggest that although twist, tilt, Φ, and Dz contribute to a certain extent to the CD of DNA, the conservative or nonconservative nature of the spectrum depends mostly on the Dx parameter.[28] A Dx value of 10 usually results in a conservative DNA CD spectrum where as a positive Dx value results in a positive nonconservative spectrum and a negative Dx value results in a negative nonconservative spectrum.[28] These results give a general guideline for relating DNA CD spectra with DNA structure. Hopefully future theoretical studies will provide a framework for these specific relationships. Finally it is important to remember that DNA may exist in a multitude of conformational forms, of which A, B, and C form DNA's are only three examples. By correlating A-, B-, and C-form DNA structures with their CD spectra, methods are developed to determine other yet unknown structures by CD spectroscopy.

Relationships between Protein CD Spectra and Protein Structure

Although a number of side chain protein groups display low intensity circular dichroism, circular dichroism in proteins is almost completely due to the dissymmetric or asymmetric arrangement of protein backbone amide carbonyl transitions.[30] The analysis of protein circular dichroism spectra has traditionally been based upon an empirical framework. Studies with synthetic polypeptide protein models have resulted in the observation of three secondary structures. Initial studies with poly-L-lysine and further studies with other polypeptides served to characterize the circular dichroism arising from a protein α-helix, protein in a β-structure and protein in a random coil configuration.[31] Shown in Fig. 13 are the CD spectra from these three geometric forms of the protein backbone.

[30] N. Greenfield and G. D. Fasman, *Biochemistry* **8**, 4108 (1969).
[31] J. Brahms and S. Brahms, *in* "Fine Structure of Proteins and Nucleic Acids" (G. D. Fasman and S. M. Timasheff, eds.), p. 256. Dekker, New York, 1970.

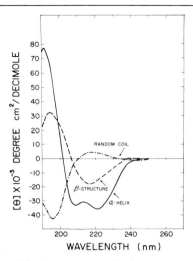

Fig. 13. Circular dichroism spectra of the α, β, and random coil structures of proteins [N. Greenfield and G. D. Fasman, *Biochemistry* **8**, 4108 (1969). Copyright by the American Chemical Society.]

Protein spectra are generally analyzed as percent α-helix, β-structure, and random coil assuming that each protein monomer is in one of these three configurations. Several computerized methods for analyzing the circular dichroism of proteins for α-helical, β-structural, and random coil contributions have been published. Two of the most useful of these methods are presented by Greenfield and Fasman[30] and more recently by Saxena and Wetlaufer.[32] These methods essentially solve three simultaneous equations each describing one of the three protein structures.

The overall approach of analyzing protein structure in terms only of α-helical regions, β-structural region, and region of random coil has been brought into serious question recently and should be used cautiously.[32] The basic flaw in this approach lies in treating the random coil as a defined single structure with a defined single circular dichroism. The circular dichroism attributed to the random coil conformation was determined by using poly-L-lysine at pH values where the polymer is highly charged and therefore in an extended conformation. This conformation is really not a random coil at all but an "extended conformation," and its use as a model for random coil regions can be misleading. Indeed, no model for the random coil configuration can be presented since random coil really means any structure which is not helical or β-structural and encompasses a wide variety of protein structures. Although these difficulties in regard to the random coil structures exist, protein CD spectra

[32] V. P. Saxena and D. B. Wetlaufer, *Proc. Nat. Acad. Sci. U.S.* **68**, 969 (1971).

are still generally analyzed in terms of α-helix, β-structure, and random coil. The advance of theoretical calculations and studies of small cyclic peptides of known precise structure promise to put the analysis of protein CD spectra on a firmer footing in the near future.[33]

Analysis of Nucleoprotein Circular Dichroism Spectra

Interpretation of nucleoprotein CD spectra requires that one deduce protein and DNA structure from both protein and DNA dissymmetric electronic transitions. This is facilitated for DNA conformational analysis due to the existence of DNA transition at energies considerably lower than normal protein carbonyl transitions. Therefore, DNA conformational analysis in nucleoprotein complexes is based upon the low energy (250 nm–300 nm) CD band and the crossover point alone since the higher energy CD band of DNA is in a region of protein transition absorption. The magnitude and position of this low energy DNA circular dichroism band may be indicative of a conservative, positive nonconservative or negative nonconservative DNA circular dichroism spectrum (see Fig. 12). The three general types of DNA spectra are indicative of certain generalized geometric characteristics of DNA (see section on relationships between DNA CD spectra and DNA structure, above).

Interpretation of the protein component of nucleoprotein CD spectra is a more difficult problem than evaluation of the DNA component. This problem becomes essentially unsolvable when the nucleoprotein complex contains more than one protein, as in studies of chromatin or isolated nucleohistone. Simple one-to-one nucleoprotein complexes may be studied by CD spectroscopy to analyze the structure of the protein component of the complex if the DNA circular dichroism of the complex is assumed to be known and can be subtracted from the nucleoprotein CD spectrum leaving the protein contribution.[34] This difference spectrum may then be analyzed as any other protein spectrum would be (see preceding section).

The approach to analyzing nucleoprotein CD spectra described in this section makes the assumption that no interaction between protein transitions and DNA transitions take place. This assumption is not a particularly strong one since some studies have suggested such interaction in nucleoprotein complexes.[35]

[33] E. R. Blout, presented at "The Advanced Study Institute on Fundamental Aspects and Recent Developments in Optical Rotatory Dispersion and Circular Dichroism," NATO Scientific Affairs Division, Tirrenia, Italy, 1971.
[34] T. Y. Shih and G. D. Fasman, *Biochemistry* **10**, 1675 (1971).
[35] M. J. Li, I. Isenberg, and W. C. Johnson, Jr., *Biochemistry* **11**, 2587 (1971).

Recently it has become more and more apparent that perhaps another contribution to nucleoprotein circular dichroism must be taken into account in analyzing the CD spectra of nucleoprotein complexes. Electronic transitions, symmetric or asymmetric, placed in an asymmetric environment display circular dichroism due to the asymmetric electric and/or magnetic fields resulting from the surrounding asymmetric environment.[36] This finding may be pertinent to interpretation of nucleoprotein spectra since DNA molecules with bound protein often associate to form macrocomplexes or bundles much as the fibers of a rope are coiled around each other. The coiled helical environment around each individual fiber may impart some circular dichroism to that already displayed by the fiber alone. Investigations of this phenomenon are just now beginning. Until these studies are complete it is unclear whether or not these macroscopic ordering effects significantly affect nucleoprotein circular dichroism.

---

[36] L. D. Haywood, presented at "The Advanced Study Institute on Fundamental Aspects and Recent Developments in Optical Rotatory Dispersion and Circular Dichroism," NATO Scientific Affairs Division, Tirrenia, Italy, 1971.

Section III
General Methods for Evaluating Hormone Effects

[18] Methods for Analysis of Enzyme Synthesis and Degradation in Animal Tissues

By ROBERT T. SCHIMKE

It is now well established that the amount of an enzyme in animal tissues results from opposing processes of continual synthesis and degradation, and that either process can be altered by hormonal, nutritional, developmental, or genetic factors.[1] There are a number of methods to assess the contributions of altered synthesis and/or degradation to changes in enzyme (protein) level, including immunologic methods, uses of tracer techniques and analysis of time courses of changes in enzyme levels. Each method embodies certain assumptions, as well as limitations, and unambiguous conclusions generally can be made only when independent estimates of rates of synthesis and degradation are made, and when more than one method is employed. These methods will be discussed largely in general terms, since the application to specific enzymes depends on the experimental system under investigation. References will be made to applications of these methods to which the reader can refer for details. Several general references giving examples include Schimke and Doyle,[1] Kenney,[2] Schimke,[3] and Rechcigl.[4]

Three fundamental questions should be asked of any experimental system in which there is an observed change in enzyme level.

Is There a Change in the Amount of Enzyme Protein?

A first approximation to this question can be made by the use of inhibitors of protein synthesis, such as puromycin or cycloheximide. Thus if the enzyme change provoked by an experimental variable, e.g., a hormone, does not occur when protein synthesis is inhibited, it can be concluded that the change results from protein synthesis,[5] as opposed to activation of a preexisting protein. It should be emphasized that such a finding does not determine whether the alteration is one affecting synthesis, as opposed to degradation, since in both cases, new protein synthe-

---

[1] R. T. Schimke and D. Doyle, *Annu. Rev. Biochem.* **39**, 929 (1970).
[2] F. T. Kenney, *in* "Mammalian Protein Metabolism" (H. N. Munro, ed.) Vol. 4, p. 131. Academic Press, New York, 1970.
[3] R. T. Schimke, *Curr. Top. Cell Regul.* **1**, 77 (1969).
[4] M. Rechcigl, ed., "Enzyme Synthesis and Degradation in Mammalian Systems." University Park Press, Baltimore, Maryland, 1971.
[5] O. Greengard, M. A. Smith, and G. Acs, *J. Biol. Chem.* **238**, 1548 (1963).

sis is required. Furthermore in such experiments the assumption is made that it is the enzyme in question whose synthesis is inhibited, rather than an enzyme (protein) which modifies the catalytic activity of the preexisting enzyme. Thus, the content or activity of the hypothetical modifying protein could be altered by the experimental variable *or* its inherent rate of turnover may be so rapid that inhibition of its synthesis, with subsequent rapid disappearance, may result in the lack of observed activation of a preexisting enzyme. In addition the use of inhibitors of protein synthesis in sufficiently large doses to completely inhibit protein synthesis generally leads to death of animals in 8–16 hours. Thus the use of these drugs is limited to systems where enzyme changes are relatively rapid. The above observations concerning interpretation of inhibitor studies hold equally well for the use of drugs that inhibit RNA synthesis.

Immunologic Techniques

For these reasons more definitive methods employing techniques that directly determine the amount of immunologically reactive protein should be employed. These techniques will be discussed in more detail.

Antigen Preparation. Obviously the antigen used for immunization should be the purest available. Ideally the protein should be purified to a homogeneous state from the organism and tissue to be studied. However in some instances there is sufficient cross-reactivity from species to species to allow use of enzyme purified from another species. This may be successful among different mammals, but often there is little or no cross-reactivity between mammals and other species.

Immunization. Approximately 1–5 mg of protein is generally used. For some proteins, successful immunization may be accomplished with as little as 100 μg of protein. The protein is suspended or dissolved in 0.5 ml of 0.15 M NaCl and mixed with 0.5 ml of Freund's complete adjuvant (Difco) and emulsified by sonication in a Branson sonifier. This can be most readily accomplished by use of a plastic 2-ml syringe into which the microtip can be introduced. In our experience the best injection site in rabbits is the toe pads[6]; approximately 0.1 ml of the emulsion can be introduced into each toe pad. This procedure has allowed for the development of antibodies which were not elicited by various series of intramuscular or subcutaneous administration. In larger animals, e.g., goats, we have routinely used intramuscular administrations using the same amount of antigen at a single injection site. After 2–3 weeks, a second administration in Freund's complete adjuvant is given intramuscularly. At 2–3 weeks after the second administration a small amount of

[6] S. Leskowitz and B. Waksman, *J. Immunol.* **84**, 58 (1960).

serum is obtained, and a preliminary test for immunologic reaction is made by a simple "ring" test in which antiserum is layered over a solution of the antigen. The appearance of a cloudy ring at the interface of the two solutions indicates a positive reaction. A sample of control serum is also used to check for possible nonspecific precipitation.

Preparation of Antibody. The animals are bled by cardiac puncture or via the marginal ear vein (rabbits), or carotid artery cannulation (goats). In inexperienced hands ear vein bleeding is simplist, in particular after wiping the ear with xylene to increase vasodilation. We routinely obtain 60–80 ml of blood from a single 4–5-kg rabbit.

The blood is allowed to clot, and the serum is collected. A γ-globulin fraction is obtained by addition of solid ammonium sulfate to 40% saturation at 4°. After allowing precipitation to proceed for 2 hours, the precipitate is centrifuged at 20,000 g and is dissolved in a volume of 10 mM potassium phosphate pH 7.4 containing 0.15 M NaCl equal to the original serum volume, and dialyzed against the same buffer to remove the ammonium sulfate. This antiserum is stored at $-20°$ in 2- to 5-ml aliquots.

Immunotitration of Enzyme Activity. Immunotitration is generally performed with antibodies that inhibit enzyme activity on reaction with it, or form a precipitate, and hence remove the enzyme from a soluble form. Two general methods can be employed: (a) the amount of enzyme or extract is varied with a constant amount of antibody, (b) the amount of enzyme (extract) is held constant and the amount of antibody is varied. The conditions for incubation of extract with antibody will vary with both the enzyme and the antiserum. Typical schedules are incubation at 30–37° for 30–60 minutes, after which the tubes are placed at 0° for periods of 1–24 hours. For rapid antigen–antibody reactions, e.g., ovalbumin, the reaction is complete in 10–20 minutes at 4°. For less rapid reactions, e.g., tryptophan oxygenase, the incubation must be at 37° for 1 hour, followed by storage at 0° for 12 hours for a visible precipitation to form. In addition the volumes necessary will also vary with the amount of extract required for enzyme assay. Following formation of the antigen–antibody precipitate and centrifugation (generally at 40,000 g for 20 minutes), aliquots of the supernatant are assayed for enzyme activity. Control antibody (animals injected with Freund's complete adjuvant only) is routinely incubated with enzyme extracts. The amounts of antibody and enzyme activity required will obviously depend on the reactivity of the antibody preparation, and must be determined empirically. Care must be taken to ensure that antigen excess does not exist, since this results in inhibition of precipitation. With unstable enzyme activities, a time and temperature must be chosen that does not result in extensive

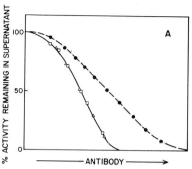 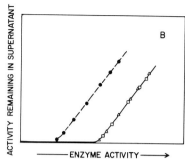

Fig. 1. Schematic representation of immunotitration techniques. (A) Constant enzyme activity and increasing antibody. The amount of enzyme activity is here indicated as percentage of activity remaining in the supernatant after removal of the immunoprecipitate. □ and △ indicate two extracts with differing specific activities, but having the same immunologic reactivity. ● shows a pattern of enzyme that is less reactive with the antibody relative to its enzymatic activity relative to □ and △. Note that at low antibody concentrations (relative to enzyme) the activity is not precipitated, since this is the area of antigen excess. This type of pattern is that in which enzyme–antibody complex is enzymatically active. (B) Constant antibody and increasing enzyme activity. As in (A), the activity remaining in the supernatant is assayed. □ and △ show the same degree of immunologic reactivity relative to enzyme activity. ● again shows a pattern in which the antibody reactivity is less relative to enzyme activity.

inactivation of enzyme activity with control serum. In some cases, the inclusion of a stabilizer, e.g., a substrate, may be of use, but this must be investigated carefully, since in certain instances substrates inhibit antigen–antibody reactions. In comparing the antigenicity of various extracts with differing enzyme activities, it is also useful to compare the antigenicity of the purified antigen (enzyme), since this may demonstrate the existence of immunologically reactive protein in extracts that is not present in the purified antigen. Such immunologically reactive, but enzymatically inactive protein, may either be an (inactive enzyme)[7] or a partial degradation product.[8]

Figure 1 indicates typical results using the two different immunotitration techniques. Figure 1A indicates the type of reaction where enzyme activity is held constant, and increasing amounts of antibody extract are added. Figure 1B depicts reactions in which antibody extract is constant and variable amounts of enzyme are added. The solid lines indicate the type of reaction in which immunologic reactivity and enzyme activity are comparable, whereas the dotted lines indicate instances in which there

[7] J. B. Li and W. E. Knox, *J. Biol. Chem.* **247**, 7546 (1972).
[8] E. Kominani, K. Kobayashi, S. Kominami, and N. Katunuma, *J. Biol. Chem.* **247**, 6848 (1972).

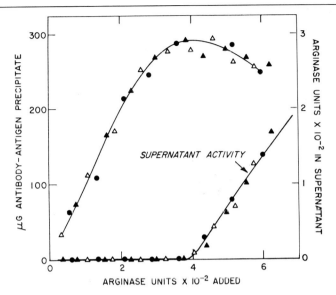

Fig. 2. Representative quantitative percipitin reactions. This result is from R. T. Schimke [*J. Biol. Chem.* **239**, 3808 (1964)] and shows the immunotitration, as well as quantitative precipitation reaction of purified arginase, and arginase from liver extracts with differing specific activities.

is a disparity between the two measures of enzyme content, specifically where enzyme activity is greater relative to immunologic reactivity. See Doyle and Schimke[9] and Ganshow and Schimke[10] for examples.

It should be noted that the immunotitration techniques described above do not require the use of a pure antigen, and hence can be employed in those instances where the purification of a homogeneous protein cannot be accomplished.

Quantitative Immunoprecipitation. Immunotitration procedures may not detect small differences in immunologic reactivity of enzyme present in two experimental states. In addition the immunotitration procedures do not give any information on the purity of the antibody, information that is necessary for use of antisera for immunoprecipitation of labeled proteins (enzymes) for various turnover studies. Quantitative precipitation employs essentially the same experimental procedures as does immunotitration (Fig. 1A), except that the protein content of the immmunoprecipitate is determined. With this technique, it is essential that nonspecific precipitation of protein is minimal, and that washing procedures are optimal. These problems are discussed below under Specificity of Immunoprecipitate. Figure 2 presents a typical result in which the anti-

[9] D. Doyle and R. T. Schimke, *J. Biol. Chem.* **244**, 5449 (1969).
[10] R. E. Ganschow and R. T. Schimke, *J. Biol. Chem.* **244**, 4669 (1969).

gens used include the antigen used for antibody formation, as well as extracts with two differing specific activities of enzyme (arginase). If the antigens are the same, and the antiserum reacts only with the specific antigen, then the precipitates should contain the same amount of protein as a function of added enzyme activity.

Use of Immunoprecipitation for the Isolation of Labeled Proteins. For the study of the rates of synthesis and degradation of specific proteins (see below), it is obviously necessary to obtain the protein (enzyme) in a pure state. Although in some cases enzyme has been obtained by standard purification procedures,[11,12] in most cases, either because of the small amount of protein obtained, or because of variability in the degree of purity obtained by standard procedures, immunoprecipitation techniques have proven invaluable as a means of rapid and reproducible isolation of specific proteins.[13-18]

The major problem involved in immunoprecipitation for tracer studies involves the specificity of the precipitation reaction. Criteria used to substantiate the specificity should include the following.

Immunodiffusion and Immunoelectrophoresis. A first approximation to the demonstration of a single reaction of antigen with antibody comes from immunodiffusion or immunoelectrophoresis techniques. Commercially available microscope slides coated with suitable agar, and appropriate well patterns, such as those obtained from Hyland Division, Travenol Laboratories, Los Angeles, California, are very convenient. Agar-coated slides can also be made readily. Five percent agar (Ionager from Oxoid, Consolidated Laboratories, Chicago, Illinois) is dissolved in Veronal buffer, pH 8.6, ionic strength 0.075. Each slide is coated with approximately 2 ml of the agar. Electrophoresis is performed at a constant current of 2.5 mA per slide for 3 hours at 4° in the above Veronal buffer. After electrophoresis of the extract, antibody is placed in the center well. The amounts of antibody and antigen used will vary, as well as the time required for full development of the precipitin lines. In practice, several concentrations of both extract and antibody should be employed, both to obtain the best definition of the precipitin lines, as well as to ensure that no minor precipitin lines appear. Generally 2–4 days

[11] Y. Kuriyama, T. Omura, P. Siekevitz, and G. E. Palade, *J. Biol. Chem.* **244**, 2017 (1969).
[12] L. W. Johnson and S. F. Velick, *J. Biol. Chem.* **247**, 4138 (1972).
[13] R. T. Schimke, *J. Biol. Chem.* **239**, 3808 (1964).
[14] R. T. Schimke, E. W. Sweeney, and C. M. Berlin, *J. Biol. Chem.* **240**, 322 (1965).
[15] F. T. Kenney, *J. Biol. Chem.* **237**, 1605 (1962).
[16] J. P. Jost, E. Z. Khairallah, and H. C. Pitot, *J. Biol. Chem.* **243**, 3057 (1968).
[17] A. G. Goodridge, *J. Biol. Chem.* **247**, 6946 (1972).
[18] P. J. Fritz, E. S. Vessell, E. L. White, and K. M. Pruitt, *Proc. Nat. Acad. Sci. U.S.* **62**, 558 (1969).

would be sufficient for the reactions to take place. The precipitation bands can be readily visualized by first washing the slides with several washes of 0.85% NaCl, and staining the precipitated protein with a dye, such a Ponceau Red, or Coomassie Blue. The slides may then be destained by immersion in 5% acetic acid.

Figure 3A depicts a typical immunodiffusion pattern with antibody introduced in the center well, and antigen preparations in the surrounding wells. Several patterns are depicted: (a) a single line of precipitation ostensibly showing a single reaction of antigen and antibody, (b) a pattern showing partial identity (cross-reactivity) of the reaction between the antigens in wells A and B, indicated by the "spur" formation, (c) a typical pattern in which the antibody reacts with multiple components in the extract, but not with components in well B. Figure 3B depicts a typical immunoelectrophoretic pattern of a single reaction of antigen and antibody (on the left), and a typical pattern of antibodies to multiple antigens in an extract (on the right).

Immunodiffusion does not necessarily prove that only a single reaction of antigen and antibody have occurred, since two antigens may diffuse through the agar at exactly the same rate. Furthermore, the finding of more than a single line of precipitation does not necessarily mean that

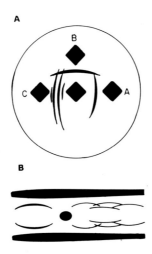

FIG. 3. Schematic representation of immunodiffusion and immunoelectrophoresis. (A) Immunodiffusion, showing: (a) a single precipitin line between the antibody (center well), and well A; (b) a single precipitin line between antibody and well B, and indication of cross reaction between the antigen in well A and well B, i.e., the "spur"; (c) multiple lines of precipitation between antibody and well C, with no bands in common between wells B and C. (B) Immunoelectrophoresis showing single (left) and multiple lines of precipitation.

the antibody is not directed to a single protein since alterations in the state of aggregation or charge (deamidation, phosphorylation, etc.) could result in multiple banding patterns by either diffusion or electrophoretic analysis.

Quantitative Precipitin Reactions. This has been discussed above (Fig. 2), and indicates that the reaction of antibody with antigen in both crude extracts and purified antigen are similar.

Electrophoresis of the Disrupted Antigen–Antibody Complex in SDS-Acrylamide Gels. This procedure is the most sensitive in indicating the specificity of the immunoprecipitation technique, once highly radioactive trace proteins may be present, but not detected by immunodiffusion or immunoelectrophoretic methods. Native enzyme (antigen) should also be subjected to the same experimental procedure so that the banding of the disrupted subunits can be compared with the radioactivity pattern obtained with the labeled precipitate.

Up to 200 μg of protein from the antigen–antibody precipitates (or highly purified antigen) are solubilized by boiling and mixing them with a glass rod for 2 minutes with 100 μl of a freshly prepared solution containing (per 2 ml) 24 mg of Tris-OH, 30 mg of DTT, 20 mg of SDS, 0.2 ml of saturated bromophenol blue in distilled water, and 0.2 ml of glycerol. The SDS dissolves the proteins and gives them a uniform negative charge such that they migrate according to their molecular weights,[19] the Tris deacylates any tRNA that is in the sample; the DTT and boiling are necessary for complete dissociation of antibody–antigen complexes; the bromophenol blue is a marker which migrates slightly faster than proteins of 10,000 daltons; and the glycerol makes the sample more dense than the overlay buffer. The sample is applied to the top of an acrylamide gel (6 × 90 mm) made by a slight modification of the procedure of Weber and Osborn.[19] The gel solution is made by mixing 10 volumes of buffer containing 50 mM sodium phosphate, pH 7.1, and 0.2% (w/v) SDS; 9 volumes of 22% (w/v) acrylamide 0.6% (w/v) of bisacrylamide, 0.03 volume N,N,N',N'-tetramethylenediamine and 1 volume of freshly prepared ammonium persulfate (5 mg/ml). The sample is overlaid with a buffer containing 25 mM sodium phosphate and 0.1% (w/v) SDS, pH 7.1 and electrophoresed in a Hoeffer (San Francisco, California) electrophoresis chamber with the same buffer in upper and lower chambers. For the first 20 minutes 2 mA per tube are applied, then the current is increased to 8 mA per tube. Total running time is about 3 hours. The gels are removed from the Plexiglas tubes and partially frozen on dry-ice. Then they are cut into disks, 2–3 mm thick, with a multiple razor blade

---

[19] K. Weber and M. Osborn, *J. Biol. Chem.* **244**, 4406 (1969).

slicer. The proteins in the gel slices are dissolved by incubating each slide with 40 µl of distilled water and 0.5 ml of NCS in tightly capped scintillation vials at 37° for 12–18 hours. They are then counted in 10 ml of a toluene-based scintillator fluid.[20]

Specificity of the Immunoprecipitate. Two major problems are encountered in obtaining highly specific immunoprecipitates. The first is that the antibody precipitates more than the desired protein (enzyme). Obviously the purer the antigen, the less the opportunity for the development of antibodies reacting with different proteins. In some instances, antibody made to a protein isolated from one tissue (or species), and which has been demonstrated to react with several proteins in extracts from that source, may be far more specific in reaction with enzyme from other tissues or species. In addition, the specificity of the antibody may be increased by attempts at adsorption of the antiserum with extracts from tissues that do not contain the enzyme activity in question, either from the same or a different species. The second major problem concerns the existence of nonspecific trapping of labeled proteins in the antigen–antibody complex. There are a number of techniques that can be used to decrease, or control, such trapping.

a. PARTIAL PURIFICATION OF EXTRACT PRIOR TO IMMUNOPRECIPITATION. In most instances it is necessary to undertake a partial purification of the protein (enzyme) to decrease the amount of nonspecific "shedding" of protein during the immunologic incubation. The specific steps to be employed will vary with the enzyme, but may include heating, salting-out procedures, or column chromatographic steps. The step(s) selected should be chosen such that the yield of active enzyme approaches 100%. Otherwise, an appropriate correction is difficult to make, since the inactivated enzyme may be or may not be precipitated by the antibody.[21]

b. MINIMIZING THE AMOUNT OF IMMUNOPRECIPITATE FORMED. With all immunoprecipitation techniques using labeled extracts, it is necessary to precipitate all the enzyme activity. Thus after such precipitations, it is routinely necessary to assay the supernatant to ensure that all enzyme activity has been removed. (In the case of nonenzymatic proteins, the assurance of precipitation can be accomplished by addition of purified protein that has been labeled with another isotope, e.g., ^{14}C versus ^{3}H amino acid(s). In addition, it is necessary to ensure that the amount of immunoprecipitate is similar in the various extracts added. Thus it is routine to add an amount of "carrier" antigen to the extracts to make constant the amount of the immunoprecipitate where tissue extracts have

---

[20] R. Palmiter, T. Oka, and R. T. Schimke, *J. Biol. Chem.* **246**, 724 (1971).
[21] R. T. Schimke, E. W. Sweeney, and C. M. Berlin, *J. Biol. Chem.* **240**, 4609 (1965).

differing amounts of enzyme. However, the amount of immunoprecipitate should be kept at some desirable minimum to decrease the amount of nonspecific trapping, and at the same time obtain sufficient immunoprecipitate for adequate handling in washing procedures. The approach must be empirical, with alterations in the proportions of added antibody and added carrier antigen to ensure optimal conditions.

c. APPROPRIATE WASHING PROCEDURES. We have found that adding small amounts of detergents, i.e., 0.5 to 1.0% of both deoxycholate and Triton X-100 during the antigen–antibody reaction markedly reduces the amount of nonspecific trapping and shedding of labeled protein.[22] This trapping can be further decreased by placing the extract containing the immunoprecipitate over a cushion of 1.0 M sucrose in 0.15 M NaCl containing the above detergents, and collecting the immunoprecipitate as a pellet at the bottom of the tube after centrifugation. This is a more efficient means of washing immunoprecipitates than repeated suspensions in saline, a procedure that may lead to variable and unpredicted loss of immunoprecipitate on the sides of the centrifuge tube. Small-volume reactions (100 μl) can be readily centrifuged in Beckman model 152 Microfuge tubes (plastic), and the bottom immersed in a dry-ice acetone bath. The frozen tip, containing the immunoprecipitate is cut off, and placed in a scintillation vial with an appropriate protein solubulizer, e.g., NCS.

Controls for Immunoprecipitation of Labeled Proteins. The controls most often used include: (a) reaction of the extract with control antibody. (b) a second precipitation of the extract (now containing no labeled enzyme) with additional (unlabeled) antigen and antibody added in amounts similar to those present in the initial immunoprecipitation. The reaction is performed as previously, and the second immunoprecipitate is collected. The radioactivity recovered in this immunoprecipitate is collected. The radioactivity recovered in this immunoprecipitate is presumed to represent nonspecific trapping of labeled proteins. If the radioactivity in the control (extract plus control antibody) and in the second immunoprecipitation is large, i.e., over 10% of the radioactivity present in the first specific immunoprecipitation, the specificity of the immunoprecipitation should be questioned, and checked routinely by SDS acrylamide gel electrophoresis of the immunoprecipitate. In some cases, it may be necessary to routinely perform gel electrophoresis to obtain sufficient specificity for meaningful results.[17]

Perhaps the best control involves use of an immunologic system in which both antibody and antigen are not reactive in the experimental system. One that can readily be developed, and is not reactive in mam-

[22] R. T. Schimke, R. E. Rhoads, and G. S. McKnight, this series, Vol. 30 [64].

malian systems is that in which ovalbumin is the antigen.[20] An amount of ovalbumin and antiovalbumin that provides a precipitate comparable to that obtained from the extract (plus added carrier antigen) and specific antibody is added to labeled extract, and carried through the same incubation and washing procedures. This technique requires information on the quantitative precipitation reaction of the experimental and control systems. It also assumes that the added immunologic system will trap protein in a manner comparable to the test immunologic system.

Is There an Altered Rate of Enzyme Synthesis?

Two general methods can be employed to this question.

Short-Term Incorporation of Isotope into Enzyme

This method involves the administration of isotope for a short period of time, and the subsequent isolation of that protein (generally using the immunologic techniques described above). In comparing two experimental states, it is obviously necessary to control for possible differences in the time-integrated specific activity of the amino acid (preferably the aminoacyl tRNA) pool from which the proteins are being synthesized. In practice this control consists of the comparison of the extent of isotope incorporation into the enzyme relative to incorporation into total cellular protein, the latter being considered to be an integration of the precursor pool. Alternative methods include measurement of the specific radioactivity of ribosome-bound nascent peptides or aminoacyl-tRNA. The major assumption of this method is that all proteins are synthesized from the same amino acid pool. There is no reason to believe that this assumption is not correct for proteins synthesized on cytoplasmic ribosomes, but may be incorrect in comparing cytoplasmic proteins with those synthesized in mitochondria.

One additional point of caution should be made regarding this method. The duration of time between isotope administration and isolation of enzyme will depend on the half-life of the enzyme. That duration of time should be optimized such that sufficient isotope has been incorporated to allow accurate counting but should not represent a significant amount of time relative to the half-life of the enzyme. For instance, with an enzyme that has a half-life of 60–90 minutes (such as tyrosine aminotransferase),[23] labeling times longer than 20 minutes should not be used; agents which may stabilize the enzyme (including that newly synthesized) would result in greater radioactivity in the isolated protein than

[23] F. T. Kenney, *Science* **156**, 525 (1967).

in the control. This finding would be then incorrectly attributed to greater synthesis, whereas the actual mechanism might involve decreased degradation.

The isotope precursor (amino acid) to be employed is not as critical in short-term pulse incorporation studies, as in either long-term incorporation or turnover studies (see discussion of these problems in later sections). Generally the least expensive isotope is preferable, as well as one that is not metabolized to other amino acids. In our laboratory, we generally use [^{3}H]- or [^{14}C]leucine or lysine since these are not metabolized to other amino acids.[13,24]

This method has the advantage that there is no need for obtaining steady state conditions (see below), but it can only provide information on relative rates of enzyme synthesis. For the majority of studies such information is sufficient. Theoretically the absolute rate of enzyme synthesis could be obtained if information were also available on the time-integrated specific activity of the precursor pool, the amount of enzyme present at the start and the completion of the experiment, and the specific activity of the amino acid in the isolated protein.[25] In practice, such measures are difficult to obtain. Therefore measures of the absolute rate of enzyme synthesis are most often obtained from knowledge of the rate of degradation of the enzyme under conditions of the steady state.

Steady State Kinetics of Enzyme Levels

As employed by various investigators,[26-28] changes in enzyme levels can be formulated by a simplified relationship between synthesis and degradation:

$$d\mathrm{E}/dt = k_s - k_d \mathrm{E} \tag{1}$$

where E is the content of enzyme expressed as units per mass, k_s is a zero-order rate constant of synthesis, expressed as units per time per mass, and k_d is a first-order rate constant for degradation, expressed as time^{-1} per mass. Generally no change in total mass of tissue occurs during the time involved, and hence no term is included for a change in total mass.

In the steady state, i.e., when $d\mathrm{E}/dt = 0$,

$$\mathrm{E} = k_s/k_d \tag{2}$$

[24] I. M. Arias, D. Doyle, and R. T. Schimke, *J. Biol. Chem.* **244**, 3303 (1969).
[25] P. J. Garlick and I. Marshall, *J. Neurochem.* **19**, 577 (1972).
[26] H. L. Segal and Y. S. Kim, *Proc. Nat. Acad. Sci. U.S.* **50**, 912 (1963).
[27] C. M. Berlin and R. T. Schimke, *Mol. Pharmacol.* **1**, 149 (1965).
[28] V. E. Price, W. R. Sterling, V. A. Tarantola, R. W. Hartley, Jr., and M. Rechcigl, Jr., *J. Biol. Chem.* **237**, 3468 (1962).

Thus, if the steady state can be assumed, the rate of synthesis, k_s can be calculated from the known k_d and E (see reference 20, for example).

Rate constants of degradation, or more simply, rates of degradation are often expressed in terms of a half-life. The half-life is given by

$$t_{1/2} = 0.693/k_d \tag{3}$$

Methods for Measurement of Specific Protein Degradation

Information concerning the control of degradation of specific proteins is only as good as the methods employed. We shall therefore review the methods available and the assumptions and limitations their use entails.

Methods Based on Isotope Incorporation and Decay

Tracer techniques are used most commonly for estimating degradation rates. Although such techniques may seem direct, they are full of pitfalls that can result in ambiguous results. Theoretical formulations of the uses of isotope techniques for measurement of protein degradation have been made by Reiner,[29] Buchanan,[30] and Koch,[31] and the reader is referred to these for mathematical analyses of the problem. Isotope methods can be divided into two general techniques: (a) single isotope administration, and (b) continuous isotope administration.

Single Isotope Administration (Pulse) Method. This method involves the single intraperitoneal, or intravenous administration of an isotopic form of a protein precursor, normally an amino acid, or exposure of cultured cells for a short time period, followed by replacement of medium. Theoretically the isotope should enter the pool from which the protein is synthesized rapidly (instantaneously), leave the pool rapidly (instantaneously), and should not return to that pool at any time after the initial isotope administration. The isotope administration, then, is a pulse, and the decay in specific radioactivity (or total radioactivity if the system is not in a steady state) of the isolated protein is a measure of the rate of degradation. This is depicted in Fig. 4. Assumptions involved in the use of this method include the following: (a) The radioactivity in the protein sample is representative of all the protein present in the isolated sample, rather than a minor contaminant. Thus, it is imperative that great care be taken to ensure absolute purity of the protein samples isolated. (b) There is no reutilization of isotope. The decay of isotope from protein is not only a function of the rate at which a labeled protein is

[29] J. M. Reiner, *Arch. Biochem.* **46**, 80 (1953).
[30] D. L. Buchanan, *Arch. Biochem.* **94**, 489 (1961).
[31] A. L. Koch, *J. Theoret. Biol.* **3**, 283 (1962).

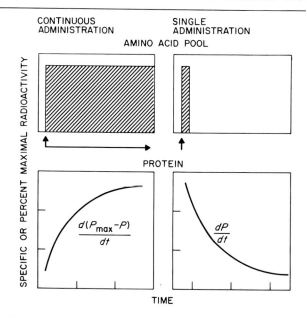

Fig. 4. Schematic representations of continuous and single administration techniques for determination of rates of degradation of proteins.

degraded to constituent amino acids, but also a function of the rate at which such isotope reenters the amino acid pool from which the protein is synthesized.[29-31] Isotope reutilization is a major limitation for obtaining accurate rates of protein degradation by the single administration method. For instance, a lack of extensive uptake of a labeled amino acid into a specific protein or cell type, and its subsequent retention in such protein, may result from a slow rate of degradation. However, it may equally well indicate, whatever the rate of intracellular protein degradation, that the amino acid is not transported into the tissue readily and, once within a cell, is not metabolized or otherwise released from the cell.

Isotope reutilization occurs in every tissue to varying degrees. Studies of Loftfield and Harris[32] and Gan and Jeffay[33] suggest that as much as 50% of the free amino acid pool of rat liver is derived from protein catabolism under normal dietary conditions. In muscle, approximately 30% of the free amino acid pool is derived from tissue catabolism.[33] Isotope reutilization is not limited to intact animals, and occurs to an extensive degree in cultured cells as well.[34]

Isotope reutilization can be minimized, theoretically, by the use of

[32] R. B. Loftfield and A. Harris, *J. Biol. Chem.* **219**, 151 (1956).
[33] J. C. Gan and H. Jeffay, *Biochim. Biophys. Acta* **148**, 448 (1967).
[34] P. Righetti, E. P. Little, and G. Wolf, *J. Biol. Chem.* **246**, 5724 (1971).

an isotope whose probability of reutilization is small. In liver of ureotelic animals, ^{14}C-labeled guanidino-L-arginine has minimal reutilization because of the large amount of arginase activity in liver.[35] Thus the administered isotope is rapidly hydrolyzed to labeled urea, which is excreted. Likewise, that labeled arginine which reenters the free amino acid pool will be hydrolyzed rapidly, and hence not be reutilized. It should be pointed out, however, that use of this isotope is limited to livers of ureotelic animals, and cannot be used in cultured cells or other tissues where little arginase activity exists. Despite the obvious advantage of guanidino-[^{14}C]arginine for obtaining more nearly "true" rates of protein degradation, a major limitation is the small extent of incorporation following its administration. The specific activity of liver protein resulting from the administration of this isotopic form of arginne is $\frac{1}{15}$ to $\frac{1}{20}$ of that when the same amount of uniformly labeled [^{14}C]arginine is administered. That difference is a crude measure of the difference in degree of reutilization, as well as initial incorporation. Thus the use of guanidino-[^{14}C]L-arginine is largely limited to proteins or cell fractions that can be obtained in insufficiently large quantities to permit accurate counting rates, i.e., about 1–2 mg.

The problem of isotope reutilization and its effect on measurable rates of turnover of proteins has been studied extensively by Poole.[36] The rate of turnover of a protein in the theoretical state (Fig. 4), where no reutilization of isotope occurs and the pulse is of short duration, is given by the formula

$$d\mathrm{E}^*/dt = k_d \mathrm{E}^* \tag{4}$$

where E* is the specific activity of the protein. The more general equation, which takes into account reutilization of the precursor is

$$d\mathrm{E}/dt = k_s F - k_d \mathrm{E}^* \tag{5}$$

where F represents a function describing the specific activity of the precursor pool. This function is determined experimentally by determining the specific activity of the free amino acid pool following a single injection of isotope.[36] Poole found that there is a rapid fall in the specific activity of the free rat liver leucine pool after intraperitoneal injection during the first day, followed by a slower decay during the subsequent 9 days.[36] Figure 5 shows Poole's computer-derived relationship between true half-lives and observed (apparent) half-lives, if leucine were used as the isotope, and using the experimentally derived F function. Several points are to be made. Proteins with short half-lives, i.e., less than 2

[35] R. W. Swick and D. T. Handa, *J. Biol. Chem.* **218**, 577 (1956).
[36] B. Poole, *J. Biol. Chem.* **246**, 6587 (1971).

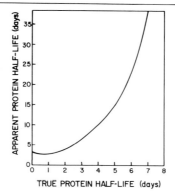

Fig. 5. Relationship between true half-lives and proteins and apparent half-lives. Data are from B. Poole [*J. Biol. Chem.* **246**, 6587 (1971)] and show the theoretical effect of the continual presence of free isotope (leucine) in the amino acid pool after single administration of leucine on half-lives of proteins.

days will all have apparent half-lives of between 3 and 4 days, and it will be difficult to determine differences in the true values in this range. Proteins with true half-lives of between 1 and 4 days can be discriminated by pulse-decay techniques, but the true half-lives will be much shorter than the apparent half-lives. For proteins with long half-lives, trace contamination with more rapidly labeled proteins, i.e., those turning over more rapidly, becomes a major problem. This may lead to an inaccurate measure of degradation rates.

Results with pulse-labeling techniques are even more difficult to interpret when comparisons of relative rates of degradation are made between the same protein in different tissues, or the same protein in a single tissue under different experimental states, where differing degrees of reutilization may occur.

Theoretically such experiments could be accomplished by determining the specific activity of the precursor pool, and correcting for the apparent versus true half-life according to Eq. (5). However, that method also carries the assumption that during isotope decay, the rate of synthesis of the protein (enzyme) remains constant [see Eq. (5)]. Under conditions where the rate of synthesis is also variable during this time of the experiment, its rate must be determined independently.

In summary, the use of reutilizable isotopes in pulse-decay experiments, although commonly employed, have severe limitation(s) in interpretation. They do not, by themselves, provide measures of true half-lives of enzymes. They can, under suitable conditions, provide information on whether two or more proteins in a similar tissue have different rates of turnover.

Another approach to the problem of isotope reutilization has been employed with those proteins which contain a heme prosthetic group. Advantage is taken of the fact that γ-aminolevulinate is rapidly incorporated into the heme moiety of heme proteins and is subsequently degraded to bilirubin. Thus, following a single administration of the heme precursor, decay of radioactivity in isolated protein is determined as a measure of the rate of degradation. The major assumption of this method is that heme does not dissociate from protein until such time as the entire protein-heme is degraded. This assumption may not always be correct, since Bunn and Jandl[37] have reported heme exchange in solutions of hemoglobin. However, Poole et al.[38] and Druyan et al.[39] have shown that the rate of degradation of catalase and microsomal cytochrome b_5 were similar when, respectively, determined either by the use of the isotopic δ-aminolevulinate decay or by other methods. This method has also been used to estimate the rates of degradation of mitochondrial cytochromes[39] and microsomal hemoprotein, presumably P-450.[40]

Continuous Isotope Administration. In this technique the animal is continuously subjected to an isotope whose specific activity is constant (Fig. 4). That isotope may be D_2O[41] or diet containing labeled protein,[42] amino acid,[13] or carbonate.[43] The rate of degradation of a specific protein, cell fraction, or tissue is determined by the rate at which incorporation approaches a maximum value, that maximal value being the specific activity of the administered precursor. Note that this method again assumes that the rate constant of synthesis is constant over the period of isotope administration. Theoretically the isotope should saturate instantaneously the free amino acid pool from which the protein(s) is synthesized. This requirement is a major limitation of this method, and hence, as with the pulse-decay method (see above), measurement of the specific radioactivity of the precursor pool must also be made for each time point on the curve. The differential equation describing the increase in specific activity of the isolated protein is given by

$$dE^*/dt = k_s(F) - kE^* \qquad (6)$$

where F is the function describing the time course of approach of the specific radioactivity of the precursor pool to a maximum value.

[37] H. F. Bunn and J. H. Jandl, *Proc. Nat. Acad. Sci. U.S.* **56**, 974 (1966).
[38] B. Poole, F. Leighton, and C. de Duve, *J. Cell Biol.* **41**, 536 (1969).
[39] R. Druyan, B. De Barnard, and M. Rabinowitz, *J. Biol. Chem.* **244**, 5874 (1969).
[40] W. Levin and R. Kuntzman, *J. Biol. Chem.* **244**, 3671 (1969).
[41] H. H. Ussing, *Acta Physiol. Scand.* **2**, 209 (1941).
[42] D. L. Buchanan, *Arch. Biochem. Biophys.* **94**, 500 (1961).
[43] R. W. Swick, A. K. Rexroth, and J. L. Stange, *J. Biol. Chem.* **243**, 3581 (1968).

The rate constant of degradation k_d is then given by

$$k_d = \log [E^*_{max} - E^*_t(F)] \tag{7}$$

Care must be taken using this method to ensure that the radioactivity in the precursor pool and that present in the isolated protein are present in the same amino acid(s). If not, then each component amino acid which is labeled must be isolated separately, since equilibration of the free amino acid pools of the various amino acids may differ. If there is incorporation into multiple amino acids, then one of the amino acids must be isolated separately, and its specific activity, both in protein and in precursor pool, determined at each time point.

The method of determining degradation rates in rat liver devised by Swick is particularly ingenious.[43] In this technique [^{14}C]carbonate of known specific activity is fed continuously to rats. The isotope is rapidly incorporated into arginine as a result of the reactions of the urea cycle. At least 90% of radioactivity in arginine is present in the guanidino carbon. The specific activity of the intracellular pool of free arginine can be easily determined from the specific activity of excreted urea, since the guanidino carbon of arginine is the carbon of urea. Thus animals are killed at varying times after onset of continuous [^{14}C]carbonate administration, liver protein samples are hydrolyzed, and the specific activity of the isolated arginine is determined, as well as that of urinary urea. The rate of degradation of the protein can be calculated from

$$A = U(1 - e^{-kt}) \tag{8}$$

where A is the specific activity of the arginine, U is the specific activity of the urea, and k is the first-order rate constant of degradation. Again it should be emphasized that this method is limited to liver, because of the large amount of arginase present.

Double Isotope Technique for Determining Heterogeneity of Degradation Rates. On many occasions it is desirable to know in steady state conditions whether proteins are turning over at the same, or dissimilar rates. To answer this question a technique has been developed in the author's laboratory which involves the combined use of ^{3}H and ^{14}C forms of the same amino acid. The method (double isotope method) is a modification of the standard pulse-labeling technique wherein the decay of specific radioactivity is determined (see above). The assumptions and limitations associated with the pulse-labeling technique are generally applicable to the double-isotope technique. Instead of establishing a number of time points along a decay curve, which involves isolation of the protein species at each time interval, the double-isotope technique allows for the identification of two time points on such a decay curve in the

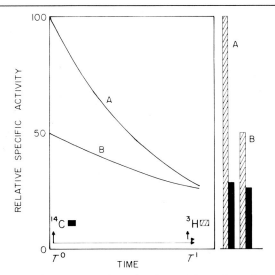

Fig. 6. Double isotope technique for determining rates of turnover of proteins. Isotope in protein at T': ^3H:^{14}C ratio, A = 3.50, B = 1.85.

same protein in the same animal. Thus one isotopic form of a [^{14}C]amino acid is administered initially, and allowed to decay a specified length of time, usually 4–6 days. This is followed by a second administration of the same [^3H]amino acid, and the animal is sacrificed a short time therafter, usually 4–6 hours. The ^3H counts represent the initial time point, and the ^{14}C counts represent the decay time point. Proteins with rapid degradation rates will have greater ^3H:^{14}C ratios. This is depicted graphically (Fig. 6) for two proteins which are present at the same concentration in a tissue, but one of which (protein A) is degraded at twice the rate of the other (protein B). The question of heterogeneity of degradation rate of proteins of a given organelle or cell fraction can be answered without knowing the specific enzymatic activity or nature of the proteins if they can be adequately separated, such as by gel electrophoresis or column chromatography. Thus, if all proteins are degraded at the same rate, the ^3H:^{14}C ratios will be the same, whereas if they are degraded at different rates, those ratios will be dissimilar. It should be clear that this method is designed largely to determine the degree of heterogeneity of turnover rates of proteins. The "apparent" rate of turnover of proteins can be calculated using this technique if an appropriate correction factor is introduced which accounts for the unequal initial incorporation of these two isotopes (^3H versus ^{14}C).[44] The correction factor

[44] K. W. Bock, P. Siekevitz, and G. E. Palade, *J. Biol. Chem.* **246**, 188 (1971).

involves the administration of both isotopic precursors simultaneously, with isolation of the protein(s) at the same time interval as the sacrifice of the animals after the second isotope administration in the decay experiments.

Thus, the degradation rate is given by

$$k_2 = 2.303/(t_2 - t_1) \log [(\text{dpm}_2/\text{dpm}_2)/(\text{dpm}_1/\text{dpm}')] \tag{9}$$

where dpm_1 and dpm_2 represent the radioactivity in the protein resulting from the initial isotope administration (t_1) (dpm_1) and the second (delayed) isotope administration (t_2) (dpm_2), and dpm' represents incorporation of the initial isotope when administered simultaneously with the second isotope (obviously a separate experiment). By the use of this correction factor, proteins with no turnover will have isotope ratio of 1.0, and hence $k_d = 0$. It should be emphasized again that these are apparent half-lives and subject to the same problems involving isotope reutilization as discussed above. Glass and Doyle[45] have compared the rates of turnover of some liver enzymes using the double isotope method, and pulse-decay experiments using guanidino-[^{14}C]L-arginine. Figure 7 shows their plot of the ratios of ^{3}H:^{14}C ratios of isolated proteins as a function of their rate constant of degradation, as determined using guanidino-[^{14}C]L-arginine.

Arias et al. (24) have used this method to demonstrate a marked heterogeneity of turnover of proteins of the endoplasmic reticulum. It has general applicability to any situation in which relative rates of degradation of proteins are to be measured. This method makes a number of specific assumptions: (a) the proteins will all follow exponential decay kinetics; (b) at the time of sacrifice all labeled proteins in the fraction under study are in a state of isotopic decay; (c) the rates of synthesis of the proteins are the same at the time of both isotope administrations; (d) the different proteins being compared are synthesized from the same amino acid pool; (e) the isotope used is not metabolized to other compounds which can enter other amino acid, carbohydrate, or lipid pools in a manner that allows them to be incorporated into the protein sample being counted.

Methods Based on Kinetics of Change in Enzyme Activity

A number of techniques have been devised during the past five years to estimate the rate of turnover of various enzymes on the basis of time courses of changes in enzyme levels. These methods have the advantage of not requiring the isolation of a specific enzyme or protein in a homoge-

[45] R. D. Glass and D. Doyle, *J. Biol. Chem.* **247**, 5234 (1972).

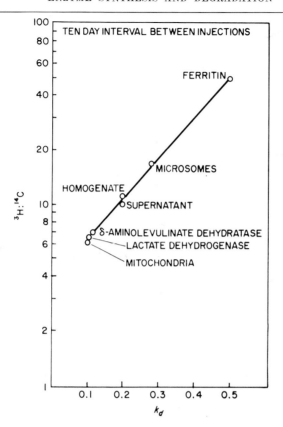

Fig. 7. Plot of rate constant of degradation of proteins determined by decay of guanidino-[^{14}C]L-arginine versus the log of the ^{3}H:^{14}C ratios, using the double isotope technique. Data are from R. D. Glass and D. Doyle, *J. Biol. Chem.* **247**, 5234 (1972).

neous state, or the manipulation of isotopes with its attendant assumptions. Such techniques, however, carry another set of assumptions and limitations.

Decay of Enzyme Activity after Elevation to High Levels by Administration of Appropriate Inducing Agents. Thus as originally used by Feigelson et al.,[46] the levels of tryptophan pyrrolase were increased by the administration of hydrocortisone and/or tryptophan, and the rate of return to a basal level was determined. The rate constant of degradation (k_d) is given by the slope of the straight line obtained from a plot on semilogarithmic paper of $(E_t) - (E_{ss,0})$ versus time where E_t is the enzyme activity at any time, t, during the decay period, and $E_{ss,0}$

[46] P. Feigelson, T. Dashman, and F. Margolis, *Arch. Biochem. Biophys.* **85**, 478 (1959).

is that enzyme activity characteristic of the basal state.[26] This technique has now been used with a number of enzymes. The assumptions of this technique include: (a) a measure of the decay of enzyme activity is a true measure of the decay of enzyme protein, and not simply a reflection of the activity. Operationally, degradation and irreversible inactivation cannot be distinguished by this method—or by any other, for that matter. (b) The stimulus of the administered agent or altered physiological status ceases relatively abruptly, relative to the rate of decay of protein under study, and the rate of synthesis of the protein immediately decreases to that characteristic of the basal state of enzyme synthesis. This method is inappropriate where the half-life of a protein is short when compared to the rate of physiologic adaptation. (c) The rate of decay of the elevated enzyme activity is similar to the rate of degradation of the enzyme in the basal state.

Decay of Activity after Inhibition of Protein Synthesis. The observed decay of enzyme activity following the inhibition of protein synthesis should reflect the inherent rate of enzyme degradation. This method carries the assumption that the inhibition of protein synthesis, such as produced by the administration of puromycin or cycloheximide, does not itself alter the rate of enzyme degradation. Although inhibitors of protein synthesis do not appear to affect the rate of degradation of γ-aminolevulinic acid synthetase[47] and tryptophan pyrrolase[14] of rat liver, puromycin, and cycloheximide diminish or abolish the breakdown (inactivation) of tyrosine transaminase in the same tissue.[23,48] These drugs exert similar effects on carbamyl phosphate synthetase[49] and glutamate dehydrogenase[50] in frog liver. In addition, the administration of actinomycin to rats inhibits the normally occurring decrease in the activity of mitochondrial α-glycerophosphate dehydrogenase that results from the cessation of thyroxine treatment in hypothyroid rats.[51] Numerous other examples could also be quoted. Therefore, the assumption that drugs will not alter degradation rates cannot be applied indiscriminately.

Time Course of Approach of an Enzyme to a New Steady State When the Rate of Synthesis of the Enzyme is Increased by Some Hormonal or Physiological Stimulus. To understand this method, we must consider further the model for steady-state enzyme levels.[27]

[47] H. S. Marver, A. Collins, D. P. Tschudy, and M. Rechcigl, Jr., *J. Biol. Chem.* **241**, 4323 (1966).
[48] A. Grossman and C. Mavrides, *J. Biol. Chem.* **242**, 1398 (1967).
[49] G. E. Shambaugh, III, J. B. Balinsky, and P. P. Cohen, *J. Biol. Chem.* **244**, 5295 (1969).
[50] J. B. Balinsky, G. E. Shambaugh, and P. P. Cohen, *J. Biol. Chem.* **245**, 128 (1970).
[51] O. Z. Sellinger, K. L. Lee, and K. W. Fesler, *Biochem. Biophys. Acta* **124**, 289 (1966).

Consider the time course of the change of E with time where the rate of synthesis is changed from k_s to k_s', and the rate constant of degradation is changed from k_d to k_d'. Then

$$dE/dt = k_s' - k_d'E$$
$$dE/(k_s' - k_d'E) = dt$$
$$\ln(k_s' - k_d'E) = k_d't + c$$

at $t = 0$, $E = E_0$, c, the constant of integration is

$$c = \ln(k_s' - k_d'E_0) = -k_dt$$
$$(k_s' - k_d'E)/(k_s' - k_d'E_0) = e^{-k_d't} \quad (10)$$
$$E/E_0 = k_s'/(k_d'E_0) - [(k_s'/k_d'E_0) - 1]e^{-k_d't}$$

This equation represents a general solution of Eq. (1). It should be emphasized that in Eq. (10), the only expression that determines the time course whereby an enzyme approaches a new steady state is the rate constant of degradation. The new steady-state level, on the other hand, is determined by the ratio of the rate constant of synthesis to the rate constant of degradation (k_s'/k_d').

It is possible to calculate, using Eq. (10), the time at which the enzyme has undergone one-half of its total change in content as follows:

Steady-state level at infinite time, $E' = k_s'/k_d'$
Initial enzyme level, $E_0 = 1$
One-half of the total change $= \frac{1}{2}(k_s'/k_d' - 1)$

This expression is substituted for the left-hand side of Eq. (10) and solved for t, the time required to attain one-half the total change.

$$\frac{1}{2}(k_s'/k_d' - 1) = (k_s'/k_d' - 1)e^{-k_d't}$$
$$\frac{1}{2} = e^{k_d't} \quad (11)$$
$$t = \ln 2/k_d'$$

The solution, i.e., $\ln 2/k_d'$, is the familiar half-life definition. Thus the time taken to increase to one-half of the final increase at the steady state is equal to the half-life of the enzyme. In this method the rate constant of degradation (k_d) can also be obtained from the slope of the straight line obtained from a plot on semilogarithmic paper of $(E_{ss'}) - (E_t)$ where $E_{ss'}$ is the new (higher) steady state level, and E_t is the enzyme activity at any time, t, during the increase of activity resulting from treatment.[26,52] It is possible, then, to estimate experimentally the half-life of any enzyme by following its time course of increase to a new steady state level. This method assumes that the rate of synthesis of the given enzyme is increased rapidly to some constant rate, and is maintained subsequently

[52] B. Szepesi and R. A. Freedland, *Arch. Biochem. Biophys.* **133**, 60 (1969).

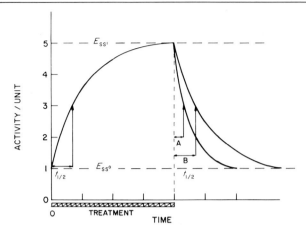

FIG. 8. Theoretical time course of change in enzyme level during and after experimental variable.

at that level during the entire experimental period. It carries the further assumption that the increase in enzyme activity does, in fact, represent *de novo* enzyme synthesis.

Figure 8 illustrates the theoretical determination of the half-life from the time course of change of enzyme level by both the decay and approach to steady state techniques. This type of analysis may be of use in determining whether an observed change in enzyme level represents an effect largely due to increased enzyme synthesis, or a result of decreased enzyme degradation. Thus, if the half-time of increase from one steady state ($E_{ss,0}$) to another (E_{ss}) is the same under treatment as it is following the cessation of such treatment (representative of decay curve B in Fig. 8) then the treatment has not altered the rate of degradation, and hence the treatment has increased the rate of synthesis only. If, on the other hand (case A), the half-life following cessation of treatment (presumed to be representative of the degradation rate in the basal state) is less than that during treatment, we can conclude that the treatment has in part retarded the rate of enzyme breakdown. Such an analysis has been applied to cortisone induction of glutamate-alanine transaminase,[26] phenobarbital induction of drug side-chain oxidation activity,[53] dietary induction of alanine and ornithine transaminase (43) and glucose-6-P dehydrogenase.[54] In all cases the major effect was on enzyme synthesis.

Time Course of Return of Enzyme Activity to the Steady State after

[53] I. M. Arias and A. De Leon, *Mol. Pharmacol.* **3**, 216 (1967).
[54] D. Rudock, E. M. Chrisholm, and P. Halter, *J. Biol. Chem.* **246**, 1249 (1971).

Irreversible Inhibition. The estimation of the rate of degradation (k_d) is similar to that described above for the time course of change from a lower to a higher steady state [see Eq. (9)]. This method has considerable potential because of the availability of suitable irreversible inhibitors of enzymes whose activities are not required for continued life. This method has been used to measure the half-life of plasma cholinesterase[55,56] and brain monoamine oxidase activity.[57] It has also been applied to rat liver catalase[28] and has been exploited by Rechcigl[58] in numerous studies on factors controlling levels of catalase in mice and rats. Catalase can be irreversibly inhibited by the administration of a single dose of 3-amino-1,2,4-triazole intraperitoneally. Catalase activity returns to half of the steady-state level in 24–30 hours in rat liver[28]; therefore the half-life of the enzyme is 24–30 hours (see Fig. 4). The major assumption of this method is that the return of enzyme activity represents new enzyme synthesis, and does not represent simply return of activity of existing enzyme. Price *et al.*[28] have shown that the incorporation of ^{59}Fe into catalase following aminotriazole administration to rats is consistent with new heme synthesis. These same workers have also shown that the repeated administration of allyliosopropylacetamide to rats results in a progressive loss of catalase activity with a half-life of 24–30 hours, and have concluded that this drug inhibits the synthesis of catalase. Consistent with this result is the finding of Schmid *et al.*[59] that the incorporation of [^{14}C]glycine into the heme portion of catalase is inhibited by allylisopropylacetamide. These results can be criticized as relating only to the heme portion of catalase. More recently, Ganschow and Schimke[10] and Poole *et al.*[38] have obtained evidence with labeling of the protein portion of catalase that is consistent with this rapid rate of degradation.

No single method for determining turnover rates is completely free of assumptions and/or limitations, nor is any single method applicable to all proteins. The most reasonable course is to be aware of such assumptions and limitations and whenever possible to use more than one method for determining degradation rates. In a few cases different techniques have been applied to the measurement of degradation rates of the same protein. The rate of degradation of tyrosine transaminase of rat liver has been measured by decay of enzyme activity after prior induction with glucocorticoid,[27,48] by decay of enzyme activity under experimental conditions

---

[55] D. Grob, J. L. Lilienthal, Jr., A. M. Harvey, and B. F. Jones, *Bull. Johns Hopkins Hosp.* **81**, 217 (1947).

[56] H. W. Neithlich, *J. Clin. Invest.* **45**, 380 (1966).

[57] S. H. Barondes, *J. Neurochem.* **13**, 721 (1966).

[58] M. Rechcigl, Jr., *Enzymologia* **34**, 23 (1968).

[59] R. Schmid, J. F. Figen, and S. Schwartz, *J. Biol. Chem.* **217**, 263 (1955).

in which enzyme synthesis has been prevented by administration of growth hormone,[23] and by determining the decay of pulse-labeled protein precipitable by a specific antibody under conditions in which a basal, steady-state enzyme level exists.[23] By all methods, the half-life of the enzyme is 1.5–2 hours. The rate of degradation of tryptophan pyrrolase of rat liver has been estimated by the decay of induced enzyme activity[46,60] and by the decay of pulse-labeled protein precipitated by a specific antibody under basal, steady-state conditions.[14] The half-life of this enzyme is about 2–4 hours. The rate of turnover of rat liver arginase has been measured by both continuous isotope and single-isotope administration methods, with resultant half-life measurements of 4–5 days.[13]

[60] R. T. Schimke, E. W. Sweeney, and C. M. Berlin, *Biochem. Biophys. Res. Commun.* **15**, 214 (1964).

[19] Enrichment of Polysomes Synthesizing a Specific Protein by Use of Affinity Chromatography

By E. Brad Thompson and J. V. Miller, Jr.

The purification of polysomes synthesizing a specific protein by affinity chromatography is based on a direct extension of the method as it is used for the purification of proteins. The adsorbent is prepared by covalently linking an appropriate ligand to an insoluble support material, and a suspension of polysomes is passed through a column of this adsorbent. In theory, those ribosomes which have attached to them nascent peptide chains sufficiently complete to interact with the ligand will be retained on the column, while the remainder pass through. The retained polysomes are then eluted by an appropriate change of buffer. After concentration by centrifugation, the purified polysomes can be used in a standard cell-free protein synthesizing system. Such purified polysomes are a potential source of enriched specific messenger RNA and possibly of special factors if any, related to the synthesis of the protein in question.

The basic principles of affinity chromatography have been treated at length elsewhere.[1,2] The essential requirement for the method to be effective is the existence of a ligand which meets the following three criteria: First, it must be reasonably specific for the protein to be purified. Ligands most commonly used are substrates, products, cofactors, or, in

[1] P. Cuatrecasas and C. B. Anfinsen, this series, Vol. 22, p. 345.
[2] P. Cuatrecasas and C. B. Anfinsen, *Annu. Rev. Biochem.* **40**, 259 (1971).

special cases, antibodies. Second, the ligand–protein association must be characterized by an association constant low enough to permit efficient binding. Third, one must be able to link the ligand to the supporting matrix. When affinity chromatography is to be used to purify polysomes synthesizing a specific protein an additional restriction placed on the ligand is that it be displaced from the peptide under fairly mild conditions so as to prevent inactivation of the polysome complex. It is an impression that all these criteria but the third are most likely to be met when the protein–ligand interaction is characterized by an association constant ranging from 10^{-5} to 10^{-9} M. The published experience in polysome purification by affinity chromatography is limited, as far as we know, to the single example discussed in this chapter.

It must be emphasized that the means by which an agarose-bound ligand selects polysomes synthesizing a specific protein is not known. We originally supposed that in order for affinity chromatography to be effective, a nearly complete nascent peptide must retain its association with tRNA and the messenger–ribosome complex while at the same time being sufficiently folded to interact appropriately with the bound ligand. Other possible means of association cannot be excluded. Thus, a free completed subunit might associate simultaneously with the ligand and an incomplete nascent peptide causing the binding of the latter to the column. Alternatively the polysome complex might for some reason have adsorbed to its surface completed proteins or subunits which are capable of interaction with the ligand.[3,4] In addition, some element of ion exchange may occur besides the affinity binding. In the absence of unequivocal proof of the mechanism involved it is impossible to predict with certainty the effectiveness of affinity chromatography in a given situation. If the criteria above seem favorable, the method can be tried.

In this chapter we will discuss some of the general methods for preparing affinity chromatography adsorbents and their application to the purification of polysomes. Our own experience in preparing adsorbents containing pyridoxamine phosphate and their use in purifying polysomes synthesizing tyrosine aminotransferase will serve as the example.

Preparation of Adsorbents

Agarose beads[5] provide a good inert support which can be readily derivatized.[6] Other supports, such as polyacrylamide gels, can also be

[3] D. Zipser, *J. Mol. Biol.* **7**, 739 (1963).
[4] Y. Kiho and A. Rich, *Proc. Nat. Acad. Sci. U.S.* **51**, 111 (1964).
[5] S. Hjertén, *Arch. Biochem. Biophys.* **99**, 446 (1962).
[6] P. Cuatrecasas, *J. Biol. Chem.* **245**, 3059 (1970).

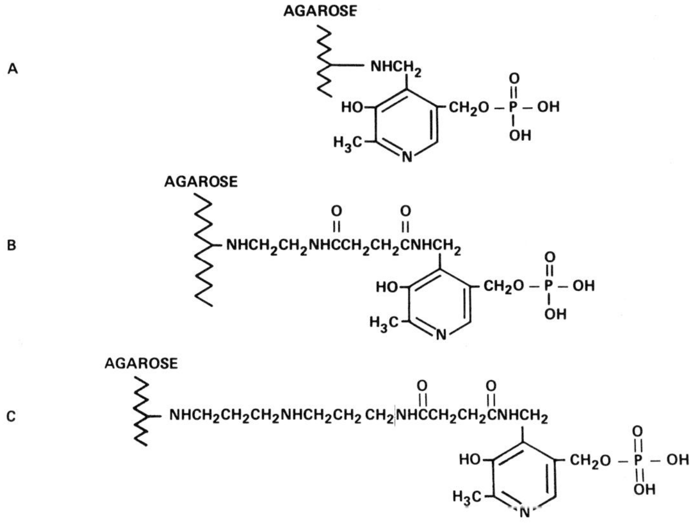

FIG. 1. Affinity chromatography adsorbents. Ligand pyridoxamine phosphate is attached to agarose beads directly (A) or by two different intervening arms, an aminoethylimino arm (B) or a longer succinyl-3,3'-diiminodipropylamine arm (C). Figure reprinted by permission of publisher, from J. V. Miller, Jr., P. Cuatrecasas, and E. B. Thompson, *Proc. Nat. Acad. Sci. U.S.* **68**, 1014 (1971).

derivatized. Preparations of agarose beads are commercially available (Sepharose from Pharmacia, Piscataway, New Jersey) as aqueous slurries which must be protected from freezing or drying. The chosen ligand can be linked directly to the agarose beads, or side chains of varying lengths can be interposed between ligand and bead (Fig. 1). Such sidearms are mandatory in some cases for the method to be effective. Presumably the displacement of the ligand from the bead favors its association with the binding site of the protein. In our experience, sidearm derivatives were essential, but only empirical testing can determine the optimal adsorbent for a new purification.

To attach a ligand by an amino group, agarose is activated by cyanogen bromide and then reacted with the ligand. Several methods, varying only slightly from one another, have been published for carrying out these reactions.[1,7,8] The procedure we have used[9] is as follows: Packed Sepharose 4B is mixed with an equal volume of water. Cyanogen bromide, 240 mg/ml of packed beads, in an amount of water sufficient to double the

[7] R. Axén, J. Porath, and S. Ernbäck, *Nature* (London) **214**, 1302 (1967).
[8] J. Porath, R. Axén, and S. Ernbäck, *Nature* (London) **215**, 1491 (1967).
[9] J. V. Miller, Jr., P. Cuatrecasas, and E. B. Thompson, *Proc. Nat. Acad. Sci. U.S.* **68**, 1014 (1971).

starting volume, is added with constant stirring (use a hood) at 20–25° while continuously monitoring the pH, which is maintained at pH 11 by the addition of 4N NaOH. The reaction is completed and no further NaOH is required after about 10 minutes. (The rate of this reaction may be slowed if desired, by carrying it out at lower temperatures.) The activated beads are then washed with 20 or more volumes of 0.1 M NaHCO$_3$ at 1–5° on a sintered-glass funnel and resuspended in an equal volume of a buffer of composition and pH (usually around 9) appropriate to the ligand being bound. Ligand in buffer is then added all at once (volume increase up to 15%), and followed as soon as possible by slow constant stirring for 24 hours at 20–25°. The derivatized agarose is then washed extensively with buffer and can be further washed with a dilute albumin or glycylglycine solution to further ensure that no reactive groups remain. If a method is available, the washes should be monitored for the presence of ligand and washing be continued until no more is detected. The quantity of ligand bound can be regulated by limiting the amount applied or by saturating the agarose as fully as possible and then diluting the derivative with unsubstituted agarose. The quantity of ligand bound can be estimated by difference if the unbound material in the washes is quantitated, or by measuring ligand released from an aliquot portion of the derivatized agarose by hydrolysis with acid, alkali, or a suitable enzyme (e.g., pronase, ribonuclease).

As already noted, the adsorbent may be more effective if the ligand is displaced from the agarose bead by a hydrocarbon side arm. Gels with sidearms of various lengths already incorporated are available commercially (Affitron Corp., El Monte, California). In addition, several methods of preparation are available. In general, cyanogen bromide-activated agarose can be reacted with aliphatic diamines, which in turn can be reacted with ligand directly or indirectly. Bromoacetamidoethyl agarose can be readily prepared and will react with primary aliphatic and aromatic amines, imidazole, and phenolic compounds. Tyrosyl agarose prepared with tyrosine as the carboxy terminal amino acid of an agarose-bound peptide can be used to couple diazotized compounds. These, and other techniques, described in detail in Volume 22 of this series[1] can provide a variety of useful sidearms for various ligands.

An example of the attachment of a ligand through a sidearm to agarose is the preparation of pyridoxamine phosphate gels.[9] Cyanogen bromide-activated agarose, prepared as described above, was allowed to react with 0.3 M ethylenediamine in 0.2 M NaHCO$_3$, pH 9.8, using the conditions already discussed for attachment of ligand. The ω-amino group was then succinylated by the addition at 4° of 1 mmole of succinic anhydride per 0.5 ml of a slurry of equal volumes of water and packed agarose. With constant stirring, the pH was adjusted and maintained at pH 6 with 5 M NaOH. After the reaction was complete and no further NaOH was required, the slurry was left to stir at 4° for an additional 5 hours. If desired, the completeness of the reaction could be determined by the sodium nitrobenzene–sulfonate color reaction.[6]

After washing the succinylated derivative with water, 20 ml of the packed succinylaminoethyl agarose in 40 ml of water was mixed with 100 mg of pyridoxamine phosphate. The pH of the slurry was adjusted to 4.7, and 1 g of 1-ethyl-3-(3-dimethylaminopropyl)carbodiimide in 3 ml of water was added rapidly at 20–25°. With constant stirring, the pH was monitored and kept at 4.7 for 2 hours, after which the reaction was complete. Stirring was continued for another 3 hours, and the final adsorbent was washed with 8 liters of 0.1 M NaCl over a 6-hour period to remove unbound reactants.

Preparation of Columns

Column dimensions and preparation of columns with the derivatized gel are not as critical as with ion-exchange or gel filtration procedures. We routinely used broken pipettes or Pasteur pipettes plugged with a small amount of loosely fitting glass wool. A slurry of the gel in buffer is poured onto the column evenly and gradually until the desired column volume is reached. The column is then washed with liberal amounts of buffer by gravity flow. After use the column can be regenerated by washing with 0.1 M NaOH followed by enough buffer to remove the NaOH and can be stored at 4° with an overlying quantity of buffer. If the column runs slowly, the width to height ratio can be increased.

Preparation of Polysomes

Polysomes can be prepared by standard methods used to obtain them for cell-free protein synthesis.[9,10] Care should be exercised to avoid exposure to RNase. If the cell source is rich in membrane-bound ribosomes, these should be removed by centrifugation over 2 M sucrose; alternatively, the membranes can be dissolved in 0.5 to 1% sodium deoxycholate. Removal of lipid-rich membranes is essential because they tend to clog the columns. Preliminary experiments are required to optimize the conditions of ribosome preparation from a given organ or cell type and to show that the unfractionated ribosomes are active in synthesizing protein. The preparation of ribosomes to be fractionated on the pyridoxamine phos-

---

[10] T. Staehelin and A. K. Falvey, this series, Vol. 20, p. 434.

phate column was done as follows: hepatoma tissue culture (HTC) cells in logarithmic growth and maximally induced for tyrosine aminotransferase synthesis were harvested by centrifugation at 600 g for 5 minutes. The cell pellet was washed twice in buffer containing 50 mM Tris pH 7.6, 25 mM KCl, and 5 mM MgCl$_2$, and the cells were suspended in a volume of 5 mM MgCl$_2$ equal to half the packed cell volume. The cells were allowed to swell at 4° for 20 minutes and then were disrupted in a Potter-Elvehjem homogenizer. Nuclei, mitochondria, and debris were removed by centrifugation at 20,000 g for 10 minutes. The ribosomes in the supernatant fraction were pelleted by centrifugation at 105,000 g for 2 hours. The ribosomes could be used immediately or stored at −90° without loss of activity. In this case, the tissue culture cells have little endoplasmic reticulum, so that removal of membranes was not necessary.

Affinity Chromatography of Polysomes Synthesizing Specific Proteins

Before any attempt is made to purify protein-synthesizing ribosomes, it is first necessary to study the conditions required for the binding and elution of the protein in question. Factors that affect binding include the type of adsorbent used (i.e., sidearms of various lengths, or none at all), and the physical state of the protein. In the case of tyrosine aminotransferase, for example, it was found that binding occurred only when the ligand was displaced from the column by a sidearm.[11] In addition, care had to be taken not to expose the enzyme to the extraneous pyridoxal phosphate normally added to stabilize the enzyme in extracts, since this resulted in formation of holoenzyme incapable of binding. Once binding is achieved, the problem of elution remains. The conditions for elution must be empirically determined. The variables that affect elution include pH, varying the concentration of salt, nonspecific protein, free competing ligand, temperature, and hydrogen-bond disrupting agents (which commonly destroy enzyme and ribosome activity). In addition, other agents not necessary for elution per se may be needed to stabilize the protein. When a set of eluting conditions is found that seems compatible with obtaining functional ribosomes, chromatography of the ribosome preparation can be attempted. It should also be noted that if fresh adsorbent is prepared, elution conditions may vary slightly and must be reestablished for the new preparation.

The empirically determined conditions that we have used to purify tyrosine aminotransferase-synthesizing ribosomes prepared from induced

---

[11] J. V. Miller, Jr., P. Cuatrecasas, and E. B. Thompson, *Biochim. Biophys. Acta* **276**, 407 (1972).

hepatoma tissue culture cells are as follows[9]: Ten to fifteen milligrams of ribosomes were suspended in 3–5 ml of 50 mM Tris·HCl buffer, pH 7.6 with 25 mM KCl and 5 mM MgCl$_2$. The solution was clarified by centrifugation at 20,000 g for 5 minutes and was then applied to a 5-ml column of pyridoxamine phosphate-substituted succinylaminoethyl-agarose kept at 4°. The column was washed with 40 ml of cold buffer and the bound ribosomes eluted with cold 50 mM Tris·HCl buffer, pH 4.0, containing 10 mM pyridoxal phosphate, 25 mM KCl, 5 mM MgCl$_2$, 0.5 M NaCl, and 0.1% bovine serum albumin (these were the same conditions that had been found previously to effectively elute tyrosine aminotransferase from this column). The first 10–12 ml of yellow eluate having an observed pH of 5.5–6.0 were collected and the ribosomes in the elute were pelleted by centrifugation. The yield of purified ribosomes was 50–100 µg of protein. By monitoring the A_{260} of the washes it could be determined that approximately 1% of the ribosomes were retained by the column. Most of the retained ribosomes were recovered upon elution.

Assay for Specific Protein Synthesis by Affinity Chromatography Purified Ribosomes

The fractionated ribosomes should be tested for their ability to incorporate radioactive amino acid into protein in a suitable cell-free system. Such systems are described in detail elsewhere.[9,12,13] If the preparation is active, analysis of the products for the specific protein in question can be carried out. This will require highly specific and sensitive techniques, the choice of which depends in large part on the protein being assayed. Techniques which have been employed include immunoassay, polyacrylamide gel electrophoresis, fingerprinting, analysis of tryptic peptides, copurification with known proteins or enzyme assay. In addition to establishing its identity, it is equally important to document that the protein in question is truly being synthesized. This can be assessed by establishing the temperature- and energy-dependence of the cell-free system and by showing that the apparent synthesis is blocked by agents, such as cycloheximide or puromycin, that are known to inhibit protein synthesis.

Continuing the example that has been cited throughout this chapter, ribosomes purified on pyridoxamine phosphate columns behaved in a cell-free protein synthesizing system in a manner very similar to the parent ribosome population. That is, the kinetics of incorporation, the ionic and energy requirements, and the sensitivity to blocking agents were essen-

[12] H. Aviv, I. Boime, and P. Leder, *Proc. Nat. Acad. Sci. U.S.* **68**, 2303 (1971).
[13] K. Moldave and L. Grossman (eds.), this series, Vol. 20, Part C of "Nucleic Acids and Protein Synthesis."

tially the same. If, however, the products of cell-free synthesis were assayed immunochemically or by enzyme activity, the purified ribosomes were found to produce substantially more tyrosine aminotransferase.[9] Thus, in this system affinity chromatography appears to be a useful tool in separating those ribosomes which are in the process of synthesizing tyrosine aminotransferase.

Published reports of polysomes containing specifically immunoreactive or enzymatically active peptide[14-19] suggest that other protein-synthesizing systems may contain nascent peptides with adequate specificity to be amenable to enrichment by affinity chromatography.

[14] K. Uenoyama and T. Ono, *J. Mol. Biol.* **65**, 75 (1972).
[15] R. Palacios, R. D. Palmiter, and R. T. Schimke, *J. Biol. Chem.* **247**, 2316 (1972).
[16] J. Hamlin and I. Zabin, *Proc. Nat. Acad. Sci. U.S.* **69**, 412 (1972).
[17] T. L. Delovitch, B. Davis, G. Holme, and A. H. Sehon, *J. Mol. Biol.* **69**, 373 (1972).
[18] R. M. Clayton, D. E. S. Truman, and J. C. Campbell, *Cell Differentiation* **1**, 25 (1972).
[19] G. Spohrand and K. Scherrer, *Cell Differentiation* **1**, 53 (1972).

[20] The Use of Inhibitors in the Study of Hormone Mechanisms in Cell Culture

By WALTER A. SCOTT and GORDON M. TOMKINS

Inhibitors of macromolecular synthesis have often been used to investigate mechanisms of hormone action. If a hormone response takes place in the presence of an inhibitor, it can be argued that the process inhibited is not essential for the response (at least during the period studied). If the response itself is inhibited, it can be argued, although with somewhat less conviction, that the inhibited process is required for the hormone action. The indirect nature of such experiments means that conclusions are seldom entirely convincing; however, the evidence obtained may lead to more direct investigations of molecular mechanisms. Furthermore, the pitfalls encountered in the use of inhibitors have become better defined as experience has been accumulated by different laboratories, so that models derived from inhibitor experiments are increasingly more respectable.

This chapter is designed to facilitate the use of inhibitors of macromolecular synthesis to analyze hormone action. Information has been drawn from studies with various types of cultured cells, and not limited to hormonal responses. The discussion is biased toward the more frequently used inhibitors since the extensive experience with these makes them more useful. We have, however, omitted discussion of inhibitors which predominantly affect mitochondrial function although they may also be useful in the study of hormone responses. Mechanisms of inhibitor action are considered only briefly since these have been covered in numerous recent reviews: (e.g., inhibitors of DNA and RNA synthesis,[1-8] protein synthesis,[6-11] and mitosis[1,12]). The chemical structures of many inhibitors can be found in a review by Kersten.[8] Simard[13] deals with the action of various drugs on nuclear cytology. Hormone antagonists are discussed in several articles and symposia.[14-17]

General Considerations

To measure inhibitor concentrations, optical methods are usually more reliable than gravimetric methods. It is also valuable to analyze visible and ultraviolet spectra and to compare these with published data (for this information, see references cited in footnotes 18–24) or to examine the compounds by chromatography.

[1] B. A. Kihlman, "Actions of Chemicals on Dividing Cells." Prentice-Hall, Englewood Cliffs, New Jersey, 1966.
[2] L. L. Bennett, Jr. and J. A. Montgomery, in "Methods in Cancer Research" (H. Busch, ed.), Vol. 3, p. 549. Academic Press, New York, 1967.
[3] P. Roy-Burman, *Recent Results Cancer Res.* **25**, 1 (1970).
[4] B. A. Kihlman, *Advan. Cell Mol. Biol.* **1**, 58 (1971).
[5] I. H. Goldberg and P. A. Friedman, *Annu. Rev. Biochem.* **40**, 775 (1971).
[6] D. Gottlieb and P. D. Shaw (eds.) "Antibiotics," Vol. I, "Mechanism of Action." Springer-Verlag, Berlin and New York, 1967.
[7] T. Bücher and H. Sies (eds.), Gesellschaft für Biologische Chemie, "Inhibitors, Tools in Cell Research." Springer-Verlag, Berlin, 1969.
[8] H. Kersten, *FEBS Lett.* **15**, 261 (1971).
[9] N. S. Beard, Jr., S. A. Armentrout, and A. S. Weisberger, *Pharmacol. Rev.* **21**, 213 (1969).
[10] S. Pestka, *Annu. Rev. Microbiol.* **25**, 487 (1971).
[11] J. Lucas-Lenard and F. Lipmann, *Annu. Rev. Biochem.* **40**, 409 (1971).
[12] G. Deysson, *Int. Rev. Cytol.* **24**, 99 (1968).
[13] R. Simard, *Int. Rev. Cytol.* **28**, 169 (1970).
[14] F. James and K. Fotherby, *Advan. Steroid Biochem.* **2**, 315 (1970).
[15] G. Raspé (ed.) "Advances in the Biosciences," Vol. 7, Schering Workshop on Steroid Hormone "Receptors." Pergamon, Oxford, 1971.
[16] H. G. Williams-Ashman and A. H. Reddi, *Annu. Rev. Physiol.* **33**, 31 (1971).
[17] E. V. Jensen and E. R. DeSombre, *Annu. Rev. Biochem.* **41**, 203 (1972).
[18] V. Ševčik, "Antibiotica aus Actinomyceten. Leitfaden für die Isolierung und Charakterizierung." Fischer, Jena, 1963.
[19] T. Korzybski, Z. Kowszyk-Gindifer, and W. Kuryłowicz, "Antibiotics. Origin, Nature and Properties" (E. Paryski, translator), 2 vols. Pergamon, Oxford, 1967. Original (in Polish) Panstwowe Wydawnictwo Naukowe, Warsaw, 1967.

The expected inhibitory activity must be confirmed under the actual conditions of experimentation. Cell lines vary considerably in permeability or sensitivity to drugs and susceptibility in a single line may be a function of growth conditions. In experiments where macromolecular synthesis is assayed by precursor incorporation,[25,26] potential inhibitor effects on the labeling of precursor pools as well as incorporation into macromolecules must be assessed.

In Tables I–VI, inhibitors are grouped according to the processes they affect most directly, but additional effects are also tabulated. Frequently, such effects are secondary to the direct action of the drugs.

Except where indicated, all inhibitors are effective in intact mammalian cells. References are, where possible, to reviews or to recent articles. We do not intend to ignore the important original observations in our selection, but rather to aid investigators who are unfamiliar with the often extensive literature on these compounds in finding all the pertinent information.

Inhibitors of RNA and DNA Synthesis

These compounds have been grouped into those that inhibit primarily DNA synthesis (mostly analogs of deoxynucleosides) (Table I), those which inhibit both RNA and DNA synthesis (usually compounds which bind to DNA) (Table II), and compounds that affect mainly RNA synthesis (analogs of ribonucleosides and inhibitors of RNA polymerases) (Table III).

Reagents that primarily affect thymidylate synthetase or ribonucleotide reductase (Table I) offer experimental advantages because they inhibit DNA synthesis specifically. Intracellular deoxyribonucleotide pools are often decreased, which may lead to an underestimate of the inhibition. Amethopterin and aminopterin affect pools of folic acid derivatives,

---

[20] H. Umezawa (ed.), "Index of Antibiotics from Actinomycetes." Univ. of Tokyo Press, Tokyo, 1967.

[21] P. G. Stecher (ed.), "The Merck Index," 8th ed. Merck and Co., Rahway, New Jersey, 1968.

[22] R. M. C. Dawson, D. C. Elliot, W. H. Elliot, and K. M. Jones (eds.) "Data for Biochemical Research," 2nd ed. Oxford Univ. Press (Clarendon) London and New York, 1969.

[23] R. Brunner, and G. Machek (eds.) "Die Antibiotica," Band III: "Die Kleinen Antibiotica." Verlag Hans Carl, Nuremberg, 1970.

[24] R. J. Suhadolnik, "Nucleoside Antibiotics." Wiley (Interscience), New York, 1970.

[25] D. M. Greenberg and M. Rothstein, this series, Vol. 4 [28].

[26] F. Bollum, this series, Vol. 12B, [106b].

TABLE I
INHIBITORS OF DNA SYNTHESIS

Inhibitor	Synonym or abbreviation	MW[a]	Solubility[b]	Actions reversible?[c]	Other effects[d]	Mechanism	Comments
Amethopterin	4-Amino-N^{10}-methylpteroyl-glutamic acid; methotrexate; methylaminopterin	454.5	Ins. H_2O, EtOH; sol. dil. NaOH	No[e]	Depletion of 1-carbon pools; chromosome breakage[f]	Inhibits dihydrofolate reductase[e, h]	Unstable to base and light (so should be neutralized); inhibition reversed by dT[i] or folinic acid[a, j]
Aminopterin (dihydrate)	4-Aminopteroyl-glutamic acid	476.4	Same as amethopterin	No[c, k]	Depletion of 1-carbon pools; chromosome breakage[f, l]	Same as amethopterin[e, g, h]	Same as amethopterin except more toxic[g]
Cytosine arabinoside (hydrochloride)	Cytarabine; ara-C; Cytostar	279.7	H_2O	Yes[m] (DNA synthesis reversed, cell viability not)	Chromosome breakage[f, l]	Inhibits DNA polymerase[m]; may also inhibit $CDP \rightarrow dCDP$[l, m]	DNA repair synthesis not inhibited[n]; inhibition reversed by dC[l, m]; RNA and protein synthesis not inhibited[m]
2'-Deoxyadenosine (monohydrate)	dA	269.2	H_2O	Yes[l]	Chromosome breakage[f, l]	Feedback inhibits ribonucleotide reductase[o]	Inhibition reversed by dG + dC[o]; inhibition requires high concentration (6 mM)
5-Fluoro-2'-deoxyuridine	FUdR, floxuridine	246.2	H_2O	Yes[l]	Chromosome breakage[f, l]	Inhibits thymidylate synthetase[l] (dU analog)	DNA repair synthesis not inhibited[n], inhibition reversed by dT or BUdR[l]
Hydroxyurea	Hydrea, $H_2NCONHOH$	76.1	H_2O	Yes[p]	Chromosome breakage[f, l]	Inhibits ribonucleotide reductase[q]	DNA repair synthesis not inhibited[n]
Thymidine	dT	242.2	H_2O	Yes[r]	May inhibit RNA synthesis, chromosome breakage[f, r]	Feedback inhibits $CDP \rightarrow dCDP^o$	Inhibition reversed by dC[o, q]; inhibition requires high concentration (2 mM)

[20] USE OF INHIBITORS WITH CULTURED CELLS 277

[a] The molecular weights are given for the forms of the compounds indicated in the first column.
[b] The solubility (sol.) is given for a suitable stock solution; ins., insoluble; dil., dilute; sl., slightly.
[c] This column indicates whether cells recover inhibited function when the inhibitor is washed out. Reversal by adding other agents is indicated under Comments.
[d] These secondary effects refer only to those reported in the literature. Absence of an entry is not meant to imply the function is known to be unaffected.
[e] R. E. Handschumacher and A. D. Welch, in "The Nucleic Acids" (E. Chargaff and J. N. Davidsen, eds.), Vol. 3, p. 453. Academic Press, New York, 1960.
[f] M. W. Shaw, Annu. Rev. Med. 21, 409 (1970).
[g] T. H. Jukes and H. P. Broquist, in "Metabolic Inhibitors" (R. M. Hochster and J. H. Quastel, eds.), Vol. 1, p. 481. Academic Press, New York, 1963.
[h] J. R. Bertina (ed.). "Folic Antagonists as Chemotherapeutic Agents," Ann. N. Y. Acad. Sci. 186, 1–519 (1971).
[i] Abbreviations: dT, dC, dA, dG = deoxynucleosides of thymine, cytosine, adenine, and guanine; T, C, A, G, = corresponding ribonucleosides; U = uridine.
[j] M. C. Berenbaum and I. N. Brown, Immunology 8, 251 (1965).
[k] G. Deysson, Int. Rev. Cytol. 24, 140 (1968).
[l] B. A. Kihlman, "Action of Chemical on Dividing Cells." Prentice-Hall, Englewood Cliffs, New Jersey, 1966.
[m] R. J. Suhadolnik, "Nucleoside Antibiotics," p. 156, Wiley (Interscience), New York, 1970. Conflicting reports. Mechanism and reversibility of ara C may depend on cell type studied.
[n] J. E. Cleaver, Advan. Radiat. Biol. 4, 1 (1974).
[o] A. Larsson and P. Reichard, Progr. Nucl. Acid Res. Mol. Biol. 7, 303 (1967).
[p] D. Gallwitz and G. C. Mueller, J. Biol. Chem. 244, 5947 (1969).
[q] B. A. Kihlman Advan. Cell Mol. Biol. 1, 58 (1971).
[r] J. M. Mitchinson, "The Biology of the Cell Cycle.", p. 26. Cambridge Univ. Press, London and New York, 1971.
[s] F. H. Kasten, F. F. Strasser, and M. Turner, Nature (London) 207, 161 (1965).

which may, in turn, influence methylation of diverse compounds. Curiously the actions of most of these inhibitors are reversible, even though they can produce extensive chromosome fragmentation.[27]

Actinomycin D (Table II) is perhaps the most popular of all the inhibitors of RNA synthesis, possibly because of the high efficiency of its inhibition. Low concentrations (0.05 μg/ml = 40 nM) preferentially inhibit synthesis of the large precursor of ribosomal RNA (rRNA); intermediate concentrations inhibit heterogeneous nuclear RNA (hnRNA) and messenger RNA (mRNA) synthesis as well; while high concentrations (5 μg/ml = 4 μM) are required to prevent transfer RNA (tRNA) or 5 S rRNA synthesis.[28–30]

In the presence of a high concentration of actinomycin D, many cell lines show a decay of general protein synthesis with a half-life of 3–5 hours which has been widely interpreted as a measure of normal mRNA degradation. However, several laboratories[31,32,33] have demonstrated, using more direct measurements, that mRNA is much more stable than this even in the presence of actinomycin D. It is likely, therefore, that some other component required for protein synthesis is depleted in the presence of the drug. This inhibition of protein synthesis is more pronounced when measured in cell-free extracts from actinomycin D-treated cells.[34,35]

A similar, but more rapid decay of protein synthesis, probably also without concomitant decay of mRNA, occurs in the presence of 5-azacytidine (Table III).[36] Camptothecin (Table II) inhibits protein synthesis in a manner that is similar to, but perhaps more gradual than, actinomycin D.[37,38]

5-Bromo-2′-deoxyuridine (Table II) is incorporated into DNA. Neither mitosis nor RNA synthesis is grossly inhibited; however, differ-

---

[27] T. C. Hsu, R. Humphrey, and C. E. Somers, *J. Nat. Cancer Inst.* **32**, 839 (1964).
[28] W. K. Roberts and J. F. E. Newman, *J. Mol. Biol.* **20**, 63 (1966).
[29] S. Penman, H. Fan, S. Perlman, M. Rosbash, R. Weinberg, and E. Zylber, *Cold Spring Harbor Symp. Quant. Biol.* **35**, 561 (1970).
[30] R. P. Perry, J. R. Greenberg, and K. D. Tartof, *Cold Spring Harbor Symp. Quant. Biol.* **35**, 577 (1970).
[31] R. H. Singer and S. Penman, *Nature (London)* **240**, 100 (1972).
[32] J. R. Greenberg, *Nature (London)* **240**, 102 (1972).
[33] W. Murphy and G. Attardi, *Proc. Nat. Acad. Sci. U.S.* **70**, 115 (1973).
[34] M. Revel, H. H. Hiatt, and J.-P. Revel, *Science* **146**, 1311 (1964).
[35] R. D. Ivarie and W. A. Scott, unpublished observations, 1972.
[36] M. Reichman, personal communication, 1972.
[37] S. B. Horwitz, C.-K. Chang, and A. P. Grollman, *Mol. Pharmacol.* **7**, 632 (1971).
[38] H. T. Abelson, personal communication, 1973.

entiated cellular functions are abolished by an unknown mechanism.[39-43]

Cordycepin (Table III) is of interest because it prevents the appearance of mRNA in the cytoplasm but does not interfere with the synthesis of hnRNA, the putative precursor of cytoplasmic mRNA. The inhibitor in its active form, cordycepin triphosphate, appears to prevent the formation of the polyadenylate sequences at the 3′-terminus of mRNA.[44] In HeLa cells, DNA and protein synthesis as well as synthesis of hnRNA are insensitive to the inhibitor, whereas rRNA and mRNA formation are prevented,[45] suggesting that the nuclear polymerase which synthesizes rRNA precursor and the enzyme catalyzing the hnRNA to mRNA conversion are particularly susceptible to the inhibitor. This conclusion is reinforced by the finding[46] that another inhibitor, 3′-deoxycytidine, inhibits only rRNA synthesis and allows normal appearance of mRNA. Cordycepin may not be as selective in other cell types,[24] however, since inhibition of the synthesis of purines and pyrimidines and of DNA and protein have been reported in Ehrlich ascites cells,[47,48] Novikoff hepatoma cells,[49] and HTC cells.[50]

Inhibitors of Mitosis and Cell Movement

The effects of Colcemid, colchicine, and vinblastine (Table IV), which inhibit chromosome movement (karyokinesis), are quite distinct from those of cytochalasin B, which inhibits cell movement and cell division (cytokinesis). The former group of inhibitors interferes with microtubule organization while the latter disorganizes microfilaments.

Inhibitors that affect microtubules alter the uptake of various compounds in a manner dependent on nutritional state. Under conditions of extended serum starvation, the transport of leucine, uridine, and deoxyglucose are dramatically stimulated by Colcemid and vinblastine.[51] In

[39] F. Stockdale, K. Okazaki, M. Nameroff, and H. Holtzer, *Science* **146**, 533 (1964).
[40] W. J. Rutter, J. D. Kemp, W. S. Bradshaw, W. R. Clark, R. A. Ronzio, and T. G. Sanders, *J. Cell Physiol.* **72**, Suppl. 1, 1 (1968).
[41] R. H. Stellwagen and G. M. Tomkins, *J. Mol. Biol.* **56**, 167 (1971).
[42] H. Weintraub, G. L. Campbell, and H. Holtzer, *J. Mol. Biol.* **70**, 337 (1972).
[43] B. L. Kotzin and R. F. Baker, *J. Cell Biol.* **55**, 74 (1972).
[44] M. Adesnik, M. Salditt, W. Thomas, and J. E. Darnell, *J. Mol. Biol.* **71**, 21 (1972).
[45] M. Siev, R. Weinberg, and S. Penman, *J. Cell Biol.* **41**, 510 (1969).
[46] H. T. Abelson and S. Penman, *Biochim. Biophys. Acta* **277**, 129 (1972).
[47] H. Klenow, *Biochim. Biophys. Acta* **76**, 354 (1963).
[48] K. Overgaard-Hansen, *Biochim. Biophys. Acta* **80**, 504 (1964).
[49] P. G. Plagemann, *Arch. Biophys. Biochem.* **144**, 401 (1971).
[50] L. A. Dethlefsen, B. B. Levinson, and G. M. Tomkins, *Abstr. 11th Annu. Meeting Amer. Soc. Cell Biol.*, p. 75 (1971).
[51] R. Kram and G. M. Tomkins, *Proc. Nat. Acad. Sci. U.S.* **70**, 1659 (1973).

TABLE II
Inhibitors of Both RNA and DNA Synthesis

Inhibitor	Synonym or abbreviation	MW[a]	Solubility[a]	Actions reversible?[a]	Other effects[a]	Mechanism	Comments
Actinomycin D (trihydrate)	Dactinomycin; Cosmegen; meractinomycin	1255.5	Sl. sol.[b] H$_2$O; sol. EtOH	Yes (slowly)	Protein synthesis gradually inhibited (see text); nucleolus distorted[d]; chromosome breakage[e]; binds thyroxine and prevents cellular uptake[f]	Binding to DNA via[g, h] —dGpdC— ⋮ ⋮ —dCpdG— (RNA ≫ DNA)[g, i]	Light sensitive and unstable in dilute base[j]; c class of RNA inhibited depends on inhibitor concentration (see text)
5-Bromo-2′-deoxyuridine	BrdU; BUdR; BrdUrd	307.1	H$_2$O	Yes[k] (slowly)	Makes DNA sensitive to visible light[l]; can induce some cells to produce virus[m, n]; chromosome breakage[f]	Incorp. into DNA in place of dT[o]	Differentiated functions uniquely sensitive (see text); negligible effect on mitosis, RNA or protein; synthesis[o]
Sodium camptothecin		388.3	H$_2$O	Yes[p, q]	Protein synthesis gradually inhibited[p]; makes DNA unstable to alkali[p, r, s] (only if cells alkali-treated while drug is present)[s]	Unknown (RNA ≃ DNA)[i, p]	Inhibitor unstable on storage as frozen aqueous solution[t]; inhibition seen only in intact cells[p, t]; no effect on 4 S or 5 S RNA synthesis[p, u] or mitochondrial nucleic acid synthesis[u, w]; synthesis of rRNA precursor and hnRNA strongly inhibited[u, v] (shortened RNA species seen)

| Daunomycin (hydrochloride) | Rubidomycin; daunorubicine; daunoblastine | 564.0 | H₂O | Chromosomal abnormalities[e,v]; nucleolus distorted[d,z] | No[x] | Binding to DNA[o,v] (DNA ≃ RNA)[i,v] or (DNA > RNA)[x,z] | Nucleolar RNA synthesis most sensitive[v,y] |
| Mitomycin C | | 334.3 | H₂O | Degradation of DNA, RNA[aa]; chromosome fragmentation[aa,bb,cc] | No[aa] | Covalent binding to DNA crosslinking DNA[cc] (RNA > DNA)[i,cc] or (RNA < DNA)[dd] | Sensitive to light and oxidation[aa], enzymatically converted to an active form intracellularly[cc] |

[a] See explanation in footnotes to Table I.
[b] Dissolves (with some patience) at 0.5 mg/ml, which is adequate for a stock solution in most cases.
[c] A. Schluederberg, R. C. Hendel, and S. Chavanich, *Science* **172**, 577 (1971). Extensive washing required; RNA made is not normal.
[d] R. Simard, *Int. Rev. Cytol.* **28**, 169 (1970).
[e] M. W. Shaw, *Annu. Rev. Med.* **21**, 409 (1970).
[f] K.-H. Kim, L. M. Blatt, and P. P. Cohen, *Science* **156**, 245 (1967).
[g] I. H. Goldberg and P. A. Friedman, *Annu. Rev. Biochem.* **40**, 775 (1971).
[h] H. M. Sobell, S. C. Jain, T. D. Sakore, and C. E. Nordman, *Nature (London) New Biol.* **231**, 200 (1971).
[i] Preference of inhibitor for DNA or RNA synthesis.
[j] T. Korzybski, Z. Kowszyk-Gindifer, and W. Kurylowicz, "Antibiotics. Origin, Nature and Properties" (E. Paryski, translator), Vol. 1, p. 1025. Pergamon, Oxford, 1967. Original (in Polish), Panstwowe Wydawnictwo Naukowe, Warsaw, 1967.
[k] J. Coleman and A. Coleman, *J. Cell Biol.* **31**, 22A (1966).
[l] B. Djordjevic and W. Szybalski, *J. Exp. Med.* **112**, 509 (1960).
[m] D. R. Lowy, W. P. Rowe, N. Teich, and J. W. Hartley, *Science* **174**, 155 (1971).
[n] S. A. Aaronson, G. J. Todaro, and E. M. Scolnick, *Science* **174**, 157 (1971).
[o] R. H. Stellwagen and G. M. Tomkins, *J. Mol. Biol.* **56**, 167 (1971).
[p] S. B. Horwitz, C.-K. Chang, and A. P. Grollman, *Mol. Pharmacol.* **7**, 632 (1971).
[q] R. S. Wu, A. Kumar, and J. R. Warner, *Proc. Nat. Acad. Sci. U.S.* **68**, 3009 (1971).
[r] M. S. Horwitz and S. B. Horwitz, *Biochem. Biophys. Res. Commun.* **45**, 723 (1971).
[s] H. T. Abelson and S. Penman, *Biochem. Biophys. Res. Commun.* **50**, 1048 (1973).
[t] H. T. Abelson, personal communication.
[u] H. T. Abelson and S. Penman, *Nature (London) New Biol.* **237**, 144 (1972).
[v] M. Oravec, A. Kumar, and R. S. Wu, *Biochim. Biophys. Acta* **272**, 607 (1972).
[w] S. Perlman, H. T. Abelson, and S. Penman, *Proc. Nat. Acad. Sci. U.S.* **70**, 350 (1973).
[x] L. E. Crook, K. R. Rees, and A. Cohen, *Biochem. Pharmacol.* **21**, 281 (1972).
[y] J. DiMarco, in "Antibiotics," Vol. I: "Mechanism of Action" (D. Gottlieb and P. D. Shaw, eds.), p. 190. Springer-Verlag, Berlin and New York, 1967.
[z] J. Bernard, R. Paul, M. Boiron, C. Jacquillat, and R. Maral, *Recent Results Cancer Res.* **20**, 28 and 52 (1969).
[aa] T. Korzybski *et al.* (*op. cit.*[j]), Vol. 1, p. 1071.
[bb] B. A. Kihlman, "Actions of Chemical on Dividing Cells." Prentice-Hall, Englewood Cliffs, New Jersey, 1966.
[cc] W. Szybalski and V. N. Iyer (*op. cit.*[v]), p. 211.
[dd] B. Peterkofsky and G. M. Tomkins, *J. Mol. Biol.* **30**, 49 (1967).

TABLE III
Inhibitors of RNA Synthesis

Inhibitor	Synonym or abbreviation	MW[a]	Solubility[a]	Actions reversible?[a]	Other effects[a]	Mechanism	Comments
α-Amanitin		916.6	Sl. sol. H_2O[b]	No[c],[d]	Nucleolus fragmented[e]	Inhibits nucleoplasmic RNA polymerase[c-g]	Not effective in many cell lines (see inhibition with isolated nuclei or polymerase); when effective in vivo, rRNA synthesis also inhibited[h],[i] complete inhibition requires about 30 min[h]
5-Azacytidine		244.2	H_2O	No[j],[k]	Mutagenic; DNA becomes unstable[j]; polysome decay; protein synthesis reduced[j],[k]; enhances incorporation of uridine into RNA[k]	Incorporated into RNA and DNA[j]	rRNA precursor breakdown[k], inhibition prevented by cytidine or uridine[j]
Cordycepin	3'-Deoxyadenosine; 3'dA	251.2	Sl. sol. H_2O; sol. dil. HCl	Yes[l] (slowly)	Inhibits pyrimidine, purine, DNA, and protein synthesis in some cell types (see text)	Incorporated into rRNA as premature 3'-terminus[l]; inhibits synthesis of polyadenylate portion of mRNA[m]	rRNA and mRNA preferentially inhibited; Hn RNA uninhibited[n]; inhibition prevented by adenosine[l]
2-Mercapto-1-(β-4-pyridethyl) benzimidazole	MPB	255.3	Ins. H_2O; sol. DMSO	Yes[o],[p]	Inhibits labeling of nucleoside pools and nucleoside phosphorylation[o-r]; inhibits phospholipid synthesis[t]	Unknown	Isolated nuclei not affected[o]

| Toyocamycin (monohydrate) | Siromycin; naritheracin; Unamycin B; Vengicide | 309.3 | Sl. sol. H₂O[b]; sol. dil. HCl | Yes[u, v] | Inhibits DNA and protein synthesis at higher concentrations[v] | Incorporated into RNA; prevents processing of rRNA precursor[u]; some inhibition of mRNA precursor processing[w] | No inhibition of tRNA synthesis or rRNA precursor synthesis[v]; DNA and protein synthesis uninhibited after 6 hours at low inhibitor concentration[u, v] |

[a] See explanations in footnotes to Table I.
[b] Solubility in water: α-amanitin, 0.5 mg/ml; cordycepin, 0.5 mg/ml; toyocamycin, 1 mg/ml. In each case this is usually adequate for a stock solution.
[c] T. Wieland and O. Wieland, in "Microbial Toxins," Vol. VIII; "Fungal Toxins" (S. Kadis, A. Ciegler, and S. J. Ajl, eds.), p. 249. Academic Press, New York, 1972.
[d] M. Meihlac, C. Kedinger, P. Chambon, H. Faulstick, and T. Wieland, FEBS Lett., 9, 258 (1970).
[e] T. J. Lindell, F. Weinberg, P. W. Morris, R. G. Roeder, and W. J. Rutter, Science 170, 447 (1970).
[f] L. Fiume and T. Wieland, FEBS Lett. 8, 1 (1970).
[g] V. L. Chan, G. F. Whitmore, and L. Siminovitch, Proc. Nat. Acad. Sci. U.S. 69, 3119 (1972).
[h] C. E. Sekeris and W. Schmid, FEBS Lett. 27, 41 (1972).
[i] J. R. Tata, M. J. Hamilton, and D. Shields, Nature (London) New Biol. 238, 161 (1972).
[j] R. J. Suhadolnik, "Nucleoside Antibiotics," p. 271ff. Wiley (Interscience), New York, 1970.
[k] M. Reichman, personal communication.
[l] R. J. Suhadolnik, (op. cit.), p. 66.
[m] M. Adesnik, M. Salditt, W. Thomas, and J. E. Darnell, J. Mol. Biol. 71, 21 (1972).
[n] S. Penman, H. Fan, S. Perlman, M. Rosbash, R. Weinberg, E. Zylber, Cold Spring Harbor Symp. Quant. Biol. 35, 561 (1970).
[o] W. P. Summers and G. C. Mueller, Biochem. Biophys. Res. Commun. 30, 350 (1968).
[p] B. B. Levinson, G. M. Tomkins, and R. H. Stellwagen, J. Biol. Chem. 246, 6297 (1971).
[q] J. J. Skehel, A. J. Hay, D. C. Burke, and L. N. Cartwright, Biochim. Biophys. Acta 142, 430 (1967)j.
[r] Y. Nakata and J. P. Bader, Biochim. Biophys. Acta 190, 250 (1969).
[s] K. R. Cutroneo and E. Bresnick, Biochem. Biophys. Res. Commun. 45, 265 (1971).
[t] R. M. Friedman and I. Pastan, Proc. Nat. Acad. Sci. U.S. 65, 104 (1970).
[u] A. Tavitian, S. C. Uretsky, and G. Acs, Biochim. Biophys. Acta 157, 33 (1968).
[v] R. J. Suhadolnik, (op. cit.), p. 330.
[w] P. M. McGuire, C. Swart, and L. D. Hodge, Proc. Nat. Acad. Sci. U.S. 69, 1578 (1972).

TABLE IV
INHIBITION OF CELL MOVEMENT AND DIVISION

Inhibitor	Synonym or abbreviation	MW[a]	Solubility[a]	Actions reversible?[a]	Functions affected	Mechanism	Comments
Colcemid	Demecolcine colchamine	371.4	H_2O	Yes[b]	Inhibits karyokinesis[c] and secretion[d, e]; inhibits or stimulates transport (see text)	Interferes with microtubule organization[f]	Uptake effects dependent on growth conditions (see text)
Colchicine		399.4	H_2O	No[g, h]	Same as Colcemid[c, d, h, i]	Same as Colcemid[j, k]	More toxic than Colcemid[b] no effect on DNA or RNA synthesis[i]; transport effects reversible[i]
Cytochalasin B		479.6	Ins. H_2O; sol. DMSO	Yes[l]	Inhibits: cytokinesis[l], cell movement[l], phagocytosis[m], pinocytosis[n], cell adhesion[o], transport (uridine, thymidine)[p], (2-deoxyglucose, galactosamine)[q, r, aa] hormone secretion[s–u]; causes disorganization of microfilaments[l, v]	Unknown (possibly interacts with an actinlike cellular component)[w]	No effect: karyokinesis[l], DNA synthesis[p], protein synthesis[l], microtubule integrity[l]
Vinblastine (monohydrate; dihydrogen sulfate)	Vinblastine sulfate, Velbe Velban vincaleukoblastine	909.0	H_2O	Yes[x] (not easily)[b]	Same as Colcemid[c–e], also RNA synthesis inhibition[y]	Interferes with microtubule organization; causes paracrystalline formation[z]	Uptake effects depend on growth conditions (see text)

[a] See explanation in footnotes to Table I.
[b] J. M. Mitchinson, "The Biology of the Cell Cycle," p. 28. Cambridge Univ. Press, London and New York, 1971.
[c] B. A. Kühlman, "Actions of Chemicals on Dividing Cells." Prentice-Hall, Englewood Cliffs, New Jersey, 1966.
[d] R. F. Diegelmann and B. Peterkofsky, *Proc. Nat. Acad. Sci. U.S.* **69**, 892 (1972).
[e] H. P. Ehrlich and P. Bornstein, *Nature (London) New Biol.* **238**, 257 (1972).
[f] B. R. Brinkley, E. Stubblefield, and T. C. Hsu, *J. Ultrastruct. Res.* **19**, 1 (1967).
[g] Often irreversible in animal cells; reversal seen in some other cell types.
[h] G. Deysson, *Int. Rev. Cytol.* **24**, 99 (1968).
[i] S. B. Mizel and L. Wilson, *Biochemistry* **11**, 2573 (1972).
[j] G. G. Borisy and E. W. Taylor, *J. Cell Biol.* **34**, 525, 535 (1967).
[k] L. Wilson and M. Friedkin, *Biochemistry* **6**, 3126 (1967).
[l] N. K. Wessells, B. S. Spooner, J. F. Ash, M. O. Bradley, M. A. Ludueña, E. L. Taylor, J. T. Wrenn, and K. M. Yamada, *Science* **171**, 135 (1971).
[m] A. T. Davis, R. Estensen, and P. G. Quie, *Proc. Soc. Exp. Med. Biol.* **137**, 161 (1971).
[n] R. Wagner, M. Rosenberg, and R. Estensen, *J. Cell Biol.* **50**, 804 (1971).
[o] J. W. Sanger and H. Holtzer, *Proc. Nat. Acad. Sci. U.S.* **69**, 253 (1972).
[p] P. G. W. Plagemann and R. D. Estensen, *J. Cell Biol.* **55**, 179 (1972).
[q] R. F. Kletzien, J. F. Perdue, and A. Springer, *J. Biol. Chem.* **247**, 2964 (1972).
[r] R. D. Estensen and P. G. W. Plagemann, *Proc. Nat. Acad. Sci. U.S.* **69**, 1430 (1972).
[s] J. A. Williams and J. Wolff, *Biochem. Biophys. Res. Commun.* **44**, 422 (1971).
[t] J. G. Schofield, *Nature (London) New Biol.* **234**, 215 (1971).
[u] N. Thoa, G. F. Wooten, J. Axelrod, and I. J. Kopin, *Proc. Nat. Acad. Sci. U.S.* **69**, 520 (1972).
[v] T. E. Schroeder, *Biol. Bull.* **137**, 413 (1969).
[w] J. A. Spudich and S. Lin, *Proc. Nat. Acad. Sci. U.S.* **69**, 442 (1972).
[x] S. E. Malawista, H. Sato, and K. G. Bensch, *Science* **160**, 770 (1968).
[y] W. Creasey, *Fed. Proc., Fed. Amer. Soc. Exp. Biol.* **27**, 760 (1968).
[z] K. G. Bensch and S. E. Malawista, *J. Cell Biol.* **40**, 95 (1969).
[aa] S. B. Mizel and L. Wilson *J. Biol. Chem.* **247**, 4102 (1972).

rapidly growing cells, these inhibitors slightly depress uptake.[51] At least in the case of nucleoside uptake, this depression is probably not mediated by microtubules.[52] This phenomenon would complicate labeling experiments by affecting precursor uptake differently depending on nutritional conditions. By contrast, cytochalasin B inhibits transport under all growth conditions.[51,53]

Inhibitors of Protein Synthesis

Because the enzymology of protein synthesis is understood in some detail, inhibitors of this process may be used as precise tools for identifying specific steps involved in control mechanisms. Information about several of these inhibitors is given in Table V. Amino acid analogs have not been included, although they are potentially very useful, because little is known about their action in cultured cells.

Compounds that specifically inhibit protein synthesis *in vitro* often have secondary effects in other cellular processes when applied to intact cells (see Table V). These may appear immediately (e.g., inhibition of DNA synthesis) or more gradually (e.g., reduced RNA synthesis). Since compounds which inhibit protein synthesis by quite different mechanisms produce the same secondary effects, it is likely that these effects are the result of interference with protein synthesis.

Inhibition of DNA synthesis by protein synthesis inhibitors is rapid and limited to semiconservative synthesis occurring during the S phase of the cell cycle. (Repair synthesis is not susceptible.[54-56]) Therefore, the blocking of a biological process in intact cells by an inhibitor of protein synthesis cannot determine whether the process depends on protein or on DNA synthesis.

Other secondary effects of protein synthesis inhibitors depend on nutritional conditions of the cells under study as well as the malignancy of the cell line. When growth is rapid, protein degradation is not influenced by cycloheximide or puromycin; but, when the cells have been starved for serum, protein degradation (which proceeds at an enhanced rate after this treatment) is inhibited by these compounds.[57] Consequently, a biological process could appear sensitive or resistant to inhibition of protein synthesis depending on whether a particular protein is

---

[52] S. B. Mizel and L. Wilson, *Biochemistry* **11**, 2573 (1972).
[53] R. D. Estensen and P. G. W. Plagemann, *Proc. Nat. Acad. Sci. U.S.* **69**, 1430 (1972).
[54] H. Weintraub and H. Holtzer, *J. Mol. Biol.* **66**, 13 (1972).
[55] J. E. Cleaver, *Advan. Rad. Biol.* **4**, 1 (1974).
[56] J. R. Gautschi, B. R. Young, and J. E. Cleaver, *Exp. Cell Res.* **76**, 87 (1973).
[57] A. Hershko and G. M. Tomkins, *J. Biol. Chem.* **246**, 710 (1971).

stabilized by the inhibitor. The magnitude of this phenomenon depends on the cell line, being more pronounced in cells with low malignancy (where the serum requirement is greatest).[58]

The uptake of small molecules (including uridine, thymidine, leucine, and 2-deoxyglucose), which may be curtailed after a period of serum starvation in certain cell lines, can be stimulated by cycloheximide or puromycin.[58-60] This is not the case in rapidly growing cells of the same lines. Such phenomena obviously complicate the interpretation of labeling experiments.

Nutrition-dependent phenomena may be related to the influence of inhibitors of protein synthesis on cyclic AMP metabolism[60] since cycloheximide reduces cyclic AMP levels in 3T3 cells.[51]

As noted in Table V, some inhibitors selectively inhibit polypeptide chain initiation (pactamycin, aurin tricarboxylic acid, sodium fluoride, polyuridylic acid) while others preferentially affect chain elongation (anisomycin, cycloheximide, emetine, fusidic acid, sparsomycin). This apparent selectivity is usually concentration dependent and, at higher concentrations, a given inhibitor may affect more than one step in protein synthesis. In this connection, it should be noted that cells concentrate some inhibitors (e.g., emetine),[61] while maintaining a barrier against others (e.g., cycloheximide, streptovitacin, fusidic acid, sparsomycin, blasticidin).[62] For these reasons, experiments designed to inhibit particular steps in protein synthesis must be done in conjunction with experiments which verify that, under the conditions being studied, the inhibitor is indeed selective.

Experiments in cell-free extracts can be especially difficult in this regard since, for unknown reasons, the inhibitor response of cell-free protein synthesis is variable and may depend on the method of preparation of the extract or on other factors. Some cell-free systems are quite sensitive to cycloheximide[62,63] while others are relatively insensitive.[64-67] This

[58] A. Hershko, P. Mamont, R. Shields, and G. M. Tomkins, *Nature (London) New Biol.* **232**, 206 (1971).
[59] W. A. Peck, K. Messinger, J. Brandt, and J. C. Carpenter, *J. Biol. Chem.* **244**, 4174 (1969).
[60] R. Kram, P. Mamont, and G. M. Tomkins, *Proc. Nat. Acad. Sci. U.S.* **70**, 1432 (1973).
[61] A. P. Grollman, *J. Biol. Chem.* **243**, 4089 (1968).
[62] H. F. Lodish, *J. Biol. Chem.* **246**, 7131 (1971).
[63] H. F. Lodish, D. Housman, and M. Jacobsen, *Biochemistry* **10**, 2348 (1971).
[64] L. L. Bennett, Jr., V. L. Ward, and R. W. Brockman, *Biochim. Biophys. Acta.* **103**, 478 (1965).
[65] S.-Y. Lin, R. D. Mosteller, and B. Hardesty, *J. Mol. Biol.* **21**, 51 (1966).
[66] B. S. Baliga, A. W. Pronczuk, and H. N. Munro, *J. Biol. Chem.* **244**, 4480 (1969).
[67] M. B. Mathews and A. Korner, *Eur. J. Biochem.* **17**, 328 (1970).

TABLE V
INHIBITOR OF PROTEIN SYNTHESIS

Inhibitor	Synonym or abbreviation	MW[a]	Solubility[a]	Actions reversible?[a]	Other effects[a]	Mechanism	Comments
Anisomycin	Flagecidin	265.3	Ins. H_2O; sol. EtOH and dil. acid	Yes[b]	Inhibits DNA synthesis[b]	Inhibits polypeptide chain elongation[c]	Alkaline labile, also slowly decomposes in acid[d], no affect on RNA synthesis[b]; at least as effective in intact reticulocytes as in lysate[b, e]
Aurin tricarboxylic acid (triammonium salt)	Aluminon	473.4	H_2O	Unknown		Inhibits mRNA binding to 40 S ribosomal particles[c, f]	No effect in intact cells; selective for chain initiation only at low concentrations[c, g]; selectively inhibits unnatural initiation[h]
Cycloheximide	Actidione; naramycin A	281.3	H_2O	Yes[c] (rapidly)	Inhibits DNA synthesis[i-k] but not DNA repair[k]; reduces cAMP level[l] (see text)	Inhibits polypeptide chain elongation[c, g] (binds to 60 S subunit)	Unstable to dilute alkali[d]
Emetine (dihydrochloride)	Cephaeline methyl ether	553.6	H_2O	No[m]	Inhibits DNA synthesis[m], slowly appearing RNA synthesis inhibition[m]	Inhibits polypeptide chain elongation[c, g]	Concentrated by living cells[c, m]
Fusidic acid	Ramycin	516.7	Ins. H_2O; sol. EtOH and $HCCl_3$	Yes[n]		Inhibits soluble elongation factor[c, o]	Reversibly binds to serum lowering effective concentration[p]
Sodium fusidate		538.7	H_2O				
Pactamycin		558.6	Ins. H_2O; sol. dil. acetic acid	Yes[q] (slowly)	Inhibits DNA synthesis and RNA synthesis[c, r]	Inhibits chain initiation[c, s] (binds to 40 S subunit)	Chain elongation inhibition at higher inhibitor concentration (irreversible)[s, t]
Polyuridylic acid	Poly(U)	—	H_2O	Unknown		Competes with natural mRNA for ribosomes[u]	No effect in intact cells; inhibits only chain initiation even at high concentrations[v]; stimulates Phe incorporation (other amino acid incorporation inhibited)

Puromycin (dihydrochloride)	Stylomycin achromycin	542.5	H₂O	Yes[w]	Inhibits DNA synthesis[w]; RNA synthesis[w]; purine synthesis[w]; amino acid uptake (kidney)[r]; cAMP phosphodiesterase[aa]	Releases incomplete polypeptide chains (analog of aminoacyl tRNA)[c, w, z]
Sodium fluoride	NaF	42.0	H₂O	Yes[bb]	Stimulates adenyl cyclase[c]; depletes ATP by inhibiting glucose oxidation[bb]	Inhibits polypeptide chain initiation[dd]
Sparsogenin	Sparsogenin	379.5	H₂O	Unknown	Inhibits DNA synthesis[c, ee] and RNA synthesis (less)[ee]	Inhibits polypeptide chain elongation (binds to 60 S subunit)[c, e, ee] No effect in intact reticulocytes, but inhibits intact L cells[ee]

[a] See explanations in footnotes to Table I.
[b] A. P. Grollman, *J. Biol. Chem.* **242**, 3226 (1967).
[c] S. Pestka, *Annu. Rev. Microbiol.* **25**, 487 (1971).
[d] R. Brunner and G. Machek, (eds.), "Die Antibiotica." Band III. "Die Kleinen Antibiotica," p. 61. Verlag Hans Carí, Nuremberg, 1970.
[e] H. F. Lodish, *J. Biol. Chem.* **246**, 7131 (1971).
[f] W. K. Roberts and W. H. Coleman, *Biochemistry* **10**, 4304 (1971).
[g] H. F. Lodish, D. Housman, and M. Jacobsen, *Biochemistry* **10**, 2348 (1971).
[h] M. B. Mathews, *FEBS Lett.* **15**, 201 (1971).
[i] R. F. Brown, T. Umeda, S.-I. Takai, and I. Lieberman, *Biochim. Biophys. Acta* **209**, 49 (1970).
[j] H. Weintraub and H. Holtzer, *J. Mol. Biol.* **66**, 13 (1972).
[k] J. R. Gautschi, B. R. Young, and J. E. Cleaver, *Exp. Cell Res.* **76**, 87 (1973).
[l] P. Kram, P. Mamont, and G. M. Tomkins, *Proc. Nat. Acad. Sci. U.S.* **70**, 1432 (1973).
[m] A. P. Grollman, *J. Biol. Chem.* **243**, 4089 (1968).
[n] C. L. Harvey, C. J. Sih, and S. G. Knight, in "Antibiotics." Vol. 1: "Mechanism of Action" (D. Gottlieb and P. D. Shaw, eds.), p. 404. Springer-Verlag, Berlin and New York, 1967.
[o] M. Malkin and F. Lipmann, *Science* **164**, 71 (1969).
[p] T. Korzybski, Z. Kowszyk-Gindifer, and W. Kurlyowicz, "Antibiotics: Origin, Nature and Properties" (translated by E. Paryski), Vol. II, p. 1361. Pergamon, Oxford, 1967. Original, Panstwowe Wydawnictwo Naukowe, Warsaw, 1967.
[q] B. Levinson, personal communication, 1972.
[r] B. K. Bhuyan (*op. cit.*[n]), p. 170.
[s] B. Colombo, L. Felicetti, and C. Baglioni, *Biochim. Biophys. Acta* **119**, 109 (1966).
[t] M. L. Stewart-Blair, I. S. Yanowitz, and I. H. Goldberg, *Biochemistry* **10**, 4198 (1971).
[u] B. Hardesty, R. Miller, and R. Sweet, *Proc. Nat. Acad. Sci. U.S.* **50**, 924 (1963).
[v] M. B. Mathews and A. Korner, *Eur. J. Biochem.* **17**, 339 (1970).
[w] D. Nathans, (*op. cit.*[n]), p. 259.
[x] R. J. Suhadolnik, "Nucleoside Antibiotics," p. 3. Wiley (Interscience), New York, 1970.
[y] H. E. Franz, M. Franz, and K. Decker, *Hoppe-Seyler's Z. Physiol. Chem.* **336**, 127 (1964).
[z] L. J. Elsas and L. E. Rosenberg, *Proc. Nat. Acad. Sci. U.S.* **57**, 371 (1967).
[aa] M. M. Appleman and R. G. Kemp, *Biochem. Biophys. Res. Commun.* **24**, 564 (1966).
[bb] P. A. Marks, E. R. Burka, F. M. Conconi, W. Perl, and R. A. Rifkind, *Proc. Nat. Acad. Sci. U.S.* **53**, 1437 (1965).
[cc] G. A. Robison, R. W. Butcher, and E. W. Sutherland, "Cyclic AMP," p. 80. Academic Press, New York, 1971.
[dd] J. M. Ravel, R. D. Mosteller, and B. Hardesty, *Proc. Nat. Acad. Sci. U.S.* **56**, 701 (1966).
[ee] I. H. Goldberg and K. Mitsugi, *Biochem. Biophys. Res. Commun.* **23**, 453 (1966).

difference may be related to the difficulty in determining which step in protein synthesis is preferentially inhibited by cycloheximide. The studies on systems that are relatively insensitive to the drug have generally shown cycloheximide to inhibit initiation preferentially. However, in intact cells and in very sensitive cell-free systems, cycloheximide, in low concentrations, preferentially inhibits elongation.

With these caveats, it is possible to obtain meaningful data using selective inhibition. A particularly useful measurement in systems that may involve translational control is the determination of a change in the concentration of growing protein chains as a result of exposure to a hormone. In principle, such a determination can distinguish between changes in specific initiation rate and changes in the rate of polypeptide chain elongation or release.[68] As an example of this approach, cells, in different states of hormonal regulation, are incubated with radioactive amino acids in the presence of a concentration of pactamycin which preferentially inhibits initiation (2 μM for our experiment with HTC cells). Amino acid incorporation continues in the presence of the inhibitor for 5–10 minutes, in which time growing polypeptide chains will incorporate radioactive amino acids and be released as completed chains. The protein under study is then isolated and its radioactivity determined. This value is proportional to the number of peptide chains being synthesized at the time of addition of the inhibitor and the label. Any hormone-induced increase in this parameter would indicate an increased concentration of specific mRNA or an increased rate of specific initiation, rather than increases in the rates of chain elongation or release.

Other types of experiments may also be done which take advantage of the selective inhibition of polypeptide chain elongation. By slowing elongation relative to initiation it is possible to "load" ribosomes onto messenger RNA making protein synthesis independent of initiation rates and determined by mRNA concentrations. This method provides an indirect *in vivo* measurement of relative specific mRNA concentrations.[31,62,69,70] As an example, cells in a desired regulatory state are incubated with submaximal concentrations of an elongation inhibitor. When a new steady state rate of synthesis has been reached (determined by the change in polysome profile), the proteins are labeled with a short pulse of radioactive acids. If the relative synthesis of different molecular species is altered by the inhibition it can be inferred that the synthesis (in unin-

---

[68] W. A. Scott, R. Shields, and G. M. Tomkins, *Proc. Nat. Acad. Sci. U.S.* **69**, 2937 (1972).
[69] H. Fan and S. Penman, *J. Mol. Biol.* **50**, 655 (1970).
[70] S. Perlman, M. Hirsch, and S. Penman, *Nature (London) New Biol.* **238**, 143 (1972).

hibited cells) is not determined only by mRNA concentrations but that translational controls also operate.

Anti-hormones

Certain compounds interfere with hormone responses by competing with normal hormones for sites on specific receptor proteins. Such inhibition causes minimal derangement of cell metabolism and singles out the processes affected by the hormone under study. Examples of steroid hormone antagonists that inhibit by this mechanism are listed in Table VI.

These compounds are insoluble in water. Stock solutions are prepared in ethanol and diluted to the low concentrations needed to antagonize a hormone response (usually less than 1 μM) at which concentration they are water soluble. The anti-glucocorticoids have been shown to be reversible in their effect on tyrosine aminotransferase induction in HTC cells.[71] Of the other compounds listed in Table VI, only spironolactone has been demonstrated to be reversible in its antagonism of mineralcorticoid response,[72] but it is quite likely that many of the others are also.

Certain anti-hormones (termed suboptimal inducers[71]) are partially active themselves, but inhibit because they compete with maximally effective compounds. Suboptimal glucocorticoids include 11β-hydroxyprogesterone, 11-deoxycortisol, and deoxycorticosterone. For a clear-cut inhibition of a hormone response, it is obviously desirable to use a competing analog with no agonist activity.

When they are tested in whole animals, several of the compounds in Table VI have other activities besides the antagonist activities listed. For example, the antiandrogens cyproterone acetate and 17-methylnortestosterone (but not cyproterone) are progestational,[73] two anti-glucocorticoids (fluoxymesterone and 17α-methyltestosterone) are androgenic, and another two (progesterone and 17α-hydroxyprogesterone) are progestational. Presumably these side effects represent an overlapping specificity of the receptor molecules. Therefore inhibition of a hormone response by an antagonist cannot conclusively identify the type of receptor molecule involved.

In summary, the use of inhibitors can be complicated due, in large part, to the sophisticated interrelationships among cellular processes. For a similar reason, inhibitors are powerful tools for biological research, since cellular events can be analyzed with much of the sophisticated structural, metabolic and regulatory apparatus intact.

[71] H. H. Samuels and G. M. Tomkins, *J. Mol. Biol.* **52**, 57 (1970).
[72] I. Edelman, personal communication (1972).
[73] F. Neumann, R. von Berswordt-Wallrabe, W. Elger, H. Steinbeck, J. D. Hahn, and M. Kramer, *Recent Progr. Horm. Res.* **26**, 337 (1970).

TABLE VI
ANTIHORMONES

Anti-glucocorticoids	Anti-mineralocorticoids	Anti-androgens	Anti-estrogens
Cortisone[a,b]	Progesterone[d,e]	Cyproterone[i-k]	17α-Estradiol[n,o]
Fluoxymesterone[a]	SC 14266[f,g] (water-soluble spirolactone)	Cyproterone acetate[i-m]	Ethamoxytriphetol[o-q] (MER-25)
17α-Hydroxyprogesterone[b,c]	SC 19886[d] (17β-OH-12α-etiojerv-4-en-3-one)	17α-Methylnortestosterone[m]	Chlomiphene citrate[o,q,r] (clomid; MLR/41)
17α-Methyltestosterone[a,b]	SC 9420[g,h] (spironolactone; Aldactone A)		Nafoxidine[o,q,s] (Upjohn 11,100 A)
Progesterone[b]			Parke Davis Cl-628[q,t]
Testosterone[a,c]			

[a] H. H. Samuels and G. M. Tomkins, *J. Mol. Biol.* **52**, 57 (1970).
[b] G. Rousseau, J. Baxter, and G. M. Tomkins, *J. Mol. Biol.* **67**, 99 (1972).
[c] J. D. Baxter, A. Harris, G. M. Tomkins, and M. Cohn, *Science* **171**, 189 (1971).
[d] G. A. Porter and J. Kimsey, *J. Steroid Biochem.* **3**, 201 (1972).
[e] I. Edelman, personal communication (1973). Demonstrated progesterone binding to type I (mineralocorticoid) receptor.
[f] G. E. Swaneck, L. L. H. Chu, and I. S. Edelman, *J. Biol. Chem.* **245**, 5382 (1970).
[g] I. Edelman, in "Modern Diuretic Therapy in the Treatment of Cardiovascular and Renal Disease" (A. F. Lant and G. M. Wilson, eds.), p. 60. Excerpta Med. Found., Amsterdam, (1973).
[h] F. C. Bartter (ed.) "The Clinical Use of Aldosterone Antagonists." Thomas, Springfield, Illinois, 1960.
[i] F. Neumann, R. von Berswordt-Wallrabe, W. Elger, H. Steinbeck, J. D. Hahn, and M. Kramer, *Recent Progr. Horm. Res.* **26**, 337 (1970).
[j] H. G. Williams-Ashman and A. H. Reddi, *Annu. Rev. Physiol.* **33**, 31 (1971).
[k] H. Jackson and A. R. Jones, *Advan. Steroid Biochem. Pharmacol.* **3**, 167 (1972).
[l] S. Fang and S. Liao, *J. Biol. Chem.* **246**, 16 (1971).
[m] K. J. Iveter, O. Unhjem, A. Attramadal, A. Aakvaag, and V. Hansson, in "Advances in the Biosciences," Vol. 7: Schering Workshop on Steroid Hormone "Receptors," p. 193. Pergamon, Oxford, 1971.
[n] K. L. Barker and J. M. Anderson, *Endocrinology* **83**, 585 (1968).
[o] F. James and K. Fotherby, *Advan. Steroid Biochem. Pharmacol.* **2**, 315 (1970).
[p] E. V. Jensen, *Recent Progr. Horm. Res.* **18**, 461 (1962).
[q] E. V. Jensen and E. R. DeSombre, *Annu. Rev. Biochem.* **41**, 203. (1972).
[r] L. Kahwanago, W. L. Heinricks, W. L. Hermann, *Endocrinology*, **86**, 1319 (1970).
[s] J. E. Martin, *Endocrinology* **91**, 594 (1972).
[t] G. Shyamala and S. Nandi, *Endocrinology* **91**, 861 (1972).

Acknowledgments

We would like to thank the numerous investigators who made unpublished information available to us. This work was supported by Grant No. GM 17239 from the National Institute of General Medical Sciences of the National Institutes of Health and Contract No. 72-3236 within the Special Virus-Cancer Program of the National Cancer Institute of the National Institutes of Health. W. A. S. is a postdoctoral fellow of the United States Public Health Service.

[21] The Design of Double Label Radioisotope Experiments

By Edwin D. Bransome, Jr.

Within the context of this volume, it seems reasonable to consider only the limited number of radioisotopes which are usually used as labels for quantities of enzyme, hormone, or nucleotide synthesized, as substrates in systems affected by hormones or cyclic nucleotides, or nucleotides. Table I provides such a limited list. The maximum energies (E_{max}) involved in their radioisotopic decay are a fair indication of whether one radioisotope in a sample may adequately be determined in the presence of another.

The gas ionization (Geiger-Müller) detectors used in "planchet" counting are not capable of resolving the energies of such isotope pairs. Filters may be inserted, however, between sample and detector, quantitatively absorbing emissions of the lower energy radioisotope. This is satisfactory only if there is a wide separation of the two energy spectra, if

TABLE I
PROPERTIES OF RADIOISOTOPES COMMONLY EMPLOYED

Isotope	Principal emission on decay	E_{max} (keV)	Relative abundance (%)	Scintillation detector
^{3}H	β	18	100	Liquid
^{14}C	β	156	100	Liquid
^{35}S	β	168	100	Liquid
^{32}P	β	1700	100	
^{125}I	γ	35	7	NaI, liquid
	κ-X-ray	27[a]	—	
^{131}I	γ	360	80	NaI, liquid
	β	610	87	

[a] Decay by electron capture.

tritium (^{3}H) with its very low β energy is not one of the isotopes, and if samples are counted at "infinite thinness"—without significant self-absorption. Indeed, until the advent of scintillation counters about twenty years ago, double-isotope experimental designs were rarely attempted.

Scintillation Counters

Birks[1] has provided a thorough and basic review of the technology involved in "gamma" counters where the detector is usually an activated NaI crystal, which responds to αs, γs, and κ-X-rays and of liquid scintillation counters in which an aromatic phosphor in solution serves as the scintillator. These respond to βs as well as αs, γs, and κ-X-rays. In both types of counter, the photoelectron spectra are proportionate to the energy spectra of the photons emitted from radioisotopic decay.

"Gamma" Counters

There is only slight degradation in the resolution of spectra in NaI crystal counters, so that isotope pairs with energies above the crystal's scintillation threshold can be easily separated for counting, with the proviso that overlap of the two spectra be ascertained with standards of the same geometry. Regoeczi and Webber have recently shown that spillover of ^{131}I into the ^{125}I spectrum varies with the well counter used and the ^{131}I count rate.[2]

For example, optimum channels in one of our automatic well counters yield an ^{125}I efficiency of 37.6% and an ^{131}I efficiency of 38.8%. The efficiencies depend not only on pulse height analyzer settings, but also upon the shielding and geometry of the crystal and of the counting chamber.) While the spillover of ^{125}I into the ^{131}I channel is inconsequential: 0.28%, the ^{131}I contribution to the ^{125}I channel is significant: 9.17%. Therefore ^{125}I cpm $= A - (0.0917 \times B)$ if A represents the cpm in the ^{125}I channel and B represents cpm in the ^{131}I channel.

Although "gamma" counting does not involve corrections for complex quenching phenomena as does liquid scintillation counting, self-absorption of emitted energy may be quite significant and may vary with the isotope being counted. The examples of ^{125}I and ^{131}I counting given in Table II should serve as a warning that standards for γ counting should be as similar in geometry to the unknowns as possible.

[1] J. B. Birks, "The Theory and Practice of Scintillation Counting." Macmillan, New York, 1964.
[2] E. Regoeczi and C. Webber, *J. Nucl. Biol. Med.* **16**, 10 (1972).

TABLE II
Effects of Sample Geometry on Relative
Detection Efficiency[a]

Isotope	Aqueous solution	Whatman No. 1 filter paper	Cellulose nitrate filter
A. "Gamma" NaI crystal scintillation counter			
^{125}I	1.00	0.92	1.11
^{131}I	1.00	0.85	0.86
B. Liquid scintillation counter			
^{125}I	1.00	0.65	0.82
^{131}I	1.00	0.96	0.98
^{14}C	1.00	0.71	0.91
^{3}H	1.00	0.09	0.42

[a] The counts per minute in a 0.1-ml aqueous solution of [^{125}I]NaI, [^{131}I]NaI, [^{14}C]adenine, or cyclic[^{3}H] AMP were taken as 1.00 and compared to replicates air dried onto discs 20 mm in diameter. Gamma counting was carried out in a Nuclear Chicago automatic gamma system with a 2-inch NaI crystal using optimum channels for each isotope. The samples in 10 × 75 mm Pyrex test tubes were inserted into cellulose nitrate counting tubes. Liquid scintillation counting was with a Beckman LS-150 system with a wide channel encompassing the total spectrum of both isotopes. The aqueous samples were counted in toluene 7.0 g/liter PPO (2,5-diphenyloxazole), 10% Biosolv BBS-3 (Beckman) solubilizer. The samples dried onto the filters were counted in toluene-PPO without Biosolv.

Liquid Scintillation Counting

In most double-label experiments, two counting channels are used and the photoelectron spectra are only partially separated. In liquid scintillation counting there is too great a loss of detection efficiency if each isotope is eliminated entirely from the other channel: the photoelectron spectra of the fluorescence emitted from the organic phosphors are too broad. Indeed, if the photon energies are not widely separated, it is statistically impractical to eliminate the spillover of either isotope from the other channel.

A decrease in detection efficiency can be produced by impurity or chemical quenching which interferes with the excitation of the scintillation phosphor by radioactive photons or by "color" quenchers which absorb the emitted fluorescence and thus interfere with the transmission of light to the photomultiplier tubes. The extent of quenching in a homogeneous solution of sample and phosphor in solvent will cause pulse height shifts of fluorescence spectra to lower photoelectron energies as an inverse function of the energy of the emitted photons. Quenching in samples will thus affect the detection efficiencies of two isotopes variably, the spectral shift of the lower energy isotope being greater. There is then no single set of optimum instrument settings for specific isotope pairs.

Methods of quench correction for homogeneous samples (samples in solution or in translucent emulsions of fine micelles) are reviewed thoroughly in recent literature and will therefore not be dwelt upon here. A reading of references cited in footnotes 3–7 should provide a thorough grounding in theory and practice. I have summarized the principal attributes of the three approaches in Table III. Correction of the overlap of radioactivities must be carried out before quench correction. The questions that deserve attention in this review are: (a) How are the cpm of one isotope mathematically best separated from the cpm of another? (b) How are the optimum instrument settings determined for a specific isotope pair? (c) What are the restrictions on sample preparation for statistically reliable double isotope counting?

Mathematics of Double-Isotope Counting. One approach to double-isotope counting is to continue to use the discriminator settings for single isotopes. Since each channel then counts contributions from both isotopes, simultaneous equations are necessary for the calculation of the absolute radioactivity (dpm) of each sample:

H: radioactivity of the isotope H of greater E_{max} in the sample (dpm)
L: radioactivity L of lesser E_{max} in the sample (dpm)
$h1$: detection efficiency of H in channel 1 (determined by sample channels ratios or external standardization, and expressed as a fraction of 1)
$h2$: detection efficiency of H in channel 2
$l1$: detection efficiency of L in channel 1
$l2$: detection efficiency of L in channel 2
N1: net cpm (cpm − background) in channel 1
N2: net cpm in channel 2

$$H = [N1 - N2(l1/l2)]/[h1 - h2(l1/l2)] \quad (1)$$
$$L = [N2 - N1(h2/h1)]/[l2 - l1(h2/h1)] \quad (2)$$

[3] J. B. Birks, in "The Current Status of Liquid Scintillation Counting" (E. D. Bransome, Jr., ed.), pp. 283–292. Grune & Stratton, New York, 1970.

[4] C. T. Peng, in "The Current Status of Liquid Scintillation Counting" (E. D. Bransome, Jr., ed.), pp. 283–292. Grune & Stratton, New York, 1970.

[5] M. P. Neary, and A. L. Budd, in "The Current Status of Liquid Scintillation Counting" (E. D. Bransome, Jr., ed.), pp. 273–282. Grune & Stratton, New York, 1970.

[6] P. D. Klein, and W. J. Eisler, in "Organic Scintillators and Liquid Scintillation Counting" (D. L. Horrocks and C. T. Peng, eds.), pp. 395–418. Academic Press, New York, 1971.

[7] F. E. L. Ten Haaf, in "Liquid Scintillation Counting" (M. A. Crook, P. Johnson, and B. Scales, eds.), Vol. II, pp. 39–48. Heyden, London, 1972.

TABLE III
METHODS OF QUENCH CORRECTION IN LIQUID SCINTILLATION COUNTING

A. *Internal Standardization ("Spiking")*

Advantages

Most accurate of any method if addition of a known standard to the sample is careful and reproducible. Differences in the effects of impurity and color quenching (which may be important with ^{14}C and more energetic isotopes) are obviated

Disadvantages

The samples must be homogeneous; the standard must be similar chemically to the sample solute if there is any question of an incomplete solution. Handling errors in addition of the standard will be large if the procedure is careless; ideally this procedure should be done in duplicate. Once a sample is "spiked," it cannot be recounted. The procedure is extremely time-consuming. In double-label experiments, it is a laborious and statistically uncertain approach

B. *Sample Channels Ratio*

Advantages

Data may be obtained in two channels simultaneously; thus, counting the sample only once is sufficient. The sample itself is not altered. Accurate for moderate and high count rates of single isotopes which are quenched slightly to moderately. Independent of sample volume over a wide range. One series of quenched standards will correct for both sorts of quenching of ^3H. Nonhomogeneous samples (in suspension, emulsion, on solid supports) may be quench corrected, *as long as standard curves are derived from known samples of the same composition and geometry*

Disadvantages

The procedure is inaccurate for highly quenched samples or for samples with low count rates because strong color quenching of ^{14}C or more energetic isotopes is not adequately corrected. With two overlapping isotopes, corrections for spillover must be made first, with a significant additional statistical error

C. *External Standard Channels Ratios*

This procedure, involving exposure of the unknown sample to an external γ source takes advantage of the secondary Compton electron generated within the sample and the resulting broad fluorescence spectrum. Modern instrumentation monitors and subtracts sample cpm from two separate channels and then ratios the Compton cpm; single channel ES is undesirable inasmuch as the dependence of cpm on sample volume is considerable

Advantages

Application to samples of low activity. Statistically the most suitable for double-isotope quench correction

Disadvantages

Least accurate for highly quenched samples in which differences between the photoelectron spectra of sample and Compton electrons are magnified. Samples must be homogeneous. This technique is not applicable to samples in suspension, on solid supports, etc.

The advantage to this procedure is that H can be counted at maximum efficiency unless there is severe quenching, but a considerable accumulation of statistical error results from the repeated application of calculated efficiencies and from the relatively complicated algebra.

Instead we sacrifice some H efficiency: counting the higher energy isotope in a channel from which L is for all practical purposes excluded. In this case:

$$H = N1/h1 \tag{3}$$
$$L = [N2 - (N1/h2)]/l2 \tag{4}$$

The propagation of error is greatly decreased by such a simplification. In our hands, analysis of the variance occasioned by the use of Eqs. (1) and (2) results in unacceptably large cumulative errors, particularly when count rates for one or both isotopes are low and the overlap is large. There have been several suggestions in the recent literature that simultaneous equations may reliably differentiate two isotopes with considerable spectral overlap. Such advice is dangerous; if simultaneous equations are employed the exclusion of isotope L from channel 1 should still be maximal. Since the effect of quenching on photoelectron spectra tends to this exclusion, such a stipulation is reasonable.

The "spillover fraction" $(N1 \times h2)$ of Eq. (4) which is a constant in gamma counting, will increase in liquid scintillation counting as the spectrum of isotope H is quenched and progressively more H cpm fall in channel 2. In Beckman scintillation counters with "automatic quench corrections (A.Q.C.)" this fraction can be held relatively constant,[8,9] but in others, a quench correction curve for this fraction must be obtained, just as it is obtained for the detection efficiencies of the two isotopes in their optimum channels. Occasional investigators have attempted to use simultaneous equations to solve for the activities of three isotopes in the same sample, using two or more channels. The statistical errors implicit in such procedures are without exception unacceptably large.

It is sometimes possible to separate one overlapping isotope from another, however, by combining different counting methods. ^{131}I may be separated from ^{14}C or ^{35}S, and ^{125}I from ^{3}H, by counting the samples in a gamma counter and then by liquid scintillation alone. It is important that ^{131}I or ^{125}I efficiency be determined in both counters so that an accurate "spillover" curve of either iodine isotope into the ^{14}C, ^{35}S, or ^{3}H channel can be calculated from standards. This approach applies Eqs. (3) and (4) to data from two instruments rather than from one.

[8] C. H. Wang, in "The Current Status of Liquid Scintillation Counting" (E. D. Bransome, Jr., ed.), pp. 305–312. Grune & Stratton, New York, 1970.

[9] M. F. Grower, and E. D. Bransome, Jr., Anal. Biochem. 31, 159 (1969).

Energetic βs may be also counted in solution in the absence of phosphors by virtue of the Cerenkov effect.[10] Of the isotopes in Table I, ^{32}P and ^{131}I can be counted at reasonable efficiencies without any impurity quenching, but Cerenkov photoelectron spectra are so poorly resolved that double isotope counting is not practical.

Instrument Settings. The method of selecting discriminator settings for the single-channel counting of a single isotope is straightforward. Since background cpm increase with channel width, the upper potentiometer setting should be decreased to the point where the count rate of a sample of the isotope begins to become significantly diminished. The lower discriminator should be set close to 0 and raised only if there is a major contribution of instrument "noise" to background in the low energy range, a problem not significant in instruments of recent manufacture.

In double-label counting there is an obvious conflict between the desires for minimum spillover and maximum efficiency. A general approach to the most effective compromise in any particular situation would obviously be of value. Klein and Eisler[11] have proposed a performance criterion, P. In the terms of Eqs. (1)–(4)

$$P = h1 \times l2 \times S \qquad (5)$$

with $h1$ and $l2$ the detection efficiencies of the two isotopes in their respective channels and S a function of the count ratios between the two channels for each isotope. Comparing a series of settings, however, may be difficult since, as is usually the case, one setting may be superior regarding only one or two of the three factors. Davies and Deterding[12] have introduced a less biased but more complex algebraic approach to find optimum settings, using a weighted sum of the activities of the two isotopes more or less independent of the contribution of background cpm, which I consider a bit too complex to review here.

Graphical approaches which are perhaps a bit easier to comprehend are represented in Figs. 1 and 2.[13]

The procedure for choosing settings for channels for double-isotope counting in an LS counter with logarithmic or pseudologarithmic amplification is illustrated in Fig. 1. The most efficient channel for each isotope is determined separately as outlined above, and the effect of progressively lowering the upper discriminator in 20-division steps is plotted as the

---

[10] R. P. Parker, and R. H. Elrick, *in* "The Current Status of Scintillation Counting" (E. D. Bransome, Jr., ed.), pp. 110–122. Grune & Statton, New York, 1970.
[11] P. D. Klein and W. J. Eisler, *Anal. Chem.* **38**, 1453 (1968).
[12] P. T. Davies and J. H. Deterding, *Int. J. Appl. Rad. Isotopes* **23**, 293 (1972).
[13] E. D. Bransome, Jr. and S. E. Sharpe, III, *Anal. Biochem.* **49**, 343 (1972).

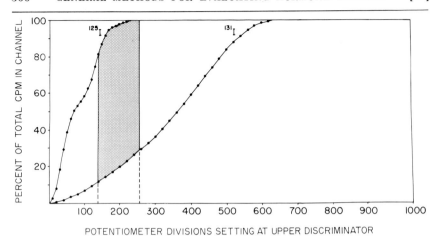

Fig. 1. Method for selecting channels for double-isotope counting in counting systems with logarithmic or pseudologarithmic amplification. The cpm in cumulative 2% (20 potentiometer divisions) narrow channels are expressed as percentage of total cpm in the most efficient wide channel. The lowest discriminator setting reaching 100% on the ordinate is the upper end-point of the isotope energy spectrum in the specific counter being examined. From E. D. Bransome, Jr. and S. E. Sharpe, III, *Anal. Biochem.* **49**, 343 (1972).

percent of total cpm in the full channel. The H channel chosen should be at the L end point so that there is no spillover of L into the channel used for H. With our Beckman LS-150 (Fig. 1), we found the ^{131}I channel to be between 260 and 660 discriminator divisions. The upper discriminator setting for the L channel can be decided by identifying the region of maximum difference between the higher and lower energy isotope plots. The shaded area in Fig. 1 indicates the area of proper choice. We chose the 0–260 channel for maximum ^{125}I. A 0–140 channel would have been a better selection if the ^{131}I activity tended to be considerably in excess of that of ^{125}I, or if the unknown samples to be counted were considerably quenched so that the ^{125}I photoelectron energy spectrum was significantly shifted to the lower end.

For "linear" LS counters, the additional requirement of gain adjustment is the most important consideration involved in selecting channels for single- or double-isotope counting. It is not possible to comprehend the complete spectra of more than one isotope by varying the discriminator settings, unless the gain settings are significantly different. The effect of grain on counting efficiency in a wide channel should therefore be systematically examined. To minimize the effect of quenching, the maximum gain for a specific level of efficiency should be chosen so that as much as possible of the spectrum fills the window.

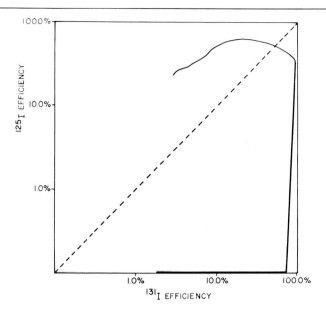

Fig. 2. Engberg plot of ^{125}I and ^{131}I in a Packard Tri-Carb 3375 LS counter. The efficiencies were determined in a fixed channel at various gain settings and plotted against each other (solid line). The best choices of isotope ratio can be identified by determining the greatest distance of the plotted curve from a 45° line drawn from the base (the dashed line). The best ^{131}I channel is obvious: the gain with maximum efficiency and no ^{125}I spillover. For ^{125}I, the best solution is a compromise between low ^{131}I spillover into the ^{125}I channel, and high ^{125}I efficiency. Maximal ^{125}I efficiency is accompanied by more than 20% spillover. If there is a substantial excess of ^{131}I dpm over ^{125}I dpm, a lower ^{125}I efficiency (at higher gain) should be offset by a further decrease in ^{131}I spillover. From E. D. Bransome, Jr. and S. E. Sharpe, III, *Anal. Biochem.* **49**, 343 (1972).

If the absolute radioactivity of the isotope(s) being counted in a sample is not known, a comparison of cpm at each gain is sufficient, but there is an advantage to knowing counting efficiencies. Kobayashi and Maudsley[14] have shown that log-log plots ("Engberg plots") of isotope efficiencies in a fixed window can be used to determine the ability of a linear LS counter to separate an isotope pair. Figure 2 shows our results counting ^{131}I and ^{125}I with a Packard 3375. Because the effect of quenching on the lower energy isotope increases the isotope separation, using only least-quenched isotope standards to set up counting channels is a sufficiently rigorous procedure.

[14] Y. Kobayashi and D. V. Maudsley, *in* "The Current Status of Liquid Scintillation Counting" (E. D. Bransome, Jr., ed.), pp. 76–85. Grune & Stratton, New York, 1970.

Sample Preparation. Most of the important considerations of sample preparation for liquid scintillation counting can be conveniently reviewed in references cited in footnotes 15 and 16. It is worth reiterating that for calculations of the radioactivity of an isotope pair to be reliable, both isotopes must be uniformly distributed in the sample. Translucent emulsions with microscopic micelles may be counted as well as true solutions if three cautions are observed: (a) If one of the isotope pair is ^{3}H or ^{125}I, it must be in the same phase as the other isotope. This might seem to be a trivial problem since emulsion counting is of aqueous samples, but it is not; there may be differential extraction of the labeled solutes into the organic solvent. (b) Absorption (of the lower energy isotope in particular) onto the vial surface may greatly alter detection efficiency.[17] (c) If the E_{max} of an isotope is equivalent to or greater than ^{14}C, and a commercial solubilizer of aqueous samples has been included in the sample, the surfactant itself may fluoresce in response to radioactivity. This renders quench correction curves derived from sealed commercial standards invalid.[18] All these proscriptions may be summarized in the cardinal, and frequently neglected, general rule for scintillation counting: standards should be of the same geometry as the unknown samples.

[15] J. C. Turner, "Sample Preparation for Liquid Scintillation Counting," Review 6. The Radiochemical Centre, Amersham, England 1967. Reprinted in "Handbook of Radioactive Nuclides" (Y. Wang, ed.) pp. 256–273. Chem. Rubber Publ. Co., Cleveland, Ohio, 1969.
[16] "The Current Status of Liquid Scintillation Counting" (E. D. Bransome, ed.), Chapters 16–27. Grune & Stratton, New York, 1970.
[17] G. J. Litt and H. Carter, *in* "The Current Status of Liquid Scintillation Counting" (E. D. Bransome, ed.), pp. 156–163. Grune & Stratton, New York, 1970.
[18] S. E. Sharpe, III and E. D. Bransome, Jr., *Anal. Biochem.* **56**, 313 (1973).

[22] Use of Antibodies to Nucleosides and Nucleotides in Studies of Nucleic Acids in Cells

By B. F. ERLANGER, W. J. KLEIN, JR., V. G. DEV, R. R. SCHRECK, and O. J. MILLER

The preparation of purine– and pyrimidine–protein conjugates and their use in eliciting base-specific antibodies that react with nucleic acids have been described in this series, Volume 12B [173]. Their reaction with nucleic acids requires that the purine or pyrimidine bases be unpaired, i.e., that the nucleic acid be denatured or, at least, have single-stranded regions. It follows, therefore, that demonstrable reaction with these antibodies is evidence for "single-stranded areas." Fluorescein-tagged antibodies have been used to study cell nuclei[1,2] and metaphase chromosomes.[2-4] With respect to the former, it was found that single strandedness could be demonstrated only during S phase of the cell cycle.[1,2] In studies with metaphase chromosomes, local areas, rich in A-T and G-C base pairs, have been localized.

Nuclear Fluorescence of Mouse L Cells

Fluorescein Conjugates. Fluorescent sheep anti-rabbit globulin and fluorescent rabbit anti-sheep globulin were obtained from Pentex Co., Kankakee, Illinois and from Nutritional Biochemicals Corp., Cleveland, Ohio. Using antinucleoside or anti-BSA antisera, globulin preparation and purification of the fluoresceinated conjugates were carried out as described by Dedmon, Holmes, and Deinhardt.[5] Crystalline fluorescein isothiocyanate (Sylvana Chemical Co., Orange, New Jersey) was conjugated to protein following the procedure of Hsu (K. Hsu, personal communication). After each step in the preparation and purification of conjugates, the antisera were checked for the presence of antibody by the gel diffusion method.

Tissue Culture. Mouse L cells (derived from strain 929, Earle) were grown in 250-ml plastic tissue culture flasks (Falcon Plastic Co., Los Angeles, California) at 37° for 4–7 days. The sheet was detached and dispersed in 0.25% trypsin in phosphate-buffered saline (PBS) with sodium ethylenediamine tetraacetate and glucose.[6]

The cells were collected by centrifugation and suspended in culture medium. (Eagle's minimum essential medium[7] with 10% fetal calf serum, heat treated at 56° for 30 minutes to inactivate possible mycoplasma contaminants; penicillin G, 15 units/ml; streptomycin, 15 μg/ml; double the concentration of glutamine, cofactors, and vitamins; and nonessential amino acids at 0.1 mM: L-alanine, L-asparagine·H_2O, L-aspartic acid, L-glutamic acid, L-proline, L-serine, and glycine. All medium constituents were obtained from Grand Island Biological Co., Grand Island, New York).

[1] W. J. Klein, Jr., S. M. Beiser, and B. F. Erlanger, *J. Exp. Med.* **125**, 61 (1967).
[2] M. V. R. Freeman, S. M. Beiser, B. F. Erlanger, and O. J. Miller, *Exp. Cell Res.* **69**, 345 (1971).
[3] V. G. Dev, D. Warburton, O. J. Miller, D. A. Miller, B. F. Erlanger, and S. M. Beiser, *Exp. Cell Res.* **74**, 288 (1972).
[4] R. R. Schreck, D. Warburton, O. J. Miller, S. M. Beiser, and B. F. Erlanger, *Proc. Nat. Acad. Sci. U.S.* **70**, 804 (1973).
[5] R. E. Dedmon, A. W. Holmes, and F. Deinhardt, *J. Bacteriol.* **89**, 734 (1965).
[6] J. Paul, "Cell and Tissue Culture," 3rd ed. Williams & Wilkins, Baltimore, Maryland, 1965.
[7] H. Eagle, *Science* **130**, 432 (1959).

Plastic flasks with 25 ml of culture medium were inoculated with 4×10^6 cells (approximately) or Leighton tubes containing cover slips were inoculated with 1.3 to 1.5×10^5 cells in 1 ml of medium. The coverslips were removed after various times of incubation and, after three washes of 5 minutes each in phosphate-buffered saline (PBS) (0.14 M NaCl, 10 mM PO$_4$ pH 7.3), were fixed by one of the following methods: (a) air-drying 1–2 hours at 37°; (b) air-drying 1–2 hours at 37°, storing at 4° for 1–21 days, and dipping in methanol at room temperature immediately prior to staining; and (c) air-drying 1–2 hours at 37°, placing at 4° for 1–21 days followed by treatment with 95% ethanol for 30 seconds at room temperature just prior to staining. Method (c) gave the most consistent results.

Fluorescent Antibody Staining. Antisera were diluted in PBS to 2–4 mg/ml total protein as measured by a hand refractometer. The coverslips were placed on a small staining rack in a humidity chamber, and 5 drops of antiserum were added onto the surface of each coverslip. The antiserum was allowed to react for 20 or 30 minutes at room temperature. Then the coverslips were drained and rinsed with a forced stream of PBS. This was followed by 2 washes in PBS for 10 and 5 minutes duration and another vigorous rinsing with PBS. The coverslips were then mounted in glycerol (1 part) : PBS (4 parts) pH 7.5 on standard microscope slides (less than 1 mm thick) and immediately examined under the UV microscope.

In "blocking" experiments, unfluoresceinated globulin was allowed to react with the cells as above for 1 hour (1 replacement with fresh globulin at 0.5 hr) at 37°. The coverslips were washed as above and then stained with the fluoresceinated antiserum.

Microscopy and Photography. A Zeiss Standard Universal Microscope fitted with an HBO 200 Osram mercury burner and a 1.2/1.4 Z dark-field condenser was employed for UV microscopy. The exciter filters were BG 12, 4 mm, and BG 39, 2.5 mm. Barrier filters were Zeiss 47 or 50. The lens system consisted of a 40 $\times$ oil immersion objective, a 10 $\times$ eyepiece (5 $\times$ eyepiece to camera) and a 2 $\times$ Optovar. Photographs were taken with a Zeiss Ikon 35mm camera (factor 0.5) using Anscochrome 200 ASA color daylight or Agfa Isopan Record (black and white) film.

Reaction of Anti-nucleoside Antibodies with Human Metaphase Chromosomes

The anti-nucleoside antibodies have been shown to bind to fixed human metaphase chromosomes only after treatment with denaturing agents such as aqueous solutions of NaOH.[2] Such treatment can produce swollen and distorted chromosomes which obscure the pattern of antibody binding to chromosomes. We therefore investigated the possibility of replacing NaOH with a saline solution of formamide, which reduces the thermal stability of DNA.[8] This method, described below, enabled the antinucleoside antibodies to bind to human chromosomes with minimal distortion.[3] We then investigated other methods for generating single-stranded DNA in chromosomes. Methylene blue-mediated photooxidation selectively destroys guanine residues in DNA in solution,[9] freeing the formerly hydrogen-bonded cytosines. We have used this procedure to attach anti-cytosine antibodies to fixed metaphase chromosomes.[4]

Preparation of Metaphase Chromosomes. Chromosome preparations were obtained from cell cultures set up from peripheral blood samples. Ten milliliters of venous blood was collected in a syringe containing 0.5 ml of an isotonic aqueous solution of sodium heparin (Organon) to prevent clotting, and allowed to sit undisturbed at room temperature until the red blood cells have settled. About 1 ml of the resulting suspension of leukocytes in plasma was transferred to sterile tissue culture flasks containing 10 ml of complete Eagle's minimal essential media (MEM) (plus penicillin–streptomycin, L-glutamine and 20% fetal calf serum) and 0.5 ml of phytohemagglutinin "M" (General Biochemicals). Cultures were grown at 37° for 3 days. After this incubation, 0.2 ml of Colcemid (10 µg/ml) in Hanks' balanced salt solution (Grand Island Biological Co.) was added to each flask, which was then reincubated at 37° for 1 hour. The culture was centrifuged (800 rpm for 8 minutes), and the pellet was suspended in hypotonic solution (0.075 M KCl). After 2 minutes, the suspension was centrifuged (800 rpm, for 8 min) and the pellet resuspended in fixative (3:1 methanol–glacial acetic acid). After three changes of fixative, slides were prepared. However, for photooxidation it proved advantageous to store the cells in fixative for several days before making slides. Slides were made by suspending the cells in fresh fixative and dropping the suspension from a Pasteur pipette onto cold, wet slides which were then air-dried. The slides were stored in the refrigerator until needed.

Chromosome preparations were also obtained from other types of cultured cells, growing either as suspension cultures or monolayers. The procedure is basically the same as for cultured leukocytes, except that cells growing attached to the glass or plastic culture vessel were first separated from the surface. Some dividing cells could be shaken off the surface of the vessel rather easily, and we prefer this method when it works. In

[8] B. L. McConaughy, C. D. Laird, and B. J. McCarthy, *Biochemisty* **8**, 3289 (1969).
[9] A. G. Garro, B. F. Erlanger, and S. M. Beiser, *in* "Nucleic Acids in Immunology" (O. J. Plescia and W. Braun eds.), pp. 47–57, Springer-Verlag, Berlin and New York, 1968.

other cases, the culture medium was removed, a trypsin–EDTA mixture (0.5 g of trypsin, 1:250, and 0.2 g of EDTA per liter of Puck's saline A which consists of 8000 mg of NaCl, 400 mg of KCl, 1000 mg of glucose, 350 mg of $NaHCO_3$, and 5 mg of phenol red per liter) was added and the cells were incubated at room temperature for a few minutes or until the cells could be removed by vigorous shaking of the vessel. The mixture was centrifuged as above, the cells were suspended in hypotonic solution (in this case 38 mM KCl). The remainder of the procedure is as already described.

Method of Denaturation

1. Formamide

Reagents

Formamide (Stabilized, Fisher Scientific Co.) pH adjusted to 7.2 with concentrated HCl
20 × SSC: 3 M sodium chloride and 0.3 M trisodium citrate
95% Formamide: 95 ml of pH 7.2 formamide + 5 ml of 20 × SSC
Ethanol, 70%, 95%, absolute
Phosphate-buffered saline (PBS): 20 g of NaCl, 85 ml of 0.25 M $NaHPO_4$, 15 ml of 0.25 M KH_2PO_4 in 2400 ml of distilled water pH 7.2–7.4

Coplin jars designed to hold micoscope slides were filled with 95% formamide in SSC and placed in a water bath maintained at 65°. The slides were heated for 1 hour in formamide. They were then rinsed twice in 70% ethanol, once in 95% ethanol and once in absolute ethanol. The slides were rehydrated for about 5 minutes in PBS before being treated with antibody.

2. Photooxidation

Reagents

Tris·HCl buffer, 0.1 M: dissolve 6 g of THAM (Fisher) in 500 ml of distilled water and adjust pH to 8.75 with concentrated HCl
Stock methylene blue (National Aniline): dissolve 0.0125 g of powdered methylene blue in 10 ml of Tris·HCl buffer
Methylene blue, 33.4 μM: dilute 0.5 ml of stock solution in 49.5 ml of cold Tris buffer

Slides were photooxidized in a Coplin jar containing a cold saturated solution of 33.4 μM methylene blue. Oxygen was bubbled into the jar, which contained the dye solution and the slides, for 10 minutes through a Pasteur pipette. As oxygen is critical to the reaction, it is imperative to seal the jar tightly after the addition of oxygen. The sealed jar was placed in a glass water bath (25°). The jar was illuminated overnight (15–18 hours) through the water bath by a 150 W Sylvania flood lamp, which was 15 cm from the jar. As the reaction progressed, the dye solution became paler and was almost colorless after 18 hours. The final temperature within the Coplin jar was usually about 1° higher than that of the water bath. The slides were rinsed briefly in PBS before antibody treatment.

Indirect Immunofluorescence. Slides treated to produce denaturation were first layered with rabbit anti-nucleoside antisera diluted 1:10 with PBS and incubated at room temperature in a moist chamber for 45 minutes. Unbound antibody was rinsed off the slides with 200 ml of PBS from a spray bottle. Indirect immunofluorescence was accomplished by incubating the slides as before with a 1:50 dilution of sheep anti-rabbit IgG tagged with fluorescein. After a second washing, the slides (cell surface) were wetted with PBS, a clean coverslip was mounted, the excess buffer blotted off and the edges sealed with clear nail polish.

Microscopy and Photography. The slides were examined with a Zeiss fluorescent microscope using light from an HBO 200 W mercury lamp, transmitted through a cardioid condenser, with a BG 12 (4 mm) exciter filter, a 530 nm barrier filter and a $100 \times$ Planapochromatic objective. Well spread metaphases were photographed on either Panatomic X or on H & W control film with 2-minute exposures.

Panatomic X film was developed with Microdol X (Kodak), and H & W film was processed with H & W control developer (H & W Co., St. Johnsbury, Vermont). The negatives were printed on Ilford paper, Grade 4.

[23] Techniques for the Study of Hormone Effect on Collagens

By DOROTHY H. HENNEMAN and GEORGE NICHOLS, JR.

I. General Precautions. 309
II. Initial Preparation of Tissues 310
   A. General 310
   B. Specific Tissues 311
   C. Defatting, Drying, and Demineralization 312
   D. Methods to Prepare Suspensions of Isolated Cells 312

III. Special Experimental Procedures 319
   A. Hormones: Preparation, Dosages, Route of Administration . . . 319
   B. Lathyrogens, Nutritional Deficiencies, Proline Analogs, Microtubular Disruptive Drugs 321
   C. Labeling of Collagen *in Vivo* 322
IV. Metabolic Studies 324
   A. Soft Tissues 324
   B. Bone 326
V. Composition of Collagen 335
   A. Purification of Collagen. 335
   B. Purification of Collagen by Digestion with Highly Purified Collagenase 340
   C. Enzymes 343
   D. Method for the Measurement of Total [^{14}C]Hydroxylysine and Glycosylated and Nonglycosylated [^{14}C]Hydroxylysine 353
   E. Chromatographic Separation of the Subunits of Several Collagens . 358
   F. Amino Acid Analyses 359
   G. Preparation of Human Tissue Obtained by Biopsy for Amino Acid Analysis 360
   H. Collagen Cross-Linking 360
VI. Measurement of DNA Content 365
VII. Measurement of Proline and Hydroxyproline and Their ^{14}C-Labeled Derivatives 366
   A. Proline 366
   B. Hydroxyproline: Stegemann Procedure 367
   C. Hydroxyproline: Prockop and Udenfriend Procedure 369
   D. [^{14}C]Hydroxyproline: Juva and Prockop Procedure without Modification 371
   E. Paper Chromatographic Separation of [^{14}C]Proline and [^{14}C]Hydroxyproline 374
VIII. Measurement of Protein with the Folin Phenol Reagent 375
   A. Original Lowry Method. 375
   B. Hartree Modification 376

Recent advances in the understanding of collagen, its synthesis, transport, maturation, degradation, and structure make studies of collagen metabolism rewarding not only in animals under experimental conditions, but also in man under the pathophysiological conditions of clinical disease. Reliable techniques are available for the study of the complex and unique biosynthetic and degradative processes involved as well as for analyzing the resultant structure of the molecule. Still others make possible the translation of observations at the cellular and molecular level into measurements of the biological properties and function of the several types of collagen present in different connective tissues—features known to be affected by phenotypic expression, stage of embryonic development, age, hormones and vitamins, trace elements and nutritional conditions, chemicals and drugs, and by disease, both genetic and acquired. Thus, basic structure can be correlated both with metabolism and function, not only in terms that the biochemist understands but which the physiologist and clinician can make useful.

It is not the purpose of this outline to describe in detail all the known techniques for such studies, nor to provide a review of collagen metabolism since several excellent ones are available.[1,2] Rather, it is designed to indicate the wide variety of avenues by which the investigator may approach the study of hormonal action on connective tissues. The procedures with which the authors have had experience are given in detail. Others, recognized to be useful and reliable, but with which we have had no experience, are discussed in a more general fashion.

The material is organized as follows: First, the general problems and procedures concerned with the choice of animal, tissue, or experimental protocol are outlined. This is followed by a brief description of the steps concerned with the excision, preparation, and aliquoting of the tissue with a brief discussion of specific procedures related to particular tissues. Methods for defatting, demineralizing, and drying tissues are described, followed by methods for isolation of connective tissue cells.

The rest of the material is divided into two major sections: The first describes the procedures available for studying active metabolic processes by long- or short-term incubation of organs, tissue fragments, or cells; the second describes procedures for studying the composition and structure of the several collagens known, including techniques for the assay of specific enzyme activities. Areas of overlap are indicated. Analytical procedures not described in the metabolic or compositional sections are described at the end.

I. General Precautions

Total body weight and food consumption must be monitored during any long-term study of any variable. Loss of weight is associated with a generalized decrease in synthesis and increase in degradation of all proteins including collagen.

Age and the stage of embryonic development must be standardized also. An excellent example of the effect of age is the decreased rate of collagen biosynthesis evident in all connective tissues. This is associated with decreased activities of glycosyltransferases and peptidyllysine and proline hydroxylases.[3,4] These changes often make measurement of the

---

[1] M. E. Grant and D. J. Prockop, *N. Engl. J. Med.* **286**, 194, 242, and 291 (1972).
[2] P. M. Gallop, O. O. Blumenfeld, and S. Seifter, *Annu. Rev. Biochem.* **41**, 617 (1972).
[3] R. G. Spiro and M. J. Spiro, *J. Biol. Chem.* **246**, 4919 (1971).
[4] J. Halme, *Biochim. Biophys. Acta* **192**, 90 (1969).

glycosylating and hydroxylating enzymes in small aliquots of tissues from older animals impossible.

In primates, including man, the rates of collagen biosynthesis and breakdown, the degree of solubility, and the turnover of collagen are all considerably less than these processes in lower mammalian species[5] at any age. Accordingly, tissue culture may be necessary to study certain aspects of metabolism in human tissue if the sample size is small. A number of additional variables must be kept in mind regardless of the tissue and animal source chosen. These include the amount and state of the collagen in the tissue, the degree to which it is soluble or insoluble, and whether or not the insoluble is mineralized. The rates of synthesis and degradation vary from tissue to tissue, from species to species, and within the same type of tissue (e.g., variations from bone to bone). The number and types of cells present also vary. Age, hormones, and nutrition all modify the sizes of these collagen pools and their rates of turnover which, in turn may vary independently. Neglect of any one of these aspects can lead to incomplete information and mistaken emphasis. This is particularly true when data derived from embryonic collagen is applied to collagen in the intact adult organism and tissue. Similarly, data derived from animals pretreated with a lathyrogenic agent (e.g., BAPN) in order to increase solubility and facilitate collagen purification may reflect the introduction of unknown variables, such as changes in the chemistry of the extracellular fluid or in the phenotypic expression of the fibroblast.[6]

II. Initial Preparation of Tissues

A. General

Tissues, reagents, and solvents should be kept chilled to reduce tissue metabolism, and the time between tissue collection and use kept minimal to avoid artifacts. Prime among these are inaccuracies in the wet weight. Whole organs as well as tissue fragments should be used to avoid sampling errors and aberrations due to cell damage. Larger, rather than smaller, aliquots are preferred. Blood and dead or unwanted adherent cells (marrow in bone, for example) should be removed by repeated shaking of tissues with buffered physiologic saline (0.15 M NaCl in 20 mM phosphate, pH 7.4) or Krebs-Ringer buffer.

Frequently, as in the measurement of metabolic phenomena *in vitro*, it is important to maximize the surface to volume ratio of the tissue sample to assure maximal exchange of substrates. Most connective tissues

[5] D. H. Henneman, *J. Med. Primatol.* **1**, 58 (1972).
[6] B. K. Hall, *Calcif. Tissue Res.* **8**, 276 (1972).

can be minced with scissors or by two scalpels cutting in opposite directions. Bone (especially adult bone) is more difficult but usually can be cut with sharp knives or orthopedic rongeurs.

Weighing, especially of fresh samples for metabolic studies, must be done as quickly as possible. A rapid-weighing torsion balance (capacity 500 mg) is useful for small samples. Rapid blotting on filter paper removes excess fluid before weighing, but tissues should be promptly reimmersed in chilled buffer (or medium) to avoid damage due to drying.

B. Specific Tissues

1. Skin

Skin should be shaved before excision. Subcutaneous fat is removed by scraping or cutting, care being taken to keep tissue moist. Special preparative steps for measurement of tensile strength are described in Section V,H,4.

2. Bone and Cartilage

The several anatomical divisions of bone and cartilage should be carefully identified and verified histologically when possible (e.g., epiphysis, metaphysis, diaphysis). In man this is not difficult. In rats and chicks failure to close the epiphyses of the long bones makes mechanical separation of cartilage and epiphysis by blunt dissection easy. In adult birds and some animals the separation is more difficult but still possible in most cases by careful sharp dissection.

3. Uterus, Prostate, and Ovary

These are best used as entire organs in most experiments. The fibrous capsule of the prostate and ovary and the cervix and oviducts of the uterus should be removed.

4. Aorta

Aorta should be stripped of surrounding supporting loose connective tissue and removed intact from the arch to the iliac bifurcation. If sections are to be studied (e.g., thoracic vs. abdominal), they should be carefully identified. All blood should be removed. If tensile strength is measured, it should be determined on the entire tube while still intact (i.e., not cut longitudinally).

C. Defatting, Drying, and Demineralization

1. Soft Tissues

A dry fat-free weight is useful as a reference base when artifactual changes in fat or water content are possible. Skin, prostate, aorta, uterus, etc., are defatted by shaking with acetone ($2\times$ 1 hour) at 4° followed by shaking in ether in the cold for 24–36 hours as needed. The fragments are dried *in vacuo* at 110° for 48 hours in tared weighing bottles before weighing.

2. Bone

Bone fragments are defatted and dried with or without prior demineralization by shaking with chloroform:methanol (2:1, v/v) followed by drying as above. Demineralization should be avoided if total hexosamine or mucopolysaccharide measurements are desired. If necessary the fragments are shaken in the cold with 10% EDTA, adjusted to pH 7.4, for 3–4 days with daily changes. When soft and easily cut with scissors, the fragments are rinsed several times with distilled water, then with acetone, before extraction with ether:absolute alcohol (1:1, v/v) for 1–3 days to remove additional lipids before drying (see below).

The procedure described here extracts fat and water together. This can also be done separately to study changes in fat or water content. Separate aliquots of tissue are analyzed for this, or defatting is carried out before drying with petroleum ether or chloroform (neither significantly miscible with water).

Fat, water, dry fat-free organic material, and mineral content are expressed as percent of the original wet weight of tissue. Other bases of reference may be used for special purposes.

D. Methods to Prepare Suspensions of Isolated Cells

The relative paucity of cells in most adult connective tissues poses a number of analytic problems to the investigator studying the cellular mechanisms involved in the metabolism of the extracellular components which make up the bulk of the tissue. Many of these can be overcome by studying isolated cells. The ideal isolation method should disperse into suspension all the cells in the sample, intact, functional, and entirely free of extracellular components. None of the methods available meet all these criteria completely, but several, described below, approach one or another well enough to be useful tools for specific studies as long as their limitations are recognized.

1. General Principles

The cells in connective tissues tend to be buried deeply in the stroma which they secrete. Therefore, to release them the tissue must be disrupted first with proteolytic enzymes (usually collagenase) mechanical procedures, or some combination of both. The cell is freed by tearing it free of the matrix mechanically or by further digestion of the sticky materials which cause the cell to adhere to its surroundings. The problems encountered center around the difficulty of disrupting the matrix completely enough to release the cells while avoiding damage to their delicate membranes and processes.

Methods for isolating cells from tendon, cartilage, lens, and bone are described below.

2. Tendon

Dehm and Prockop,[7] with trypsin and collagenase, have prepared cells from leg tendons of chick embryos which are free of extracellular matrix and which retain the ability to synthesize collagen at a rapid rate for several hours. The yield of cells appears to approximate the total present initially in the tissue, and 84–92% are viable after 2 hours at 37°.

MATERIALS

17-Day-old chick embryos incubated in a moist atmosphere at 37°
Bacterial collagenase, Type I (Sigma Chemical Corp.)
Trypsin solution: 2.5% in physiological saline (Gibco Corp.)
Eagle's minimum essential medium with glutamine
Trypan blue: 0.4% in Hanks' balance salt medium
Soybean trypsin inhibitor I (Miles Laboratories)
Modified Krebs Medium II
Small Swinnex filters (Millipore Corp.)

PROCEDURE. The legs are removed from chick embryos (10–15) and placed in modified Krebs medium in a petri dish. The tendons are isolated under a dissecting microscope by pulling on the toes. Small amounts of adherent nontendonous tissue are readily washed away before the tendons are placed in 3.0 ml of Eagle's minimum essential medium containing 0.33 ml of trypsin solution and 3.3 mg of collagenase. The system is incubated at 37° for 40 minutes with shaking (80–100 oscillations per minute) in an atmosphere of 95% O_2–5% CO_2 to disrupt the tissue. The samples are filtered through lens paper in a Swinnex filter, and the cells are pelleted by centrifugation at 600 g at room temperature for 3 minutes.

The supernatant is discarded, and the pellet of cells is gently resus-

[7] P. Dehm and D. J. Prockop, *Biochim. Biophys. Acta* **240**, 358 (1971).

pended (a Pasteur pipette with a wide mouth is good for the purpose) in 5 ml of 0.25% soybean trypsin inhibitor dissolved in Krebs medium. The cells are recovered again by centrifugation as above, and the washing with inhibitor is repeated 2 further times. Finally the cells are resuspended in 3 ml of Krebs medium for counting in a hemacytometer. Eight to 12×10^6 cells in a final volume of 2.5 ml of medium is a convenient aliquot of these cells for many studies of collagen synthesis.

Cells prepared in this manner are sufficiently active and free of extracellular collagen to permit the measurement of net synthesis of new collagen and the accumulation of specific molecular forms in sufficient quantity for identification. The metabolic activity and simplicity of the system suggest it to be suited for several types of study.

3. Lens

The lens of the chick embryo eye offers the special advantage of providing cells that synthesize the type of collagen characteristic of basement membranes. By adapting the technique described above, Grant et al.[8] have obtained approximately 75% of the cells present in the tissue sample, 90% of which are viable.

MATERIALS. As for tendon except
 Eagle's medium without glutamine
 Fetal calf serum

PROCEDURE. Identical to that for tendon cells except as follows: (1) the lenses from 2–3 dozen chick embryos (19 to 20 days old) are dissected free of adherent tissue and punctured by gentle squeezing with dissecting forceps before placement in the medium containing enzymes; (2) incubation is continued for 60 minutes; (3) cells are rinsed after harvesting in medium containing 10% fetal calf serum and no trypsin inhibitor.

Fitton-Jackson[9] has used a modification of her method for cartilage and bone to isolate cells from cornea and lens. Her method and the one described here differ only in detail.

4. Cartilage

A number of workers have isolated intact functioning cells from cartilage with a variety of proteolytic enzymes. Collagenase has also been used. Two methods, both of which give satisfactory yields of single cells in uniform suspension, which seem to maintain differentiated functions such as growth and matrix biosynthesis (in one instance through several generations), are described. Others can be found in the literature.

[8] M. E. Grant, N. A. Kefalides, and D. J. Prockop, *J. Biol. Chem.* **247**, 3539 (1972).
[9] S. Fitton-Jackson, personal communication, 1972.

a. Horwitz and Dorfman Method[10,11]

MATERIALS

Tibial and femoral epiphyses from 13-day-old chick embryos
Crystalline trypsin, 1%, in Ca-free, 0.15 M NaCl buffered with 10 mM phosphate to pH 7.4
Loose-fitting glass homogenizer
F-12 nutrient medium (Gibco Corp.)
Fetal calf serum

PROCEDURE. Cartilaginous ends of leg bones of 13-day-old chick embryos are dissected under sterile conditions and incubated for 45–60 minutes at 37° in buffered isotonic Ca-free saline (pH 7.4) containing 1% trypsin. The tissue is then disrupted gently with a loose glass homogenizer (or some equally gentle means) and large particles are removed by centrifugation at 500 g for 1 minute. The cells in the supernatant are harvested by centrifugation at 1000 g for 10 minutes and washed twice in F-12 medium containing 10% fetal calf serum to remove all traces of trypsin. After resuspension, any further clumps are removed by centrifugation at 100 g for 30 seconds to yield a uniform single-cell suspension in the supernatant.

b. Fitton-Jackson Method[12]

MATERIALS

Tibial and femoral epiphyses from 9- to 14-day-old chick embryos or minced articular cartilage from young pigs or other mammals
Crystalline Trypsin (Sigma Corp. or Armour Corp.)
Bacterial collagenase (Sigma, Type I)
Ca-Mg-free Tyrode's solution
Tyrode's solution (with Ca and Mg present)
Swinnex filters
Nutrient medium—BGJ, F_{12}, or other (Gibco Corp.)

PROCEDURE. Femoral and tibial epiphyses 30–40 embryos or slices of fresh articular cartilage are harvested (under sterile conditions for culture) into Tyrode's solution in glass concave slides or petri dishes and carefully cleaned of adherent tendon and loose connective tissue by scraping and washing with Ca–Mg-free Tyrode's solution at room temperature. The tissue is minced with scalpels in 0.3% trypsin dissolved in 20 ml of Ca–Mg-free Tyrode's 1–2 ml being added to each concave slide. All the tissue is combined in a 150–200-ml Erlenmyer flask with the rest of

---

[10] A. L. Horwitz and A. Dorfman, *J. Cell. Biol.* **45**, 434 (1970).
[11] Z. Nero, A. L. Horwitz, and A. Dorfman, *Develop. Biol.* **28**, 219 (1972).
[12] S. Fitton-Jackson, personal communication, 1972.

the trypsin solution and incubated at 37° with shaking at 75–100 oscillations per minute. Incubation is continued until the tissue fragments begin to stick together in a sticky rope (20–90 minutes depending on the age of the embryo and size of the pieces). Ten milliliters of Tyrode's solution (or other Ca-containing nutrient medium) containing enough collagenase to bring the final concentration in the 30 ml to 0.06% (i.e., 0.18%) is added to the flask and the incubation is continued until the tissue fragments have largely broken up (1–3 hours). The process is accelerated by sucking the fragments up and down repeatedly in Pasteur pipettes of progressively smaller mouth diameter. The solution is filtered through a Swinnex filter, and the cells are collected by centrifugation at 500 g for 2 minutes. The cells are washed 3 times by suspension in 1–3 ml of Tyrode's solution for the first wash and nutrient medium for the second and third washes. The final cell pellet gives a uniform suspension of single cells which are 90% or better viable and capable of synthesizing all components of the cartilage matrix.

5. Bone

Two approaches have been used to mobilize cells from bone: enzymatic digestion and mechanical disruption. Both suffer to a greater or lesser degree, because of the nature of the tissue, from low cell yield (with variable proportions of different cell types as a result) and cell damage with decreased cell survival in culture. Despite these limitations cells useful for a variety of studies can be obtained by proper selection of method, bone, and animal age.

a. Enzymatic Digestion

The majority of workers have used this approach. Two methods are described. Others have been published but differ only in detail.

i. Peck Method[13]

MATERIALS

Frontal and parietal bones from 17- to 21-day-old rat fetuses

NaCl solution, 0.15 M, buffered with Tris 10 mM, pH 7.4, containing glucose (5 μmoles/ml) and penicillin and streptomycin 50 units/ml each

Crude bacterial collagenase (Sigma, type 1)

PROCEDURE. Frontal and parietal bones are obtained under sterile conditions from 17- to 21-day-old fetal rats and carefully cleaned of super-

---

[13] W. A. Peck, S. J. Birge, Jr., and S. A. Fedak, *Science* **146**, 1476 (1964).

ficial periosteum, dura, and cartilage along major suture lines. About 10 calvaria (100 mg) are minced with a scalpel or scissors, placed in a siliconized 25-ml Erlenmyer flask containing 4.0 ml of the Tris-buffered saline with 1–3 mg of collagenase per milliliter and incubated for 90 minutes at 37° with shaking at 90 oscillations per minute. The supernatant medium is decanted into siliconized centrifuge tubes and centrifuged at 400 g for 1–3 minutes. The pellet of cells obtained is washed free of enzyme by gently resuspending the cells in fresh medium and recentrifuging three times. This procedure yields 20,000–50,000 cells per milligram of bone, of which over 95% are viable as judged by vital dye staining.

Cells prepared in this manner synthesize collagen and mucoprotein in culture, and are sensitive to parathyroid as well as other hormones. Unfortunately, cell yield drops abruptly as the animal increases in age, with an especially sharp fall near the time of birth (or hatching). Moreover, the cells from older animals have lost their ability to multiply in culture.

ii. Fitton-Jackson Method[14]

MATERIALS. Same as for cartilage (see Section II,D,4 above)

PROCEDURE. Under sterile conditions frontal bones from chick embryos (9 to 14 days old) are removed and placed in 1–2 ml of Tyrode's solution in the concavity of a glass slide (8–10 bones per slide). From then on, cell isolation procedure is the same as for cartilage except that: (1) 20 ml of trypsin solution are used for each lot of 80 bones; (2) duration of incubation with enzymes is shorter (especially with younger bone).

The yield of cells in uniform single cell suspension by this technique is about 0.4 to 0.6 $\times$ 10^6 cells per embryo. They are capable of collagen and mucoprotein synthesis, growth, and the *in vitro* morphogenesis of new bone under proper conditions. As with Peck's method, cell yields and the ability of these to grow in culture decline abruptly as the animal matures.

b. Mechanical Disruption[15,16]

MATERIALS

Bone, trabecular, cortical, or mixed, free of adherent soft tissue from any age or skeletal site. Rat, guinea pig, chick, and human bone have been used.

---

[14] S. Fitton-Jackson, personal communication, 1972.
[15] G. Nichols, Jr. and P. Rogers, *Calcif. Tissue Res.* **9**, 80 (1972).
[16] J. F. Woods and G. Nichols, Jr., *Science* **142**, 386 (1963).

"Grinding medium": 0.15 M NaCl containing Ca^{2+} 1.25 mM and $H_2PO_4^-$ 1.25 mM. Adjust pH with 0.1 N NaOH to 7.4 (The Ca^{2+} and $H_2PO_4^-$ may not be of critical importance. Other buffered isotonic media may also be used.)

Krebs-Ringer bicarbonate buffer, pH 7.4, or other nutrient medium

Porcelain mortar and pestle. The grinding surfaces of both must be smooth and well worn

30% (w/v) Dextran solution: 30 g Dextran, MW 200,000–300,000, in 100 ml of 0.15 M NaCl with Ca^{2+} and $H_2PO_4^-$ (1.25 mM each). The density of this solution should be 1.11 or slightly greater.

PROCEDURE. Finely minced washed bone fragments are placed in the mortar with about 5–10 times their volume of "grinding medium" and ground for 3 minutes with gentle pressure. After allowing a few seconds for the bone fragments to settle the supernatant medium containing the cells that have floated free is decanted into a large centrifuge tube (40 ml). The grinding procedure is then repeated 2–3 times or until the residual bone fragments have become very finely ground and decreased in volume. The supernatants from the several grinds are pooled and allowed to stand for 30 minutes, during which the small fragments of calcified collagen settle to the bottom of the tube.

The supernatant cell suspensions are collected and layered over 1–2 ml of Dextran. The system is centrifuged at 700 g for 5 minutes, separating the cell suspension into 3 fractions: a supernatant medium, a layer of cells packed on top of the Dextran, and a variable amount of heavier debris (calcified bone fragments and broken cell remnants) under the Dextran. The supernatant and material under the Dextran are discarded and the cell layer is harvested from over the Dextran by gentle resuspension in a suitable volume of nutrient medium using a Pasteur pipette. The sedimentation of cell on top of the Dextran is repeated once more after which they are washed twice in medium to remove any adherent Dextran from their surfaces.

The yield of cells is about 30–40% of the total present in the bone sample. This is increased materially by resuspending the calcified collagen, which settles out in the first 30 minutes, allowing it to settle a second time and harvesting the cells which were originally trapped in it. Some 75–85% of the cells are viable, and perform a variety of metabolic functions, including protein, mucoprotein, and nucleic acid synthesis under suitable conditions. They can be used to study amino acid, tetracycline, and calcium uptake and release. Collagen synthesis is, however, reduced, and the capacity to grow in culture apparently is lost. Some of these functions are known to reflect hormone activity *in vivo* and (probably) *in vitro*.

III. Special Experimental Procedures

A. Hormones: Preparation, Dosages, Route of Administration

1. Parathyroid Hormone (PTH)

a. Parathyroid Extract (Eli Lilly)

Subcutaneous or intraperitoneal injection of at least 1 unit/g per day are needed. Higher doses are recommended if only 1–2 days of treatment are planned; 1–2 days of treatment produce a decrease in bone collagen biosynthesis; 4–8 days produce an increase in both biosynthesis and bone resorption. *In vitro* effects on bone resorption in tissue culture are evident with 1–3 μg/ml of culture medium. Ovariectomized or thyroparathyroidectomized weanling rats are more sensitive. Old rats (more than 350 g) are relatively insensitive unless ovariectomized when weaned. Store extract refrigerated; do not freeze. Check expiration date.

b. Wilson Purified Hormone (about 200 USP units/mg)

Material varies in potency and is unstable in solution. It should be prepared daily. Stock solutions which keep for 1 month can be prepared by dissolving 20 mg in 20 ml, pH 4, buffer (citric acid) and kept frozen. Stability for longer periods is possible in a vehicle containing 2.5% gelatin and 50 mM cysteine hydrochloride. All preparations at pH 4 are painful.

c. Parathyroid Hormone Peptides (Bovine)

These peptides are available from Beckman Instruments, Inc., Spinco Division.

2. Calcitonin

Several species are available and active in *in vitro* systems in doses of 1–10 MRC milliunits per milliliter of culture medium. Inhibition by calcitonin of bone resorption produced by PTH, vitamin A, etc., results in an indirect effect on bone matrix. No direct effect on collagen metabolism has been demonstrated.

3. 17-Hydroxycorticosteroids

Cortisone modifies collagen biosynthesis and solubility in connective tissues in dosages of 3–5 mg/kg per day for 3 days. Hydrocortisone and prednisolone are active in doses which are 2/3 and 1/10, respectively, that of cortisone. Doses in man are only slightly less (e.g., cortisone 1–3 mg/kg per day $\times$ 3). *In vitro* effects are seen with 2–5 µg of cortisone per milliliter of culture medium (2–10 µM).

4. Estrogenic Steroids

Synthetic and naturally occurring estrogens modify collagen metabolism in several connective tissues. Subcutaneous administration of 50–150 µg/kg per day of propylene glycol or cottonseed oil suspension of estradiol-17β over a 3–6-week period will produce changes in several species of rodents. In man the active dose is 1/10 that in rodents. *In vitro* studies demonstrate activity with as little as 0.5–1.0 µg/ml of medium (1–2 µM).

5. Other Steroids

Testosterone (0.1 mg/kg per day), dehydrotestosterone (5 mg/kg per day), and progesterone (10 mg/kg per day) administered subcutaneously will produce changes in collagen biosynthesis in the uterus, skin, and bone of ovariectomized rats. Testosterone propionate (10 mg/kg per day for 2–3 weeks) will modify collagen biosynthesis in skin and bone of intact adult rats. *In vitro* concentrations of 50 mM to 50 µM testosterone have demonstrated activity.

6. Growth Hormone

Bovine growth hormone, 0.5 mg/kg per day, administered subcutaneously or intramuscularly to weanling or hypophysectomized rats and adult dogs of both sexes produces significant changes in collagen metabolism. Human growth hormone in man at 5–10 mg/kg per day, produces changes in urinary hydroxyproline. Studies in man of bone and collagen biosynthesis are limited. *In vitro* effects in rat tissues are produced with 5–10 µg/ml medium. Lower concentrations have failed to produce significant changes.

7. Insulin

In vitro effects in rat bone organ cultures (0.01 unit/ml culture medium) and on chick skin (5 units/ml medium) have been reported. Effects produced by the *in vivo* administration have not been studied extensively.

8. Thyroid Hormone

In vivo studies in rats have demonstrated increased degradation and decreased biosynthesis of collagen in the hyperthyroid state (1-thyroxine sodium, 10–40 μg per rat per day for 8 days). *In vitro* concentrations of 0.01–1.0 μg of triiodothyronine per milliliter of medium (1 μM to 10 mM) are effective in organ cultures of embryonic ulnas.

B. Lathyrogens, Nutritional Deficiencies, Proline Analogs, Microtubular Disruptive Drugs

1. Lathyrogenic Compounds

Nitriles [β-aminopropionitrile (BAPN) and aminoacetonitrile (AAN)] in the diet (0.05–0.10% of the diet by weight for rodents, birds, and primates) (baboon), produce marked skeletal and vascular deformities with increased solubility of collagen. The abnormalities are the consequence of an inhibition of lysyl oxidase, the enzyme responsible a primary step in collagen cross-linking. BAPN added directly to the medium (50 μg/ml) inhibits activity during assay of lysyl oxidase activity (see Section V,C).

D-Penicillamine (30 mg/kg per day for 3–4 weeks in rodents) produces increased collagen solubility of soft tissues by several mechanisms, the primary one being the blocking of the condensation step between two allysine residues—an important step in collagen cross-linking. In man, the same daily dose given over a period of 3–10 years produces changes in skin collagen solubility.

2. Nutritional Deficiencies

Copper deficiency in birds and rodents can be produced within 3–4 weeks. It is accompanied by vascular lesions and bone deformities with changes in collagen solubility and cross-linking.

Sulfate deficiency in birds and rodents produces the same general changes as copper deficiency.

Calcium deficiency produces changes which have not been determined definitively. Collagen biosynthesis in metaphyseal bone fragments from calcium–deficient rats, guinea pigs, and mice is decreased.[17]

Vitamin D deficiency, phosphorus deficiency, and magnesium deficiency are all known to modify calcium and bone metabolism. Recent work has demonstrated that rachitic chicks synthesize a collagen which

---

[17] D. H. Henneman, *Endocrinology* **87**, 456 (1970).

has an increased content of hydroxylysine and abnormalities in its aldehydric residues[18,19]

Vitamin C deficiency is well known to produce abnormalities in connective tissue metabolism. Its specific role in collagen biosynthesis appears to be related to the activity of peptidylproline hydroxylase.

The effects of nutritional deficiencies are beyond the scope of this section. The reader is urged to investigate particular conditions of interest by study of the literature.

3. Proline Analogs[20,21]

COMPOUNDS. L-Azetidine-2-carboxylic acid, cis-4-fluoro-L-proline, 3,4-dehydroproline, cis-hydroxy-L-proline, and trans-4,5-dehydrolysine, and their ^{14}C-labeled derivatives are available from Calbiochem.

EFFECTS AND DOSAGES. In vitro concentrations of 40–100 μg/ml incubation or organ culture medium embryos L-[^{14}C]proline incorporation into collagen by chick and mouse embryos. L-Azetidine-2-carboxylic acid and cis-4-fluoro-L-proline also inhibit the extent to which the incorporated L-[^{14}C]proline is hydroxylated. The in vivo administration of 500 μg per day for 5 days to chick embryos produces embryo fragility and decrease in tissue collagen content.

4. Microtubular Disruptive Drugs[21,22]

COMPOUNDS. Colchicine and vinblastine

EFFECTS AND DOSES. Procollagen secretion by chick embryonic bone and mouse fibroblasts is inhibited by both compounds. There is no effect on the percentage of proline hydroxylation. Concentrations as low as 50 μM are effective in organ culture systems.

In vivo doses of 0.05–0.2 mg/100 g body weight produce significant hypocalcemia in rodents, probably by blocking bone resorption.

C. Labeling of Collagen *in Vivo*

L-[^{14}C]Proline can be administered to animals at different stages of embryonic development, growth, and under a variety of experimental

---

[18] B. P. Toole, A. H. Kang, R. L. Trelstad, and J. Gross, *Biochem. J.* **127**, 715 (1972).
[19] G. L. Mechanic, S. U. Toverud, and W. K. Ramp, *Biochem. Biophys. Res. Commun.* **47**, 760 (1972).
[20] T. Takeuchi and D. J. Prockop, *Biochim. Biophys. Acta* **175**, 142 (1969).
[21] J. M. Lane, P. Dehm, and D. J. Prockop, *Biochim. Biophys. Acta* **236**, 517 (1971).
[22] R. F. Diegelmann and B. Peterkofsky, *Proc. Nat. Acad. Sci. U.S.* **69**, 892 (1972).

conditions with the production of labeled collagen whose specific activity (i.e., [^{14}C]hydroxyproline) is sufficiently high for reliable measurement. The effect of hormones and drugs on *in vivo* collagen biosynthesis, accumulation, and turnover in the intact organism can be examined with this procedure. The time, amount, and duration of isotope administration, the amount of nonlabeled proline in the diet, and the time interval between the last injection of ^{14}C and sacrifice of the animal are all important variables, which must be considered before interpretation of the data is possible. Short-term isotope administration (1–3 injections of 10–25 μCi per day) with sacrifice of the animal 4–144 hours later has been useful[23,24] for studying rapidly exchangeable [^{14}C]hydroxyproline pools in tissues or organisms with a high metabolic turnover or rate of growth. Pulse labeling *in vivo*, is limited, however, particularly in higher organisms where there is anatomic complexity and metabolic heterogeneity of tissues. These result in a slow equilibrium of metabolic pools, reversibility of reactions, and recycling of metabolites. Still further difficulties relate to the comparison of synthetic rates without consideration of the pool size of preexisting molecules. This problem is particularly true in bone, where the existing collagen pools are nonmineralized, partially mineralized, or fully mineralized.

Many of these problems can be circumvented by the administration of larger amounts of label (10 μCi per day of L-[^{14}C]proline for 3–4 weeks) during the phase of rapid growth (e.g., rats 50 g, guinea pigs 100 g, newly hatched chicks) in order to label the collagen in the whole animal.[25,26] The animal is sacrificed 8–10 weeks after the last injection of isotope to ensure equilibration of label and permit release of radioactivity from rapidly metabolizing tissues. Carrier-protein (e.g., unlabeled L-proline) is administered during the equilibration period (10% of protein in diet, about 0.5–1.0 g per day) or at least for the last 3 weeks before sacrifice to assure equal distribution of label. After equilibration and before sacrifice a variety of experimental conditions (e.g., hormones, drugs, nutritional deficiencies, etc.) can be studied. The effect of those on the redistribution of labeled collagen to newly synthesized collagen can also be examined if after equilibration newly synthesized collagen is stimulated in the form of a carrageenin granuloma (Section IV,A), a fracture callus, or a cotton pellet implant.[27]

Long-term prelabeling is particularly useful in experiments designed

[23] D. A. Heath, J. S. Palmer, and G. D. Aurbach, *Endocrinology* **90**, 1589 (1972).
[24] D. H. Henneman, *Endocrinology* **83**, 678 (1968).
[25] G. M. Fischer, *Endocrinology* **91**, 1227 (1972).
[26] L. Klein, *Proc. Nat. Acad. Sci. U.S.* **62**, 920 (1969).
[27] D. H. Henneman, *Biochem. Biophys. Res. Commun.* **44**, 326 (1971).

to mimic human metabolic connective tissue diseases. Too frequently insufficient attention is given to the fact that skeletal, vascular, and dermal lesions in man are of long-standing duration and that observed changes in collagen metabolism are secondary as well as primary.

IV. Metabolic Studies

A. Soft Tissues

Collagen metabolism can be studied in cartilage, skin, aorta, carrageenin[28] and gauze-pellet[29] granulomata, prostate, uterus, etc., by *in vitro* methods.

1. Incubation Procedures

Tissue samples are minced, rinsed, blotted, weighed, and placed in incubation flasks containing at least 10 times their weight (w/v) of incubation medium. The use of a bicarbonate buffer is recommended (see below for details). The flasks are incubated at 37° for 2–4 hours (or other suitable period) in a metabolic shaker (100 oscillations per minute) under the desired atmosphere (for optimum synthetic conditions 95% O_2–5% CO_2 is preferred). A period of preincubation of 10–30 minutes is often desirable to allow equilibration of tissue with the medium. After this, the medium can be changed, and substrate, isotope, or test substances added unless these have been administered to the intact animal *in vivo* (see Section III).

It is important that metabolism cease completely at the end of the period of observation and not continue while medium and sample are being separated. Chilling in ice-water mixture, heat, and the addition of metabolic inhibitors all accomplish this. Nonetheless, immediate separation in the cold is strongly recommended.

2. Preparation of Extracts after Incubation

After medium is withdrawn, the tissue fragments are rinsed with chilled buffer or equivalent and the rinse added to the original medium.

---

[28] λ-Carrageen (purified polysaccharide from carrageenin) is injected subcutaneously (4 ml of 1% saline solution) and a granuloma of newly synthesized fibrous tissue forms within 5–10 days. It is resolved completely within 30 days. The material is available as "Sea Kem," Marine Colloids.

[29] A gauze sponge (±2 × 2 cm) is implanted subcutaneously. Within 1–2 weeks a granuloma forms which goes on to become a permanent dense, fibrous tissue mass.

They are next rinsed several times with chilled saline and the rinses discarded. Next the tissue can be defatted and dried (as described in Section II,C) or extracted serially with a variety of solvents, if soluble collagen fractions are to be studied. The most useful solvents are 0.45 M NaCl in 20 mM phosphate buffer, pH 7.0 and 0.5 M citrate, pH 3.6 (or 3.8). Acetic acid (0.5 M), NaCl (1 M), and cysteamine (0.5 M) are useful for certain occasions.

If the tissue is to be extracted, it is shaken in the cold with each solvent (neutral salt solvents of increasing molarity, followed by acid solvents, and lastly by cysteamine) for 3 days with daily changes of solvent. The three extracts with each solvent are pooled and centrifuged 40,000 g at 4° for 1 hour). At this stage the tissue extract can either be purified if the original weight of tissue is large (15–20 g) (see Section V,A) or dialyzed exhaustively against distilled water if the incubated aliquot is small (200–500 mg). The former procedure can of course be done with tissue which has not been incubated and hence is not labeled or which has been prelabeled *in vivo* (see Section III, C). Dialysis rids the extract of nonincorporated L-[^{14}C]proline; it does not purify. In addition, it measures collagen peptides larger than 8000–10,000 molecular weight only, since free amino acids (including hydroxyproline and hydroxylysine) and small peptides are lost. Free hydroxyproline represents approximately one-third of the total hydroxyproline which is soluble in the medium (Krebs-Ringer is a 0.15 M neutral-salt solution). In subsequent extracts (0.45–1.0 M NaCl, acid, cysteamine) there is essentially no free hydroxyproline. After dialysis, the extracts are evaporated to dryness, hydrolyzed in sealed tubes with 6 N HCl at 115° for 16 hours, and the hydrolyzate are analyzed for hydroxyproline, [^{14}C]hydroxyproline, total ^{14}C, total nitrogen, and, if [^{14}C]lysine is present in the original incubation medium, for [^{14}C]hydroxylysine (see Sections V,D and VII).

The determination of total nitrogen (micro-Kjeldahl procedure with Nessler's reagent[30,31]) and of total hydroxyproline in the acid hydrolyzate allows for the calculation of noncollagenous protein.[32]

Radioactive collagen in the extract, including the medium, can be separated also from noncollagenous protein by digestion of the collagen under optimum conditions with highly purified collagenase. The collagen

[30] F. C. Koch and T. L. McMeekin, *J. Amer. Chem. Soc.* **46**, 2066 (1924).
[31] R. Ballentine, this series, Vol 3 [145].
[32] One milligram of collagen contains 13.3% hydroxyproline, or approximately 1 μmole, and 18.6% nitrogen, or 13.3 μmoles. One milligram of noncollagenous protein contains 11.4 μmoles of nitrogen. Total nitrogen in a tissue expressed as micromoles (minus the 6% due to nonprotein nitrogen) minus collagenous nitrogen in micromoles and divided by 11.4 equals an approximation of the milligrams of noncollagenous protein present in a given tissue.

and noncollagenous protein are first incubated with RNase to degrade aminoacyl tRNA, and then precipitated with 5% TCA (see Section V,B).

Since all metabolism is best expressed in terms of DNA, aliquots of the tissue need to be set aside (either prior to or after incubation) for measurement of DNA (see Section VI).

B. Bone

Collagen metabolism in bone can be studied *in vitro* with surviving tissue fragments or organ culture. Both biochemical and morphometric techniques are available to assess results.

1. Tissue Fragments

This approach is applicable to bone from any skeletal site, species, or age and involves the smallest number of assumptions. Its disadvantages (apart from those already attendant upon the use of any tissue incubated *in vitro* as a model) are the result of the relative paucity of cells and the abundance of extracellular material in the samples. This necessitates relatively long incubation periods (4–7 hours) and relatively large samples of viable bone (200–500 mg wet weight). This is particularly true of adult human bone which when compared to that of children or other mammals has considerably fewer cells and a lower rate of collagen turnover.[5,33]

a. General Principles of Incubation

Finely minced bone (1–2 mm fragments), thoroughly rinsed to remove as much marrow as possible, or carefully cleaned calvaria (see Section II) are placed in incubation vessels with a volume of medium of at least 10 times their weight and incubated as described (see Section IV,A) for soft tissues. Times of incubation depend upon the information desired. Generally, metabolism is stopped by chilling the flasks in ice water and promptly separating the medium from the fragments. These are rinsed twice with buffer or medium and the rinses are added to the separated medium. Since cells, some of which have been metabolically active, float off the fragments during incubation, the medium should be centrifuged at 600 g for 20 minutes if only the pellet is to be used for measurement of DNA or 40,000 g for 1 hour at 4° if the supernatant is to be analyzed for hydroxyproline (see Section IV,A). The pellet (containing cells) is

---

[33] B. Flanagan and G. Nichols, Jr., *J. Clin. Invest.* **44**, 1788 (1965).

added to the fragments which are to provide the major source of DNA (see below, under specific studies, and Section VI).

b. Incubation Medium[33]

The best, simple, chemically defined medium is a Krebs-Ringer bicarbonate buffer solution, pH 7.4.[34] Both authors have made the observation that bone fragments incubated in bicarbonate buffer have the highest rate of collagen biosynthesis and that Tris, PO_4, and Veronal buffers specifically lower several aspects of bone metabolism.

The buffer contains dextrose (13.3 mM), proline (1.73 mM), lysine (1.37 mM), glycine (2.67 mM), ascorbic acid (2.27 mM), α-ketoglutarate (0.14 mM), and calcium gluconate (0.11 mM).

STOCK SOLUTIONS. (1) 0.90% NaCl, (2) 1.15% KCl, (3) 2.11% KH_2PO_4, (4) 3.82 $MgSO_4 \cdot 7\ H_2O$, (5) 1.30% $NaHCO_3$ gassed for 1 hour with CO_2, (6) 1.22% $CaCl_2$.

To prepare: Mix 100 parts of solution (1), 4 parts solution (2), 1 part solution (3), 1 part solution (4), 21 parts solution (5), and 3 parts solution (6). After mixing, gas for 10 minutes with 5% CO_2 (95% O_2 5% CO_2) and store in glass-stoppered vessel in the cold until needed (fresh each day of use). Solutions (1)–(4) made in concentrations 5 times those listed can be stored in the cold for months.

SUBSTRATES AND COFACTORS. These are weighed out and stored in the dry state until needed. One or more of them can be combined: (A) 50 mg each of proline, lysine, and glycine; (B) 600 mg of dextrose, 100 mg of ascorbic acid, and 5 mg of α-ketoglutarate.

ISOTOPES. At least 1 μCi/ml of medium of L-[^{14}C]proline or 2 μCi/ml of medium of L-[^{14}C]lysine are added as needed. [^{14}C]Proline analogs can also be added if desired (Section III, B). If proline substrate and isotope are added in the concentrations described the specific activity of the proline is about 1.1 to 1.3×10^3 dpm/nmole of proline.

Care must be taken to keep the molarity of the amino acid carrier substrate (i.e., lysine or proline) between 1 mM and 1.75 mM so that their incorporation rate is constant over a 4–7-hour period.[35] In addition, when comparing experimental groups it is best to keep a particular molarity of proline or lysine from study to study and to include all the cofactors (e.g., α-ketoglutarate, ascorbic acid). Variations in the calcium or phosphorus content should also be kept to a minimum unless particular effects are being investigated.

---

[34] H. F. DeLuca and P. P. Cohen, in "Manometric Techniques" (W. W. Umbeit, R. H. Burris, and J. F. Stauffer, eds.), p. 132. Burgess Publ., Minneapolis, Minnesota, 1964.

[35] B. Flanagan and G. Nichols, Jr., *J. Clin. Invest.* **48**, 595 (1969).

c. Specific Studies

Many aspects of bone metabolism can be measured in this system. The age of the animal, the species, the size of the original samples, the skeletal site and type of bone are of importance in selection of the size of aliquot to be incubated, and the measurements that can be made.

Results are best expressed in terms of DNA content (mg/g). The same factors listed above determine the size aliquot required to measure DNA. This can be done by extracting the bone with 0.1 N NaOH for 16 hours with shaking at room temperature. DNA can be measured, also, in an aliquot of bone which has been incubated if the medium and fragments (see above, under General Principles of Incubation) are both extracted with 0.1 N NaOH or if the pellet after the medium has been centrifuged is added to the fragments and also extracted. For DNA method see Section VI.

OXYGEN CONSUMPTION. This can be measured either by the change in O_2 content of the medium in a sealed system with an O_2 electrode (Model 53 biological oxygen monitor, Yellow Springs Instrument Co.) or by manometry. Aliquots of 200 mg of mixed trabecular and cortical bone generally give reliable results.

ACID PRODUCTION. Accumulation of H^+, lactate, or citrate ions in the medium can be measured as a function of time: H^+ by change in pH or by manometry; lactate by the method of Barker and Summerson,[36] and citrate by the method of Natelson et al.[37] All these techniques are reliable and give results that are reproducible within 5–8%.

COLLAGEN BIOSYNTHESIS. Various steps in biosynthesis can be measured by adding L[U-^{14}C]proline to the medium and tracing its rate of incorporation into various fractions of the system. L-[^{14}C]lysine, L-[^{14}C]glycine, and ^{14}C-labeled proline analogs are useful.

(a) Intracellular synthesis is studied by extracting the cellular materials from the incubated bone fragments and the cells present in the medium with 0.1 N NaOH (2 ml/100 mg wet weight of bone) overnight with shaking at room temperature. Total incorporation of ^{14}C into this extract is measured directly by counting an aliquot of the alkaline extract in Hyamine-toluene scintillation solution (10 ml of toluene base solution[38] plus 1.5 ml Hyamine) in a liquid scintillation counter. The collagen present in the extract can be purified by digestion with highly purified collagenase (see Section V,B) or the extract itself can be dialyzed and analyzed for [^{14}C]proline and [^{14}C]hydroxyproline as described above

---

[33] S. B. Barker and W. H. Summerson, *J. Biol. Chem.* **138**, 535 (1941).
[37] S. Natelson, J. B. Pincus, and J. K. Luguvoy, *J. Biol. Chem.* **175**, 745 (1948).
[38] Four grams of POP and 50 mg POPOP to 1 liter with toluene.

(Section IV,A). If other isotopes are present, these can be measured by amino acid analysis of the acid hydrolyzates of either purified or dialyzed peptides (see Section V,F or, if [^{14}C]lysine is included in the medium the percentage glycosylation of [^{14}C]hydroxylysine can be determined by several means (see Sections V,D and V,F).

(b) Incorporation of label into soluble collagens is measured by the same procedures described above for soft tissues. Determination of this in the medium is useful in studies of intracellular collagen-transport and extracellular extrusion. It must be realized, however, that the hydroxyproline present includes not only recently synthesized and extruded collagen, but also that which was synthesized and soluble prior to incubation plus that which is resorbed during incubation. Collagen which is soluble in the NaOH extract as prepared for DNA (see above) can be measured and frequently is useful if it is assumed that the peptides present are derived from the intracellular pool of collagen. Increased, or abnormal, solubility in weak alkali may, however, reflect greater susceptibility of matrix collagen to NaOH extraction.

Bone, like soft tissues, can be extracted with neutral salts (0.45 M NaCl in 20 mM phosphate buffer, pH 7.0) and acid (0.5 M sodium citrate, pH 3.6 or 3.8). These, as well as the medium and NaOH extract, can be purified by the Peterkofsky and Diegelmann method (see Section V,B) or dialyzed, evaporated, and hydrolyzed as described above for soft tissues (Section IV,A). Analyses of total ^{14}C, [^{14}C]hydroxyproline, or [^{14}C]hydroxylysine, and their specific activities can be measured by any of the several techniques described in later sections (Sections V,D, V,F, and VII).

Acid-soluble collagen in adult bone deserves special comment. If 0.5 M sodium citrate, pH 3.6–3.8, is the solvent only a small amount (less than 0.1% of the total collagen) is extracted under normal circumstances. However, under conditions of decreased or abnormal mineralization the amount that is soluble is increased.

If soluble collagens are to be studied, a minimum of 250–500 mg wet weight of cancellous bone is required. To purify these for analytical compositional studies, 20–25 g wet weight is needed (Section V, A).

(c) Incorporation of label into insoluble (mineralized) collagen: The biosynthesis of this fraction of collagen can be determined with a high degree of accuracy since it constitutes the bulk of bone collagen. If the fragments are first "cleaned up" as described in Section V, G, the collagen can be separated into peptide subunits and hydrolyzed for amino acid analyses by a variety of techniques (see Sections V,D–V,F). Radioactivity in any of the amino acid residues (e.g., proline, lysine, hydroxyproline, and hydroxylysine) can be determined also. Collagen purification by the Peterkofsky and Diegelmann procedure (see Section V,B) is recommended if medium collagen has been purified similarly.

FORMATION RATE[35]. The total rate of proline or lysine incorporation into mature, mineralized matrix collagen also can be studied without prior preparative procedures necessary for amino acid analyses. After incubation the bone fragments should be rinsed free of contaminating medium (with Krebs-Ringer buffer, pH 7.4), and extracted with weak alkali (0.1 N NaOH) for 16 hours with shaking at room temperature. The alkali extract is withdrawn, the fragments are rinsed three times with distilled water followed by demineralization, defatting, and drying as already described (see Section II,C).

The dried insoluble matrix is hydrolyzed in 6 N HCl at 115° for 16 hours. Total radioactivity in terms of the millimicromoles of [^{14}C]proline incorporated from the original medium (see Section IV,B) is calculated and expressed as nanomoles of [^{14}]proline per hour per milligram of DNA. Formation rate is linear between 1 and 7 hours.[35]

Dried, demineralized fragments can be dissolved in 0.5–1.0 ml of 98% formic acid and 0.1 ml of this counted in a Hyamine-toluene base scintillation solution. Again, this procedure measures total protein synthesis only.

RESORPTION RATE.[35] It is difficult to quantitate this aspect of bone metabolism. Nonetheless, the resorption both of collagen that is formed during incubation and that existing in the bone sample prior to incubation can be distinguished and quantitated separately by analysis of the total hydroxyproline (free and peptide bound) released into the medium during incubation. Medium hydroxyproline is derived from several sources, but in practice only three can be distinguished specifically: collagen synthesized and extruded during incubation, soluble collagen formed before incubation, and resorption of performed and newly formed insoluble collagen. These three fractions are measured as follows: bone fragments are incubated in a medium containing L-[^{14}C]proline for 4 hours. After this, the medium is removed, the fragments are rinsed, and the rinse is added to the medium, which is saved. Fresh medium is added to the fragments and the incubation continued an additional 3 hours before the second medium is separated and analyzed separately.

Total and ^{14}C-labeled hydroxyproline are measured in both the 0–4-hour and the 4–7-hour media. The values obtained are taken to represent the resorption of collagen from each pool as follows: (a) Resorption of newly synthesized collagen is estimated from the accumulation of [^{14}C]hydroxyproline in the 4–7 hour medium since, at this time, the precursor pool of proline can be shown to be of the same specific activity as that of the incubation medium and the accumulation of labeled hydroxyproline is linear over that same period. (b) Resorption of performed insoluble collagen is measured by the accumulation of unlabeled hydroxyproline in the 4–7-hour medium. This is calculated by subtracting the labeled hydroxyproline (resorption of newly synthesized collagen) from the total hydroxyproline present. The portion of the total that is newly synthesized is usually less than 10% of the total. In adult human bone it is so small that generally it can be ignored. (c) Resorption of preformed soluble collagen is estimated from the 0–4-hour medium. However, since this will contain whole soluble collagen molecules as well as the breakdown products of resorptive activity, these must be separated out either by dialysis or gel filtration on Sephadex. The pool size of soluble collagen present initially in the bone sample can be determined by incubating bone samples inactivated either by heating for 1 minute at 95° or by three times freezing and thawing. However, both these procedures denature collagen, at least in part, thereby probably modifying its solubility properties and introducing possible artifacts.

"Resorption rate" as reported generally represents the rate of breakdown of preformed insoluble collagen (plus the breakdown of newly synthesized collagen in most cases) calculated in terms of nanomoles of hydroxyproline released per hour per milligram of DNA.

2. Organ Culture

Organ culture systems can be used to study both bone formation and resorption with advantages when longer periods of observation are needed to examine slower phenomena not demonstrable in short-term incubations. Disadvantages include the need to use very young bone if bone formation or growth are to be studied.

a. Goldhaber Method[39,40]

Others have described similar techniques.[41]

MATERIALS
  5-Day old mice
  Glass coverslips
  11–13-Day chick embryo extract
  Chicken plasma
  Leighton tubes

---

[39] P. Goldhaber, *J. Dent. Res.* **45**, 490 (1966).

[40] E. F. Voelkel, A. H. Tashjian, Jr., and P. Goldhaber, in "Calcium, Parathyroids, Hormones, and Calcitonins" (R. V. Talmage and P. Munson, eds.), p. 478. Excerpta Med. Found., Amsterdam, 1972.

[41] P. J. Gaillard, *Exp. Cell. Res.*, Suppl. **3**, 154 (1955).

Gey's balanced salt solution (Gibco Corp).
Heated horse serum
Roller drum 1 rev/5 minutes in 37° incubator
50% O_2–50% N_2

PROCEDURE. Calvaria are removed aseptically, the occipital bone is discarded, and the remaining portion (frontal and parietal bones) is placed on a rectangular glass coverslip and covered with a thin mixture of chicken plasma and embryo extract. After clotting, each coverslip with its calvarium is inserted into the flat well of a Leighton tube and covered with 2 ml of medium, the air in the tube being displaced with 50% O_2–50% N_2 before closing it. The tubes are placed in the roller drum and incubated at 37° for 2–3 weeks replacing medium and regassing every 2–3 days. The medium is Gey's balanced salt solution, heated horse serum, and chicken embryo extract (6:2:2) with 100 units of penicillin and 100 µg of streptomycin added per milliliter.

Cultures are inspected under the microscope for loss of substance and new osteoid formation at regular intervals or when the changes sought have appeared. At the end of the culture period the tissue is fixed in 10% neutral formalin, embedded in paraffin, and sectioned for staining by Von Kossa's method for phosphate, the later facilitating the visual estimation of bone loss.

Bone resorption usually begins in the sagittal suture area between the frontal bones at 2–3 days. By 5–6 days this area is largely destroyed and bone loss is extending centrifugally. Omission of embryo extract from the medium significantly reduces bone resorption while inclusion of 0.5 U/ml of parathyroid extract (Eli Lilly) greatly enhances it. Degrees of resorption can be estimated semiquantitatively by scoring sections (or transilluminated whole calvaria) on an arbitrary scale of 0–6. New osteoid and bone formation can be observed either in hematoxylin–eosin-stained sections or by noting the appearance of new areas of poorly transilluminating bony material in areas where resorption has taken place.

This technique has the advantage of measuring both mineral and matrix changes, and although it is dramatic because it is visual, it is insensitive, slow, and difficult to quantitate. Its greatest usefulness has been to screen *in vitro* the activity of various factors that might influence bone resorption *in vivo*.

Two laboratories have used organ culture chiefly to measure bone resorption. While they have concentrated on changes in bone mineral and generally assumed that matrix was resorbed also, collagen breakdown products in the medium can be measured if desired. Both approaches are similar, but details of methodology and, in some areas, results are sufficiently different to warrant description of both.

b. Raisz Method[42,43]

MATERIALS

Sterile $^{45}CaCl_2$

BGJ medium modified to contain: sodium acetate, 50 µg/ml; bovine serum albumin, 1 mg/ml; 1-arginine-HCl, 175 mg/liter; 1-serine, 200 mg/liter; 1-aspartic acid, 150 mg/liter; 1-alanine, 250 mg/liter; 1-proline, 400 mg/liter; glycine, 800 mg/liter; Na, 150 mM; K, 5 mM; SO_4^{2-}, 0.7 mM; Ca, P, Mg—as needed; streptomycin, 50 µg/ml; penicillin, 50,000 units/ml

Pregnant rats on day 18 of gestation

Small-embryo watch glasses—8 enclosed in a 100 × 15 mm petri dish—equipped with stainless steel screens so cut that each supports a bone at the surface of 0.5 ml of medium.

Incubators at 37.5° with facilities to maintain a suitable gas phase.

PROCEDURE. Calcium-45, 500 µCi as sterile $CaCl_2$ solution, is given subcutaneously to pregnant rats on day 18 of gestation. On day 19 the rats are sacrificed, the embryos are removed aseptically (all subsequent steps are under sterile conditions), and the forelimbs are separated and kept under medium in a 150-mm petri dish at 5–10°. The radius and ulna are dissected under a binocular dissection microscope, cleaned of adherent tissue, and the cartilaginous ends are removed. The remaining thin tube of calcified bone matrix surrounding a core of calcified cartilage is washed 3 times with medium (to remove any traces of developing marrow) and transferred to a culture vessel.

The bones are explanted in pairs (radius and ulna together) into small watch glasses on squares of Millipore filter which are on top of the wire screens and at the surface of 0.5 ml of medium. The medium fills the watch glass and contains any desired test substance. Eight watch glasses are placed in a petri dish on a layer of filter paper moistened with medium containing Phenol Red. Before the bones are explanted, medium and test substances are equilibrated with an atmosphere of 5% CO_2–20% O_2–75% N_2. A change in color of the Phenol Red in the filter paper indicates that equilibration of that petri dish is complete. Cultures are maintained at 37.5° under the same atmosphere for 48 hours without change in medium. If longer culture periods are desired, the time can be extended by changing the medium. Explants are made in pairs, the bone from one side of the embryo serving as the control.

At the end of incubation the fluid is removed and analyzed for total Ca, radioactivity, P, and hydroxyproline. Bones are placed in 0.5 ml of

[42] L. G. Raisz, *J. Clin. Invest.* **44**, 103 (1965).
[43] L. Raisz and I. Neimann, *Endocrinology* **85**, 446 (1969).

1 N HCl to extract the mineral for measurement of radioactivity and total Ca, P, and, if desired, Mg.

Paired bones from the same animal do not differ by more than 5% in initial ^{45}Ca content. Results may be expresed as counts or hydroxyproline released, as change in medium Ca concentration, as percent of the total label or sample collagen released, or as treated to control ratios in any of the variables.

c. Reynolds Method[44]

This differs from the Raisz method in the composition of the medium, the source of bone, and a few additional minor details. These factors may account for some of the differences in results observed by these workers.

MATERIALS

2-Day-old mice

Pregnant rats on day 16 of gestation

Medium containing the components (mg/liter) tabulated below.

l-Lysine-HCl, 240.0	Ca lactate, 555.0	Nicotinic acid, 20.0
l-Histidine-HCl · H$_2$O, 150.0	KCl, 350.0	Thiamine-HCl, 2.0
l-Arginine-HCl, 7.5	KH$_2$PO$_4$, 35.0	Ca pantothenate, 0.5
l-Threonine, 75.0	Na$_2$PO$_4$, 140.0	Riboflavin, 0.2
l-Valine, 65.0	NaCl, 6,200.0	Pyridoxal-PO$_4$, 0.2
l-Leucine, 50.0	MgSO$_4$ · 7H$_2$O, 200.0	Folic acid, 0.2
l-Isoleucine, 30.0	NaHCO$_3$, 2200.0	Biotin, 0.2
l-Tryptophan, 40.0		p-Aminobenzoic acid, 2.0
l-Tyrosine, 40.0		α-Tocopherol-PO$_4$ (sodium salt), 1.0
l-Cysteine-HCl · H$_2$O, 90.0		Choline-Cl, 50.0
l-Glutamine, 200.0		m-Inositol, 0.2
		Vitamin B$_{12}$, 0.04
		Glucose, 5,000.0
		Sodium acetate, 50.0
		Streptomycin, 50.0
		Penicillin-G, 50,000 U/ml

Heat-inactivated rabbit serum (60° for 30 minutes)

Two round, flat-bottom dishes (3 cm in diameter and 1 cm high) containing a square table of stainless steel wire mesh (2 mm high and 2 cm across) placed in a 100 × 15 mm petri dish on a layer of cotton wool or filter paper wet with 0.15 M NaCl.

Incubator and gas supply and ^{45}Ca as in Raisz method

[44] J. J. Reynolds and J. T. Dingle, *Calcif. Tissue Res.* **4**, 339 (1970).

PROCEDURE. Mouse calvaria are labeled by injecting 12.5 μCi of sterile $^{45}CaCl_2$ subcutaneously on day 2 of life, and fetal rat calvaria by injection of the mother with 200 μCi on day 16 of gestation.

Calvaria (frontal and parietal bones) are harvested aseptically from mice on day 6 of life and from rat fetuses on day 19 of gestation. Under a dissecting microscope, they are carefully cleaned of adherent skin and suture line cartilage leaving the dura and periosteum, intact. Each is divided in half along the suture line, one half being explanted onto the steel grid in each of the two culture dishes contained in a petri dish. Medium, 1.5 ml, containing 5% heat-inactivated rabbit serum (just enough to wet surface of the grid) is placed in each culture dish, the petri dishes are closed and incubated at 37° under an atmosphere of 5% CO_2–20% O_2–75% N_2 for 48 hours. The time course of ^{45}Ca (and hydroxyproline) release (or ^{32}P if P is to be studied) is followed by removing 25 μl aliquots of medium aseptically with micropipettes (disposable ones are convenient) for counting in a liquid scintillation system (see Section IV,B,1). Radioactivity remaining in the bone at the end of the experiment is determined by counting an aliquot of a solution of the bone in 0.5 ml of 98% formic acid.

In this system one half of each calvarium serves as the test object, the other as the control. Differences in radioactivity and tissue composition are no more than 5% between the two bone halves. Results may be expressed as in the Raisz method.

Equating bone resorption with the rate of release *in vitro* of ^{45}Ca or ^{45}Ca previously incorporated *in vivo* presupposes that mineral and matrix are being resorbed at the same rate, a notion that, while generally true, is not universally so. Because of this it is desirable to study the release of hydroxyproline at the same time. Uniform distribution of the label, needed for sensitivity and ease of quantitation, must also be assumed, and this requires young rapidly growing animals or, preferably, that the label be administered to the mother before birth or to the embryo before hatching. Therefore, like the morphologic approach it is applicable only to very young animals.

V. Composition of Collagen

A. Purification of Collagens

1. Introduction

Purification of collagen is essential in the study of the transcriptional and posttranscriptional stages of collagen biosynthesis. Several genetically distinct collagens have been described among the more abundant "interstitial collagens" (e.g., differences between cartilage and skin collagens[45]), and primary structural differences between basement membrane collagens and interstitial collagens are known.[46] In addition, there exists a collagen precursor, procollagen which is probably concerned with the extracellular transport of collagen[47] and contains additional sequences in its constituent pro-α_1 and pro-α_2 peptide chains (with unique amino acid residues). The methods for studying the amino acid sequence in constituent peptide chains, the residues concerned with cross-linking, and the cross-links themselves all require pure starting material. These methods are described in Sections V,E–V,H.

2. General Comments

All purification steps are carried out in the cold (0–4°) with prechilled solvents. Attention to pH and ionic strength of solutions is essential. In general about 10 ml per gram of original tissue is required for any given solvent. This applies also to the resolubilization steps of the precipitated and partially purified collagens.

3. Neutral-Salt-Soluble Collagen (NSS)[48-50]

a. Extraction of NSS

Cut and mince, or chip, tissue into small fragments. Homogenize with a Virtis set set at $\frac{1}{2}-\frac{2}{3}$ maximum speed approximately 30 minutes at 4° in 0.15 M NaCl in 20 mM phosphate buffer, pH 7.0. Extract overnight by shaking gently in the same buffer. Centrifuge (40,000 g) for 1 hour at 4°. Discard the supernatant. The concentration of hydroxyproline per gram wet weight can be measured in the supernatant if desired.

Rehomogenize the residue in 0.45 M NaCl in 20 mM phosphate buffer, pH 7.0, and extract with gentle shaking for 48 hours in the same solution. Decant the supernatant, save, and reextract residue an additional 48 hours with the same buffer, pH 7.0. Pool the extracts. Centrifuge for 1 hour (40,000 g) at 4°. Measure the volume of the supernatant and take

---

[45] E. J. Miller, E. H. Epstein, Jr., and K. A. Piez, *Biochem. Biophys. Res. Commun.* **42**, 1024 (1971).
[46] N. A. Kefalides, *Int. Rev. Exp. Pathol.* **10**, 1 (1971).
[47] H. P. Ehrlich and P. Bornstein, *Nature (London) New Biol.* **238**, 257 (1972).
[48] A. Deshmukh, K. Deshmukh, and M. E. Nimni, *Biochemistry* **10**, 2337 (1971).
[49] K. Deshmukh and M. E. Nimni, *Biochem. J.* **112**, 397 (1969).
[50] K. Deshmukh and M. E. Nimni, *Biochim. Biophys. Acta* **154**, 258 (1968).

an aliquot for determination of the total collagen soluble in neutral salt (micromoles of hydroxyproline per gram wet weight).

b. Purification of NSS in Supernatant

Precipitate the collagen in the 0.45 M NaCl extract by dialyzing the supernatant for 1 week against dibasic phosphate (0.2 M Na_2HPO_4). Change the dibasic phosphate daily. Centrifuge contents of the bag for 1 hour (40,000 g) at 4°, discard the supernatant.

Resuspend the precipitate in 0.5 M acetic acid until fully dispersed in solution. Salt-out with 5% (w/v) NaCl. Let sit in cold overnight to assure complete precipitation. Centrifuge 1 hour (40,000 g) at 4°, discard the supernatant, and suspend the precipitate in 0.45 M NaCl in 20 mM phosphate buffer, pH 7.0. Add enough buffer to fully solubilize the collagen. Put this solution in a dialysis bag and dialyze exhaustively (daily changes) against the same 0.45 M NaCl phosphate buffer, pH 7.0. After 1 week the material should be fully solubilized. If not, centrifuge and save the supernatant.

To contents of bag (or supernatant if centrifuged after dialysis) add NaCl to 20% at neutral pH. *Be sure it is neutral.* To assume complete precipitation of collagen let sit overnight in the cold. Be sure collagen is well dispersed and in solution when 20% NaCl is added. Centrifuge 1 hour (40,000 g) at 4°. Discard the supernatant and save the precipitate. Redissolve in 0.45 M NaCl, phosphate buffer, pH 7.0 and dialyze 2–3 days vs. 0.45 M NaCl phosphate buffer, pH 7.0. Contents of the bag are lyophilized and stored at −20° for future use.

c. Possible Uses of ^{14}C-Labeled purified NSS

　i. As substrate for lysyl oxidase activity
　ii. For gel formation to be used as substrate for collagenase activity
　iii. If labeled with [^{14}C]lysine in the presence of α-α'dipyridyl, it can be used as the protocollagen substrate for lysyl protocollagen hydroxylase activity
　iv. If labeled with [^{14}C]proline in presence of α-α'-dipyridyl, it can be used as the protocollagen substrate for prolyl protocollagen hydroxylase activity

4. Cysteamine-Soluble Colagen[48-50]

a. Extraction of NSS and Acid-Soluble Collagen

After neutral-salt-soluble (NSS) and acid-soluble collagens (see Section V, A,5) are extracted, wash the tissue with 0.45 M NaCl in 20 mM

Na$_2$HPO$_4$ buffer, pH 7.0. If only NSS has been extracted, wash with 0.5 M citrate buffer, pH 3.6 and then the 0.45 M NaCl solution.

b. *Cysteamine Extraction*

Extract the tissue with 0.2 M cysteamine (also in 0.45 M NaCl in 20 mM Na$_2$HPO$_4$ buffer pH 7.0) for 72 hours in the cold with shaking. Decant extract and repeat procedure with the same cysteamine solution for another 48 hours. Pool the extracts and centrifuge (40,000 g) for 1 hour.

Dialyze the supernatant vs. distilled water to precipitate the collagen. Centrifuge contents of bag at 40,000 g for 1 hour. Redissolve the precipitate in 0.45 M NaCl in 20 mM Na$_2$HPO$_4$ buffer, pH 7.0. Redialyze vs distilled water, reprecipitate, recentrifuge, and redissolve in the 0.45 M NaCl phosphate buffer. Redialyze vs distilled water, centrifuge the precipitate at 105,000 g for 2 hours, lyophilize, and store at $-20°$.

5. *Acid-Soluble Collagen (Soft Tissue)*[51-53]

a. *Preliminary Extraction of Tissue with Neutral Salts*

Extract finely minced tissue (granuloma, blood vessel, skin, uterus, etc.) overnight in 0.15 M NaCl in 20 mM phosphate buffer, pH 7.0. The following morning, decant and discard the 0.15 M NaCl extract and add 0.45 M NaCl in 20 mM phosphate buffer, pH 7.0. Extract for 24 hours also with gentle shaking. Decant and discard the 0.45 M NaCl extract (unless purification of NSS is needed) and rinse fragments twice with 0.5 M acetic acid; discard these rinses.

b. *Extraction with 0.5 M Acetic Acid (or 0.5 M Sodium Citrate, pH 3.6)*

Add sufficient 0.5 M acetic acid to disperse the tissue fragments (about 10–15 ml per gram of tissue) and extract a total of 72 hours with daily changes of 0.5 M acetic acid. Do not shake vigorously or material will become gelatinous. Decant the extracts each day and pool. Centrifuge for 1 hour (40,000 g) at 4° and discard pellet after centrifugation. The remaining residue can be extracted with cysteamine (skin) or guanidine (bone) if needed (see below).[54]

---

[51] P. Bornstein and K. Piez, *J. Clin. Invest.* **43**, 1813 (1964).
[52] A. H. Kang, Y. Nagai, K. Piez, and J. Gross, *Biochemistry* **5**, 509 (1966).
[53] P. Bornstein and K. Piez, *Biochemistry* **5**, 3460 (1966).
[54] The extraction of the NSS fraction (0.45 M NaCl in 20 mM phosphate buffer, pH 7.0) can precede the extraction of the acid-soluble collagen in any tissue includ-

c. Purification of Supernatant (Acid-Soluble Collagen)

Add NaCl (5%, w/v) to the pooled 0.5 M acetic acid-soluble extracts in order to precipitate the collagen. Let the solution sit overnight in the cold. Centrifuge (40,000 g) for 1 hour at 4°.

Dissolve the precipitate by the addition of water, suspend this solution in a dialysis bag, and dialyze exhaustively against 0.15 M acetic acid. When completely solubilized, or nearly so, centrifuge solution in bag (40,000 g); discard the precipitate and save the supernatant.

Neutralize the supernatant by adding 0.2 Na_3PO_4 to pH 7.0 and precipitate collagen by addition of NaCl to 20% (w/v). Let the solution sit overnight in the cold to salt out the collagen completely. Centrifuge and suspend the precipitate in 0.15 M acetic acid. Dialyze the solution vs 0.15 M acetic acid exhaustively. Lyophilize the solution in a bag. Store at −20°.

6. Extraction of Chick Bone Collagen[55]

Note: This method recovers about 5–10% of the total collagen in normal bone and about 78% from lathyritic bone.

a. Decalcification and Extraction with Acid

Put the bone in cellulose dialysis tubing containing 0.5 M acetic acid (200 ml/70 bone cylinders) and dialyze it against 6-liter volumes of 0.5 M acetic acid. Change the 0.5 M acetic acid twice daily. Continue dialysis for 3 weeks. Save the acetic acid solution from the dialysis bag.

The insoluble residue is next extracted for another week with constant stirring in 200 ml of fresh 0.5 M acetic acid. Combine the extract with the acetic acid solution saved after dialysis. Repeat this extraction. Pool all extracts. Clarify all extracts by centrifugation at 100,000 g for 2 hours in ultracentrifuge.

b. Purification of Pooled Acid Extracts

The collagen is precipitated by dialyzing the extracts against large volumes of 20 mM disodium phosphate. Centrifuge the solution (40,000

---

ing bone. If both fractions are desired, the tissue is first extracted with 0.15 M NaCl overnight (discard), then 48 hours with 0.45 M NaCl in 20 mM phosphate buffer, pH 7.0 (see NSS procedure) then 0.5 M acetic acid for 72 hours (see acid-soluble collagen procedure). The amount of collagen (micromoles of hydroxyproline per gram wet weight) present in each extract can be measured without purification.

[55] E. J. Miller, G. R. Martin, K. A. Piez, and M. J. Powers, J. Biol. Chem. **242**, 5481 (1967).

g) for 30 minutes. Dissolve the precipitate in 0.5 M acetic acid, and precipitate the collagen again with the addition of sufficient NaCl to give a concentration of 5%. Collect the precipitate by centrifuging at 40,000 *g*. Redissolve the pellet in 0.5 M acetic acid and dialyze the solution against a large volume of 0.5 M acetic acid. Lyophilize the contents of the bag and store at $-4°$ to $0°$.

c. Guanidine Extract[48-50]

If guanidine extract is desired, extract the fragments (after acid extraction) with 200 ml of 5 M guanidine-HCl for 1 week with constant stirring. Dialyze the extract exhaustively against distilled water and lyophilize. When dialyzing against distilled water fill the dialysis bags only half full with the guanidine extract.

7. Reconstitution of Purified, Lyophilized Collagen

a. Method I

Shake aliquots of the purified, lyophilized collagen in phosphate buffer 20 mM pH 7.6) (about 400 mg/100 ml) overnight at $0–4°$. On the following morning dialyze the solution against 50 mM Tris buffer, pH 7.6, containing 0.4 M NaCl and 3 mM CaCl$_2$ for 24 hours at $4°$. Centrifuge the contents of the bag to remove any undissolved collagen after dialysis. Dilute the highly viscous material with equal volumes 50 mM Tris buffer, pH 7.6 containing 1 mM CaCl$_2$. Measure either the protein content (milligrams of protein per milliliter of purified collagen in solution) or the hydroxyproline content per milliliter of solution.

b. Method II

Dissolve the lyophilized material in cold 50 mM acetic acid or 0.1 M citrate buffer, pH 4.3. Stir gently overnight at $4°$. Remove the undissolved particles by centrifugation (100,000 g for 1 hour) in a Spinco Model L ultracentrifuge.

B. Purification of Collagen by Digestion with Highly Purified Collagenase[22,56]

The study of collagen biosynthesis *in vitro* with either tissue fragments or isolated cells is handicapped by the fact that the several soluble collagen fractions (e.g., the newly synthesized, recently extruded peptides which are soluble in the incubating medium or in neutral salt, 0.45 M

---

[56] B. Peterkofsky and R. Diegelmann, *Biochemistry* **10**, 988 (1971).

NaCl) are present in such small amounts that their purification by the usual procedures (see Section V,A) is impossible. Under these conditions several novel approaches have been described, one of which although not fully established, is nonetheless worthy of mention. The procedure has been useful in the study of the effect of microtubular disruptive drugs on collagen secretion. A brief description of this method follows.

CHEMICAL PRINCIPLE. Radioactive collagen is measured in the presence of large amounts of other proteins. Bacterial collagenase purchased as "purified" is further purified and used to cleave the collagen into acid-soluble peptides. Radioactive protein is precipitated from a homogenate with trichloroacetic acid after degrading aminoacyl RNA with RNase. The precipitate is redissolved and digested under optimal conditions with highly purified collagenase for 90 minutes.

MATERIALS

    a. Chromatographically "pure" collagenase (Worthington)
    b. Protease-free RNase, Type XI-A
    c. Bovine tendon collagen
    d. Tannic acid
    e. Sephadex G-200 gel
    f. N-Ethylmaleimide (NEM)
    g. Hydroxyethylpiperazine-N-2-ethanesulfonic acid (HEPES buffer) and azocoll

Components b–g are available from Sigma Chemical Corp.

1. Purification of Collagenase

The procedure is similar to that of Keller and Mandl[57] except that the elution buffer is 50 mM Tris·HCl (pH 7.6) 5 mM CaCl$_2$. A 1.6 × 30 cm column of Sephadex G-200 is equilibrated with the above buffer for several days before the collagenase is applied. A 1.0-ml sample of collagenase (30–50 mg) is layered on the top of the gel column followed by 1.0 ml of buffer. After this has passed into the gel, a 5.0-ml head of buffer is added and elution is started. The flow rate is 6.5 ml per hour; 2.7 ml fractions are collected at room temperature, and their collagenolytic activity is determined. In general, fractions 52–62 contain the peak activities. These are assayed against native tendon collagen, azocoll, and radioactive chick embryo protein.[56] A mixture of the most active fractions are combined and stored at −20°. Activity is retained for as long as one year. Modifications of the original purification procedure have been made by both the original authors[58] as well as by others.[59] The assay

[57] S. Keller and I. Mandl, *Arch. Biochem. Biophys.* **101**, 81 (1963).
[58] P. Benya, K. Berger, M. Golditch, and M. Schneir, *Anal. Biochem.* **53**, 313 (1973).
[59] R. F. Diegelmann and B. Peterkovsky, *Develop. Biol.* **28**, 443 (1972).

procedure and reagent mixtures have been modified also depending upon the tissue studied (see above).[58]

2. Preparation of Radioactive Protein Substrate

Substrate is prepared from radioactive tissue homogenates, *in vitro* incubation media, organ culture, or cell-culture media.

Soft tissue fragments (or chick embryos) are washed several times with cold 0.15 M NaCl, blotted, weighed, and homogenized with two volumes of 50 mM Tris·HCl buffer (pH 7.6). Embryonic bones are homogenized and then sonicated for 20 seconds with a Branson sonifier (3 A); adult bones are first frozen in dry-ice and then pulverized in a Wigl-Bug Shaker (Fisher or VWR). RNase (20 μg/ml final volume) is added, and the homogenate is incubated at 37° for 5 minutes in order to degrade the aminoacyl-tRNA. Trichloroacetic acid (TCA) is added to give a final concentration of 5%. This mixture is kept at 0° for 5 minutes, and the resultant precipitate is collected by centrifugation (400 g for 5 minutes). The precipitate is resuspended in 5% TCA and recentrifuged. This washing procedure is repeated twice more. Finally the precipitate is washed twice with ethanol:ether (3:1, v/v), once with absolute ether, and dried slowly to yield a gray powder.

3. Assay of Radioactive Collagen in Presence of Other Proteins by the Use of Purified Collagenase

The dry radioactive protein (see 2 above) is dissolved in 0.1 N NaOH (2.5–5.0 mg/ml) by warming at 37° for 5 minutes with occasional shaking. The suspension can be gently homogenized if necessary. The protein solution is chilled at 0°.

Each assay reaction mixture of 0.5 ml contains (1) 0.20 ml of the protein solution, (2) 0.20 ml of 0.08 N HCl (added to neutralize the NaOH), (3) 60 μmoles HEPES buffer (pH 7.2), (4) 1.25 μmoles of NEM, (5) 0.25 μmoles CaCl$_2$, and (6) 12–26 μg of Sephadex G-200 purified collagenase. The total volume is 0.5 ml. HEPES buffer, NEM, and CaCl$_2$ are combined into one solution such that when 0.1 ml is added to the 0.2 ml of protein of 0.2 ml of 0.08 N HCl the correct final molarities are obtained.

The reaction mixture is prepared in 2-ml conical centrifuge tubes, and the tubes are incubated with the collagenase at 37° for 90 minutes with shaking at 100 oscillations per minute; 0.5 ml of 10% TCA containing 0.5% tannic acid is added to stop the reaction. The mixture is kept at

0° for 5 minutes. The tubes are centrifuged at 400 g for 5 minutes at 4°, and the supernatants are transferred to counting vials. The sediment is resuspended in 0.5 ml of 5% TCA with 0.25% tannic acid and recentrifuged; the second supernatant is added to the first already in a counting vial. Ten milliliters of Triton–Omnifluor (2:1, v/v) scintillation fluid is added, and the mixture is counted in a liquid scintillation counter. Suitable standardized quench curves are used to determine the percent efficiency (about 80% for ^{14}C and 20% for ^{3}H).

If reaction mixtures are larger than 0.5 ml, the protein solution is incubated with the collagenase in larger tubes, but after incubation 0.5-ml aliquots are transferred to smaller tubes, 0.5 ml of the TCA-tannic acid is added to each aliquot, and the counts from each aliquot are combined only after each has been run through the remainder of the procedure separately (i.e, centrifugation, resuspension of the precipitate, pooling of the supernatants, and counting).

Further details for the measurement of recoveries of [^{14}C]proline and [^{14}C]hydroxyproline, for the purity of collagenase, and for the specific activity of the enzyme preparations have been published.[56]

C. Enzymes

1. Peptidylproline and Peptidyllysine Hydroxylase Activity

The hydroxylation of proline to hydroxyproline and of lysine to hydroxylysine can be assayed *in vitro* with a proline-rich (hydroxyproline-poor) protocollagen substrate or a lysine-rich (hydroxylysine-poor) protocollagen substrate and either purified enzyme preparations or the supernatant of tissue homogenates containing enzyme protein. The basic conditions of assay and reaction cofactors are the same when the hydroxylation of either proline or lysine is measured.

a. Preparation of Enzyme

TISSUE HOMOGENATE.[4,60] Samples of embryos or tissues are homogenized in cold 0.1 M KCl in 50 mM Tris·HCl buffer, pH 7.6 (10 ml/g) (some methods use 10 mM Tris·HCl buffer, pH 7.6) with a Teflon and glass homogenizer. Skin and bone must be minced manually, frozen in liquid nitrogen or carbon-dioxide ice, and then pulverized by grinding in a mortar or an instrument such as the Wigl-Bug Shaker (VWR Scientific). The frozen powder is solubilized in cold 0.1 M KCl in 50 mM Tris·HCl buffer, pH 7.6. The resultant homogenates are centrifuged at

---

[60] E. Mussini, J. J. Hutton, Jr., and S. Udenfriend, *Science* **157**, 927 (1967).

15,000 g for 30 minutes, and aliquots of the supernatant (1–2 mg protein) are stored at −20° for future assay. Storage time is limited since enzyme activity deteriorates rapidly (reason, unknown[61]), hence it is best to assay activity immediately with preprepared substrate and enzyme standards.

PURIFIED ENZYME. Preparations of peptidylproline hydroxylase which are either homogeneous in the analytical ultracentrifuge and over 90–95% pure on disc electrophoresis,[62] or which are carried through the initial purification steps,[63,64] can be isolated if specific degrees of purity are needed.

b. Preparation of Substrate

SOURCE

"Carrageenin" granuloma: 10 g of minced 9-day-old granuloma produced in young guinea pigs (3–6 months) by the subcutaneous injection of purified carrageen polysaccharide (Sea Kem, Marine Colloids, Springfield, New Jersey). Guinea pigs should be given 50 mg of ascorbic per day in a 5% dextrose solution by direct oral administration for 10 days before the injection of "carrageenin" (4 ml of 1% solution, subcutaneously) and until sacrifice 9 days after injection.

Abdominal skin: 10 g of minced, shaven abdominal skin from young rats (100 g) or guinea pigs (1–3 months)

Embryonic cartilage, bone, or aorta

INCUBATION OF TISSUE. The minced tissue (0.5 g/3 ml of medium) is incubated *in vitro* for 4 hours at 37.8° with an atmosphere of 95% O_2–5% CO_2 in a Dubnoff metabolic shaker (100 oscillations per minute). The incubation medium is the same as that described in Section IV, B,1; 5–10 μCi L-[U-^{14}C]proline or 15–20 μCi of L-[U-^{14}C]lysine per milliliter of final medium are added, and enough α,α'-dipyridyl to make to a final concentration of 1.5 mM.

EXTRACTION OF COLLAGENOUS PEPTIDYL ^{14}C-LABELED SUBSTRATE. The contents of the flask (medium and fragments) are homogenized at 4° for 45 seconds at two-thirds the maximum speed of a Virtis homogenizer and centrifuged at 100,000 g for 1 hour in a Spinco L model ultracentrifuge. The supernatant is dialyzed exhaustively against 1 M KCl and 50 mM Tris·HCl, pH 7.6. The contents of the dialysis bag are boiled

[61] D. J. Prockop, personal communication.
[62] M. Pankalainen, H. Aro, K. Simons, and K. I. Kivirikko. *Biochim. Biophys. Acta* **221**, 559 (1970).
[63] K. I. Kivirikko, H. J. Bright, and D. J. Prockop, *Biochim. Biophys. Acta* **151**, 558 (1968).
[64] J. Halme and K. I. Kivirikko, *FEBS Lett.* **1**, 223 (1968).

for 5 minutes (to inactivate any residual enzymes), divided into 1–2-ml aliquots containing ± 50,000 dpm, and stored at −20°. Aliquots stored thus will give reproducible results over a period of 4–6 weeks. The maximal degree of [^{14}C]proline hydroxylation varies between 15 and 23% with different protocollagen preparations. The variations are explained by the presence of varying amounts of noncollagenous protein proline in the biologically synthesized substrates. Hence, comparison of enzyme activities is possible only within those measurements made with the same substrate. The theoretical maximal hydroxylation of pure collagen is about 43%.

Preparation of the appropriate substrate and measurement of peptidyllysine hydroxylase activity has been described with only a few minor modifications of the procedures provided here.[65] Details of the specific oxidation of hydroxylysine by periodate and its subsequent measurement are provided in Sections V,D, V,F, and V,G.

c. *Assay System*[4,60,66,67]

REACTION MIXTURE (4 ml)

0.1–1.0 ml (0.1–0.5 mg protein) of the 15,000 g supernatant of the tissue homogenate containing enzyme activity (see Hartree modification of Lowry protein method, Section VIII, B)

[^{14}C]Peptidylproline (or lysine) collagenous substrate (1–2 ml, containing ± 50,000 dpm). Determine percentage of total ^{14}C that is [^{14}C]hydroxyproline, or [^{14}C]hydroxylysine, despite the presence of 1.5 mM α,α′-dipyridyl during synthesis (see Section b above).

2.0 ml of 50 mM Tris·HCl buffer, pH 7.6, at 24° to which is added, fresh each day, 4.45 mg FeSO$_4$, 70.4 mg ascorbic acid, and 14.6 mg α-ketoglutarate per 100 ml. Final concentration per 4 ml of reaction mixture is 80 mM FeSO$_4$, 2 mM ascorbic acid, and 0.5 mM α-ketoglutarate.

Crystalline bovine serum albumin, 2 mg, and

Catalase (Sigma), 0.2 mg, both added directly to reagent mixture

SERIES OF FLASKS. Each series should include: (1) duplicate unknown enzyme solutions (supernatant of tissue homogenates); (2) two reagent blanks (no enzyme) with determination of dpm present as [^{14}C]hydroxyproline, or [^{14}C]hydroxylysine per milligram of substrate protein before and after incubation; (3) one of each unknown tissue homogenate with

[65] S. M. Krane, S. R. Pinnell, R. W. Erbe, *Proc. Nat. Acad. Sci. U.S.* **69**, 2899 (1972).
[66] K. I. Kivirikko and D. J. Prockop, *Arch. Biochem. Biophys.* **118**, 611 (1967).
[67] J. Halme, K. I. Kivirikko, and K. S. Simons, *Biochim. Biophys. Acta* **198**, 460 (1970).

added α,α'-dipyridyl (final molarity 1.5 mM) to determine specificity of hydroxylation.

d. Measurement of Total ^{14}C, [^{14}C]Hydroxyproline, and [^{14}C]Hydroxylysine

After incubation, 4.0 ml of concentrated HCl is added to each flask, and the contents are hydrolyzed in sealed tubes at 115° for 16 hours.

Total ^{14}C is counted in the hydrolyzate: 0.05–0.10 ml added to 10 ml of Hyamine-toluene base scintillation solution (1.5 ml Hyamine/10 ml of solution containing 4 g of POP and 50 mg of POPOP in 1 liter of toluene). Appropriate internal standards or suitable externally standardized quench curves are used to determine counting efficiency.

[^{14}C]Hydroxyproline is measured either by paper chromatography or silicic acid column chromatography (see Section VII).

[^{14}C]Hydroxylysine is measured by the periodate oxidation technique (see Section V,D) or by amino acid analysis (see Sections V,D and V,F).

Aliquots of the 15,000 g supernatant of the tissue homogenate are hydrolyzed with 6 N HCl (115° for 16–19 hours) in sealed tubes for measurement of protein content (see Hartree modification of Lowry protein technique, Section VII,B). Total nitrogen content can also be determined as a reference base (see Section IV,A).

e. Calculations

The change in percent total activity in the collagenous substrate that is hydroxyproline or hydroxylysine before and after incubation with the enzyme represents the activity of the enzyme. Alternatively, this can be expressed as disintegrations per minute per milligram (dpm/mg) of protein (or nitrogen). Since the maximal percentage of hydroxylation of the proline in a given substrate is 15–23% (see above), the activity can be described also as a percentage of the maximal hydroxylation possible.

f. Peptidylproline Hydroxylase Activity

This activity can also be measured by the tritium release assay of Hutton et al.[68,69] In this technique collagenous substrate containing [3,4-^{3}H]peptidylproline is prepared by extracting (see Section V,A) newly synthesized acid-soluble collagen (0.5 M acetic acid) from rapidly growing connective tissues (see Section b above) labeled in vitro with L-[3, 4-^{3}H]proline (20 μCi/ml) in the presence of 1.5 mM α,α'-dipyridyl. The assay mixture contains enzyme protein (partially purified protein

---

[68] J. J. Hutton, A. L. Tappel, and S. Undenfriend, Anal. Biochem. 16, 384 (1966).
[69] R. E. Rhoads, N. E. Roberts, and S. Udenfriend, this series, Vol. 17B [176].

or supernatant of tissue homogenate), substrate, and the cofactors described above (see Section c, above). After incubation at 37° for 60 minutes, the water in the same is isolated by distillation and 1 ml is counted in a suitable phosphor solution.

2. Lysyl Oxidase Activity[70,71]

The cross-linking of collagen involves the conversion of certain lysine and hydroxylysine residues to α-aminodipic-δ-semialdehyde (allysine), and α-amino-δ-hydroxyadipic-δ-semialdehyde (hydroxyallysine), respectively. The formation of the actual cross-links occurs by spontaneous condensation of the aldehyde derivatives. The conversion of lysyl and hydroxylysyl to the aldehydes, however, is carried out by the enzyme lysyl oxidase.

Lysyl oxidase activity can be assayed in the pooled media from flasks of cultured fibroblasts, from the cultured fibroblasts[72] themselves (including those obtained from skin biopsy specimens of patients with Marfans and Ehlers-Danlos syndromes), and from extracts of chick cartilage and other embryonic connective tissues.

a. Preparation of Enzyme

Enzyme activity in fibroblast medium is concentrated by adding an equal volume of cold saturated $(NH_4)_2SO_4$ solution adjusted to pH 7.4 with NH_4OH in the cold (4°) for 20 minutes. The precipitate that forms is collected by centrifugation, resuspended in 2 ml of 0.1 M Na_2HPO_4 with 0.15 M NaCl, pH 7.5, and dialyzed against the same buffer at 4° for 4–6 hours. The contents of the dialysis bag are centrifuged at 10,000 g for 10 minutes, and the concentration of NaCl is adjusted to 0.35 M by the addition of 20 ml of 1.75 M NaCl per 100 ml.

Tissue extracts are prepared by homogenizing cells or tissues in 2 ml of 0.1 M Na_2HPO_4 buffer with 0.15 M NaCl, 7.5, and centrifuging at 17,000 g for 10 minutes. The supernatant is assayed directly, or it can be concentrated as already described for culture medium (see above). The final enzyme preparation should be about 7–10 mg of protein per milliliter (see Section VIII,B) with a collagen content of less than 0.5 mg/ml.

b. Preparation of Collagen Substrates

Activity of the supernatant is determined by the tritium-release assay of Siegel and Martin[70,71] with [6-³H]lysine-labeled collagen or elastin

---

[70] S. R. Pinnell and G. R. Martin, *Proc. Nat. Acad. Sci. U.S.* **61**, 708 (1968).
[71] R. C. Siegel and G. R. Martin, *J. Biol. Chem.* **245**, 1653 (1970).
[72] D. L. Layman, A. S. Narayanan, and G. R. Martin, *Arch. Biochem. Biophys.* **149**, 97 (1972).

substrate. The substrate is prepared from aorta elastin, parietal bones, or cartilaginous tibias and femurs. Bones from 15 to 20-day-old chick embryos are incubated in 10 ml of Eagle's minimum essential medium minus lysine and glutamine. The incubation medium described in Section IV,B,1 *in vitro* can also be used, again minus lysine. β-Aminopropionitrile fumarate (50 μg/ml), ascorbic acid (50 μg/ml), penicilin G (2000 units/ml), and either L-[U-^{14}C]lysine (5 μCi/ml) or DL-[6-^{3}H]lysine (20 μCi/ml) are added to the 10 ml of medium. By inhibiting allysine production and collagen cross-linking with BAPN, increased amounts of labeled soluble collagen with a low aldehyde content are extracted.

The bones, or aortas, are incubated for 24 hours with shaking at 37° in an atmosphere of 95% oxygen—5% CO_2. After this they are removed, rinsed well with distilled water, and homogenized in cold 1 M NaCl and 50 mM Tris·HCl buffer, pH 7.4; the homogenate is allowed to stand 1–3 hours at 4°. The mixture is centrifuged (17,000 g for 10 minutes), and the supernatant is dialyzed with three changes against 2 liters of 0.15 M NaCl in 0.1 M Na_2HPO_4, pH 7.4 (some methods call for pH 7.5).

c. *Assay System*

REACTION MIXTURE. The reaction mixture consists of 1 ml of the tissue supernatant enzyme and 0.5 ml of a suspension of the substrate; 1 mM lysine as carrier is added. BAPN (150 μg/ml) is added to some of the flasks containing either unknown enzyme extract or chick cartilage standard enzyme. By inhibiting lysyl oxidase activity (i.e., with BAPN), the specificity of the activity can be verified. The final volume is adjusted to 1.7 ml with 0.16 M NaCl in 0.1 M phosphate buffer, pH 7.4. Any convenient volume can be chosen. Of importance is the need for the final solution to be 0.15 M NaCl in 0.1 M phosphate buffer, pH 7.4. Toluene is added to inhibit bacterial growth. The flasks are incubated at 37° for 8–12 hours with shaking in an atmosphere of air. The reaction is stopped by freezing the samples.

SERIES OF FLASKS. Each series of assays should include flasks with: (1) unknown enzyme (tissue supernatant) in duplicate; (2) unknown enzyme with added BAPN. (3) known chick cartilage enzyme (5–10 mg of protein per milliliter) as standard; (4) standard chick cartilage enzyme with added BAPN; (5) reagent blank (in duplicate)—no added enzyme (only substrate present).

d. *Measurement of Activity*

[6-^{3}H]LYSINE-LABELED SUBSTRATE. After incubation, the water in the sample is isolated by distillation,[68,69] and 1 ml is counted in a suitable phosphor solution e.g., Bray's Hyamine-toluene, or the phosphor solutions described by Juva and Prockop in Section VII.

[^{14}C] LYSINE LABELED SUBSTRATE. After incubation, the pellets are reisolated by centrifugation (105,000 g for 1 hour at 5°) and oxidized with performic acid to convert [^{14}C]allysine to [^{14}C]α-aminoadipic acid. Samples are next hydrolyzed with 6 N HCl at 107° for 72 hours. The amino acids are separated on an automatic amino acid analyzer with the radioactivity of the column effluent monitored continuously. For amino acid analytical techniques see (Section V,G).

3. Collagen Degradation: Collagenase and Lysosomal Enzymes

The effect of hormones (*in vivo* and *in vitro*) on collagen degradation has generally been studied by measurement of the collagenolytic activity in pooled media (with or without partial purification) from cultures of tissue fragments. The concentrated media are assayed by measurement of the release of soluble ^{14}C-containing peptides from gels of reconstituted [^{14}C]glycine-labeled collagen fibrils. This and other assay procedures, such as viscometry and gel electrophoresis, have been described and recently reviewed.[73,74] Several new advances, since these reviews have expanded the available procedures to include (a) methods for the study of the release and activation of a collagenase zymogen,[75] (b) methods for the activation of a bone and skin procollagenase,[76] and (c) methods for the separation of inhibitory, serum α-globulins from collagenases present in tissue homogenates and supernatants.[77] A brief description of the methodology required for these observations follows the more general procedures of assay.

a. Methods for Tissue Culture, Enzyme Purification, Substrate Preparation, and Assay

CULTURE OF TISSUE EXPLANT. The culture of fragments of uterus, skin, cartilage, and embryonic bone with partial purification of the collagenase present in pooled culture media is the most useful system to date. Tissue fragments (from animals which are normal or treated *in vivo* with one or more hormones) are removed under sterile conditions, washed in Krebs-Ringer bicarbonate buffer, pH 7.4 containing 200 units/ml of peni-

---

[73] S. Seifter and E. Harper, this series, Vol. 19 [44].
[74] S. Seifter and E. Harper, *in* "The Enzymes" (P. D. Boyer, ed), Vol 3, p. 649. Academic Press, New York, 1971.
[75] E. Harper, K. J. Block, and J. Gross, *Biochemistry* **10**, 3035 (1971).
[76] G. Vaes, *Biochem. J.* **126**, 275 (1972).
[77] A. Z. Eisen, E. A. Bauer, and J. J. Jeffrey, *Proc. Nat. Acad. Sci. U.S.* **68**, 248 (1971).

cillin and streptomycin, cut into uniform explants (about 2 mm^2), and placed in a series of disposable culture flasks (Falcon Plastics, Richmond, California) containing 2.5 ml of Dulbecco's modified Eagle's medium. The flasks are gassed with 95% O_2–5% CO_2 and incubated at 37°. The *in vitro* effects of hormones on collagenase activity can be studied by adding known amounts directly to the culture medium (e.g., 50 mM progesterone to cultures of uterine tissue[78]).

Culture medium is changed daily, centrifuged, augmented with 1/10 volume of 1 M Tris·HCl (pH 6.5), and stored frozen until the end of the experiment, (5–10 days). The media are then pooled, dialyzed in the cold vs distilled water, and lyophilized. The crude enzyme powder is redissolved in 50 mM Tris·HCl (pH 7.5) containing 5 mM $CaCl_2$ to a final concentration of 10 mg of protein per milliliter (protein is measured by the Hartree modification of the Lowry, Folin-reagent method, see Section VIII,B). Partial enzyme purification can be accomplished by fractional salt precipitation with ammonium sulfate [25–50% $(NH_4)_2SO_4$].

PARTIAL PURIFICATION OF COLLAGENASE. The pooled culture medium is measured and sufficient ammonium sulfate added to effect 20% saturation of the chilled solution. The mixture is agitated gently and allowed to stand for 1 hour in the cold (4°). If a precipitate forms, the mixture is centrifuged at 48,000 g for 1 hour at 4°, and the sediment is discarded. Ammonium sulfate is then added to effect 50–60% saturation. After standing for 1 hour in the cold, the mixture is centrifuged at 48,000 g for 1 hour at 4°. The sediment is dissolved in 1–2% of the initial volume of culture fluid with cold 50 mM Tris·HCl buffer (pH 7.5) containing 5 mM $CaCl_2$ and dialyzed overnight against the same buffer. After dialysis, the material is assayed immediately for collagenase activity or it is lyophilized and stored at −20°.

PREPARATION OF [^{14}C]GLYCINE-LABELED COLLAGEN. The method for preparation of the [^{14}C]glycine-labeled collagen substrate has been published in an earlier volume of this series.[73] [^{14}C]Glycine, 150 μCi, is injected intraperitoneally 8 hours before sacrifice in young guinea pigs. Neutral-salt-soluble collagen is purified by one of several methods, lyophilized, and stored at −20°. It is reconstituted to a concentration of about 30,000 cpm/mg protein (see Section V,A).

ASSAY PROCEDURE. Activity is determined by the release of soluble [^{14}C]glycine-containing peptides from native, reconstituted guinea pig fibrils. A typical reaction mixture (200 μl) contains: 200 μg of [^{14}C]glycine-labeled collagen fibrils (see above) containing approximately 4000

---

[78] J. J. Jeffrey, R. J. Coffey, and A. Z. Eisen, *Biochim. Biophys. Acta* **252**, 136, 143 (1971).

cpm; 100 μl of 50 mM Tris·HCl, 5 mM $CaCl_2$, pH 7.5; 50 μl of partially purified enzyme solution.

The mixture is incubated at 37° for 45 minutes to 3 hours. The reaction is stopped by chilling on ice, and the tubes are centrifuged (12,000 g for 15 minutes). The supernatant is suspended in phosphor solution (Hyamine-toluene base, or Bray's solution containing 4% Cab-o-Sil) and counted in a liquid scintillation counter.

Protein concentrations of enzyme solutions are determined (see Hartree modification of Lowry procedure, Section VIII,B), and the enzyme-specific activities are expressed as micrograms of collagen solubilized per minute per milligram of protein. The amount of collagen present in the pooled media and substrate is measured as hydroxyproline (see Section VII) present in the acid-hydrolyzates of the medium and substrate (6 N HCl at 115° for 16 hours).

Detection and measurement of collagenolytic activity in tissue fragments can also be determined by the direct sterile incubation of tissue fragments on gels of [^{14}C]glycine-labeled collagen. The tissue is incubated on 0.5 ml of reconstituted ^{14}C-labeled substrate in Leighton tubes for 72 hours at 37°. Activity is assessed by centrifuging the contents of the tubes and determining the radioactivity in the supernatant. Activity is expressed as the milligrams of collagen solubilized (cpm/μmole of collagen) per minute per milligram of enzyme protein.

Owing to variations in the specific activity (i.e., [^{14}C]glycine per micromole of hydroxyproline of substrate) or different substrate preparations no standard unit of activity has been defined.

Hormones can be added directly to culture media in either of the above systems (i.e., culture of tissue with pooling of media and purification of collagenase or culture of tissue directly on ^{14}C-labeled gels). Progesterone in concentrations of 50 μM abolishes activity while 10 mM is 32% inhibitory.[78] Estradiol (1 μM to 10 μM) and testosterone (1 μM) are not effective *in vitro*.[78]

b. New Advances

THE ZYMOGEN OF TADPOLE COLLAGENASE.[75] A precursor of tadpole collagenase which is enzymatically inactive and incapable of binding collagen can be isolated. It is prepared from extracts of tadpole tail-fin tissue using the same techniques as those used to isolate the active enzyme from the media of tail-fin cultures.[73] It is activated when incubated with collagenase-free tail-fin culture medium, but not when incubated with trypsin or chymotrypsin (in contrast to the "procollagenase" described by Vaes, see next section). Both the inactive zymogen and the active enzyme have been purified by ammonium sulfate precipitation and agarose chromatography. Their estimated molecular weights are 115,000 and 104,000, respectively. The activator in the collagenase-free culture medium is heat labile and nondialyzable. The technique for the preparation of the zymogen is described by Harper et al.[75]

PROCOLLAGENASE OF MOUSE BONE AND SKIN.[76] A latent collagenase (procollagenase) has been demonstrated in culture media which is activated by trypsin and partially by chymotrypsin. They are not activated by trypsin-activated collagenase, but are by, an as yet, unidentified, thermolabile agent present in culture medium or by purified liver lysosomes, pH 5.5–7.4. Activation by the liver lysosome is of interest in view of the report that lysosomal enzymes, particularly cathepsin D, are implicated in the extracellular breakdown of the protein-polysaccharide component of cartilage (most sensitive at pH 5).[77]

The methodology for the culture of bone and skin fragments, for the preparation of the [^{14}C]collagen substrate, and for the assay of collagenase are described in detail by Vaes.[76] These are not significantly different from those already described.[73] Of significance, however, are the details for trypsin activation which follow.

The pooled culture media are buffered by the addition of 1 M Tris·HCl buffer adjusted to pH 7.5 with HCl to a final concentration of 50 mM. The buffered pooled media are filtered on 0.45 μm Millipore filters to remove floating debris, and assayed immediately or stored at $-20°$. If stored, they are first diluted with 50 mM Tris·HCl buffer, pH 7.5, containing 1 mM CaCl$_2$.

If activation of the media is required prior to assay of collagenase activity, the pooled, buffered media are preincubated with trypsin (2 μg/ml) for 10 minutes at 25°. Soybean trypsin inhibitor (8 μg/ml) is added and the culture is incubated a further 10 minutes at 25° to block the activity of trypsin before assay of collagenase activity (see Section a above).

SEPARATION OF COLLAGENASE FROM INHIBITORY SERUM ANTI-PROTEASES.[79] The role of the serum α-globulins in the control of collagenase activity can be studied both *in vivo* and *in vitro*. Specific antibodies to collagenase (e.g., human skin) have identified the presence of inactive collagenase, which can be activated by the chromatographic separation of the enzyme from the serum anti-proteases. To obtain separation, gel filtration is performed at 4° using reverse flow on a column of Sephadex GV-150 equilibrated with 50 mM Tris·HCl buffer (pH 7.5) containing 5 mM CaCl$_2$ and 0.2 M NaCl. Crude enzyme powder from one day of culture medium is dissolved in the Tris·HCl buffer (pH 7.5) (about 25 mg/ml). Individual column fractions are assayed for collagenase activity and for immu-

[79] J. T. Dingle, A. J. Barrett, and P. D. Weston, *Biochem. J.* **123**, 1 (1971).

nologic evidence of anti-trypsin and α_2-macroglobulin.[79] Antiserum to human α_1-anti-trypsin and α_2-macroglobulin are obtained from Hoechst and Hyland Laboratories, respectively.

Tissue homogenates can be assayed also. About 8–10 g of tissue (e.g., human skin) is trimmed of fat, minced, and homogenized in 4.0 ml of 50 m Tris·HCl (pH 7.5) buffer containing 5 mM CaCl$_2$. The homogenates are centrifuged at 10,000 g for 15 minutes at 4°. The supernatant is assayed for activity and for immunoreactive collagenase.

OTHER HYDROLYTIC ENZYMES. A liver lysosomal cathepsin D, active maximally at pH 5.0, has been implicated in the extracellular breakdown of the protein-polysaccharide component of cartilage and may also be of importance in activating the procollagenase systems.[79] Similarly, intracellular collagenolytic activity has been ascribed to a cathepsin maximally active at pH 3.5. This particular enzyme is present in all tissues and may be of importance in those conditions where active phagocytosis is taking place.[80] Other lysosomal acid hydrolytic enzymes have been implicated in bone collagen resorption.[81] The methods for their measurement have been well-described.[79,80,82]

4. UDP-Glucose: Galactosylhydroxylysine-Collagen (Basement Membrane) Glucosyltransferase

Carbohydrate units linked to hydroxylysine are a characteristic structural feature of collagens and basement membranes. Their biosynthesis proceeds through the action of two specific glucosyltransferases, which have been found to occur in many tissues. The preparation and assay of these enzymes has been described recently.[83]

D. Method for Measurement of Total [^{14}C]Hydroxylysine and Glycosylated and Nonglycosylated [^{14}C]Hydroxylysine (Blumenkrantz and Prockop[84])

1. Preparation of Tissues

INCUBATION *in vitro* IN A CHEMICALLY DEFINED MEDIUM CONTAINING L-[U-^{14}C]LYSINE. Fragments of tissue (minced skin, aorta, or uterus, etc., bone chips, or chick embryos) which have been rinsed free of blood, and marrow, blotted, and weighed are incubated *in vitro* for 4–7 hours

[80] D. J. Etherington, *Biochem. J.* **127**, 685 (1972).
[81] G. Vaes, *J. Cell Biol.* **39**, 676 (1968).
[82] G. Vaes and P. Jacques, *Biochem. J.* **97**, 389 (1965).
[83] R. G. Spiro and M. J. Spiro, this series, Vol. 28 [84].
[84] N. Blumenkrantz and D. J. Prockop, *Anal. Biochem.* **30**, 377 (1969).

in a synthetic medium in a Dubnoff metabolic shaker, in an atmosphere of 95% O_2–5% CO_2, at 37.8° and 100 oscillations per minute. The flasks are equilibrated for 10 minutes under these conditions prior to the start of the time period. The medium is prepared fresh for each use and equilibrated with the appropriate gas mixture immediately prior to use (see preparation of tissue for incubation *in vitro*, Section IV,B). The time between sacrifice of the animal, excision of the tissue, mincing, and the start of the incubation period should not exceed 1 hour. All materials are kept cold (on ice), and the tissue samples are kept moist with cold Krebs-Ringer buffer at all times. A minimum of 100 mg wet weight of cancellous bone, 200 mg of skin or other soft tissue, or 30–50 mg of embryonic connective tissue are required per incubation flask which contains 3 ml of medium.

DIALYSIS. After incubation, the flasks are placed in ice, and 1 ml of 1.33 mM α,α'-dipyridyl per 3 ml of medium are added in order to inhibit further hydroxylation of lysine. Fragments and medium are transferred with rinsing (distilled water) to dialysis bags (dialysis tubing, Fisher, 0.984-inch flat width) and dialyzed, first, overnight against running tap water; second, 48 hours against distilled water at 4° (changing twice daily); third, 3 hours against 1 mM $CaCl_2$; and fourth, 48 hours against 50 mM Tris·HCl buffer with 1 mM $CaCl_2$, pH 7.6 (changing buffer twice daily) also at 4°. Buffer pH should be adjusted at 4° to ensure proper pH.

COLLAGENASE DIGESTION. After dialysis the contents of the bag are transferred with rinsing (50 mM HCl·Tris buffer with 1 mM $CaCl_2$, pH 7.6) to an Erlenmeyer flask. Approximately 1 mg of purified bacterial collagenase (Worthington Biochemical Corp. or Sigma) per 100 mg of tissue is added, and the flask is incubated at 37.8° for 48 hours in an atmosphere of air. Additional 1.0-mg amounts of collagenase are added after 24 and 36 hours of incubation. The original method[84] was set up for embryonic cartilage. For more adult connective tissues, especially bone, larger amounts of collagenase, and longer incubation periods are needed.

After digestion with collagenase the supernatant is withdrawn, the fragments are rinsed twice with 1–3 ml of 50 mM Tris·HCl buffer pH 7.6 with 1 mM $CaCl_2$. The rinse is added to the supernatant, and the entire volume is heated for 15–20 minutes at 60°. The insoluble proteins are removed by centrifuging for 1 hour at 4° (36,000 g). After centrifugation the supernatant is brought to a known volume.

Since newly synthesized collagen (i.e., that collagen produced during the period of hormonal treatment *in vivo* or during the incubation period per se) or collagen not yet polymerized might differ from older, more insoluble collagen, two periods of collagenase treatment are recommended, particularly when dealing with human bone. Accordingly, the rinsed fragments are demineralized over a 4-day period with daily changes of 10% EDTA, pH 7.4 with shaking at 2–4°. After demineralization, the fragments are rinsed with water, Tris·HCl buffer with $CaCl_2$ is added and collagenase incubation is repeated as already described. The resultant supernatant is treated as described for the supernatant after the first incubation with collagenase (see below). Generally, soft tissues do not require two periods of treatment with collagenase.

2. Assay

CHEMICAL PRINCIPLE. The specific periodate oxidation of hydroxylysine gives rise to ammonia and formaldehyde.[85] The assay[84] depends on the formation of a [^{14}C]formaldemethone complex derived from the released [^{14}C]formaldehyde and dimedon. The [^{14}C]formaldemethone is then extracted into toluene. Only the ϵ-carbon of the uniformly labeled [^{14}C]hydroxylysine is measured as [^{14}C]formaldehyde in the assay. The recovery of the ϵ-carbon of hydroxylysine is about 75%. The method distinguishes between glycosylated and nonglycosylated [^{14}C]hydroxylysine since hydrolysis under alkaline conditions splits the peptide bonds but does not remove the ortho-glycosidic substituents. Acid hydrolysis breaks both bonds. Hence the percentage of total (acid hydrolysis) [^{14}C]hydroxylysine that is glycosylated can be calculated as the difference between the radioactive hydroxylysine present in the acid hydrolyzate and that present in the alkaline hydrolyzate. Nonglycosylated [^{14}C]hydroxylysine can also be measured directly in the collagenase-soluble peptides without prior alkaline hydrolysis.

MATERIALS

Citrate–phosphate buffer: Mix 154 ml of 0.15 M citric acid with 346 ml of 0.3 M dibasic sodium phosphate. The pH of the final solution is 6.4.

Carrier hydroxylysine solution: δ-Hydroxylysine-HCl (mixture of DL and DL-allo, Sigma Chemical Corp.) at a concentration of 30 mg/ml

Periodate solution: a 0.3 M solution of sodium metaperiodate (Fisher Scientific Co.) in water is prepared, and stored in the dark in a brown bottle covered with aluminum foil for up to 2 weeks.

Dimedon solution: Dimedon (5,5-dimethyl-1,3-cyclohexanedione) (Mann Research Laboratories) (2.8 g) is dissolved in 50 ml of 50% ethanol and water. The solution is stable for 4 weeks when stored in a brown bottle.

Phosphor solution: 2,5-Diphenyloxazole (PPO, Packard Instrument

[85] R. B. Aronson, F. M. Sinex, C. Franzblau, and D. D. Van Slyke, *J. Biol. Chem.* **242**, 809 (1967).

Co.) 6 g, and 20 mg of 1,4-bis[2-(4-methyl-5-phenyloxazolyl]benzene (POPOP, Packard Instrument Co.) are dissolved in 1000 ml of toluene, and 600 ml of ethylene glycol monomethyl ether (methyl Cellosolve, Fisher) is added.

Stock Tris·HCl buffer: $CaCl_2$, 1 mM in 50 mM Tris·HCl buffer, pH 7.6. To make 2 liters of stock solution (10 mM $CaCl_2$ in 500 mM Tris) dissolve 2.2198 g of $CaCl_2$ in 500 ml of Tris (2 M) and 650 ml of 1 N HCl. Check the pH and adjust to 7.6 by adding more HCl or Tris). Then dilute to 2 liters. Working solution: dilute stock $\times$ 10. Prepare in cold room.

DIRECT PROCEDURE FOR NONGLYCOSYLATED [^{14}C]HYDROXYLYSINE. The centrifuged supernatant containing the collagenase-soluble peptides is brought to a known volume (about 5.0 ml), and 1–3 ml are pipetted into a screw-capped tube (Teflon-lined cap, 200 $\times$ 25 mm Kimax). Then 10 ml of citrate-phosphate buffer, pH 6.4 is added, followed by 15 ml of toluene. The mixture is shaken on a Burrell wrist shaker in the horizontal position for 15 minutes. The toluene is removed with suction and discarded. Another 15 ml of toluene is added, and mixture is shaken a second time. The toluene is again removed and discarded. Extraction with toluene is necessary in order to remove [^{14}C]lysine present in proteins other than collagen which have been solubilized by collagenase. After shaking, the mixture may require centrifugation for 5–10 minutes at 10,000 g for good separation of the layers.

After toluene extraction, add 0.1 ml of hydroxylysine carrier solution, and 2.0 ml of 0.3 M sodium metaperiodate. Mix and allow to stand 1–2 minutes. Add 3 ml of dimedon solution and mix on a Vortex for about 1 minute. Add exactly 15 ml of toluene and shake the mixture for 30 minutes on a Burrell shaker in the horizontal position. Transfer exactly 10 ml of the upper toluene phase to a counting vial containing 10 ml of phosphor solution and count in a liquid scintillation counter (Packard Model 3375 or its equivalent). Suitable quench curves or internal standards should be established for the particular system being used (e.g., the phosphor solution described, Bray's solution).

The aliquots of the collagenase-soluble peptides that are measured should be concentrated sufficiently to give at least 200 cpm above background. In human tissue this requires incubation of at least 150 mg wet weight of cancellous bone and 200 mg wet weight of skin.

Nonglycosylated [^{14}C]hydroxylysine in the isolated peptides can also be assayed after hydrolysis under alkaline conditions which splits the peptide bonds but does not remove the *ortho*-glycosidic substituents. For the technique of alkaline hydrolysis, see Blumenkrantz and Prockop.[84] Values obtained on the unhydrolyzed peptides with this rapid assay are comparable to those obtained after alkaline hydrolysis with either the assay or the chromatographic procedures described by others. A rapid chromatographic method for the simultaneous quantitative measurement of [^{14}C]lysine, hydroxylysine, and hydroxylysine glycosides from alkaline hydrolyzates of biological tissues has been described.[86]

MEASUREMENT OF TOTAL [^{14}C]HYDROXYLYSINE (GLYCOSYLATED PLUS NONGLYCOSYLATED). The remainder of the centrifuged supernatant containing the collagenase-soluble peptides is hydrolyzed in a final normality of 6 N HCl, for 16 hours, at 115° in sealed tubes. After hydrolysis, aliquots are set aside for measurement of hydroxyproline, total protein, total nitrogen, and, if desired, for separation of [^{14}C]lysine and [^{14}C]hydroxylysine (see Amino Acid Analyses). The remainder of the hydrolyzate is evaporated to dryness, resolubilized in water, reevaporated, and brought to 2–3 ml with water. The now neutral sample is buffered with citrate–phosphate buffer, pH 6.4 and extracted twice with toluene (discarding the toluene layer each time) as described for nonglycosylated [^{14}C]hydroxylysine. Hydroxylysine carrier, periodate, and dimedon are added and mixed; the solution is extracted with 15 ml of toluene. Of this final toluene extract, 10 ml is counted in 10 ml of phosphor solution. Since acid hydrolysis breaks the *ortho*-glycosidic linkages as well as the peptide bonds the counts recovered represent the label in the glycosylated plus the nonglycosylated hydroxylysine.

CALCULATIONS. Periodate oxidation of hydroxylysine gives rise to ammonia and formaldehyde. The assay presented depends on the formation of a [^{14}C]formaldemethone complex from the released [^{14}C]formaldehyde and dimedon. The [^{14}C]formaldemethone complex is then extracted into toluene. The recovery of the ϵ-carbon is about 75% and only the ϵ-carbon of the uniformly labeled [^{14}C]hydroxylysine is measured as [^{14}C]formaldehyde in the assay. Hence, factors of 6 and of 100/75 are present in the calculations. Of the 15 ml of toluene added to extract the [^{14}C]formaldemethone complex, only 10 ml are removed for counting. Hence an additional factor of 15/10 is present. The formula for calculating the total dpm in the aliquot measured is

$$[(\text{cpm}) - (\text{background})]/\text{efficiency} \times 100 \times 6 \times 100/75 \times 15/10 = \text{dpm}$$

From this one can calculate the dpm/g wet weight, or dpm/mg DNA, or dpm/μmole hydroxyproline present in the same collagenase soluble extract.

3. Separation of [^{14}C]lysine and [^{14}C]Hydroxylysine

Samples labeled with [^{14}C]lysine and [^{14}C]hydroxylysine can be hydrolyzed in three times distilled 6 N HCl, evaporated under reduced

[86] R. S. Askenasi and N. A. Kefalides, *Anal. Biochem.* **47**, 67 (1972).

pressure, and the [^{14}C]hydroxylysine and [^{14}C]lysine separated by chromatography at pH 5.25 on the short column of a Beckman Model 116 amino acid analyzer. Aliquots of the column eluate are counted in a liquid scintillation counter. Approximately 90% of the total ^{14}C is recovered in peaks corresponding to hydroxylysine and lysine.[87] This method and the one referred to already[86] are useful when purified collagens are being analyzed.

Recent improvements for the more rapid determination of hydroxylysine making use of the same basic principles described above[84] have been published[88] as well as other new chromatographic techniques.[89]

4. Separation of Glycosylated Hydroxylysines[90]

Since the O-glycosidic linkage of Hyl(GlcGal) and Hyl(Gal) is alkali-stable, purified skin and bone collagens can be hydrolyzed under N_2 in alkali-resistant glass tubes in 2 M KOH at 108° for 20 hours. The hydrolyzates are neutralized with 1 M HClO$_4$, and the resultant precipitate is separated by filtration. Aliquots of the hydrolyzate are placed on a column (2 × 98 cm) of Bio-Gel P-2 (200–400 mesh) and eluted with 0.1 M acetic acid. The column is previously calibrated with purified Hyl(GlcGal), Hyl(Gal) and hydroxylysine, thereby permitting the isolation and identification of these amino acids in the test hydrolyzate. The appropriate fractions containing Hyl(GlcGal), Hyl(Gal) and hydroxylysine are analyzed directly by amino acid analysis using the conditions of Miller and Piez.[91]

Recent improved and rapid methods for the assay of hydroxylysine and its glycosylated residues are described in detail in the references listed in Section 3 above. Determinations of glycosylated hydroxylysine residues are also possible in urine.[92]

E. Chromatographic Separation of the Subunits of Several Collagens

These techniques are adequately described in the references listed. In general, the procedures are modifications of the original technique de-

---

[87] N. Blumenkrantz, J. Rosenbloom, and D. J. Prockop, *Biochim. Biophys. Acta* **192**, 81 (1969).
[88] N. Blumenkrantz and G. Asboe-Hansen, *Anal. Biochem.* **56**, 10 (1973).
[89] R. Askenasi, *Biochim. Biophys. Acta* **304**, 375 (1973).
[90] S. R. Pinnell, R. Fox, and S. M. Krane, *Biochim. Biophys. Acta* **229**, 119 (1971).
[91] E. J. Miller and K. A. Piez, *Anal. Biochem.* **16**, 320 (1966).
[92] L. W. Cunningham, J. D. Ford, and J. P. Segrest, *J. Biol. Chem.* **242**, 2570 (1967).

scribed by Piez, Eigner, and Lewis[93] and by Bornstein and Piez.[53] The method has been further modified by Kang et al.[94] for rat tail-tendon collagen and by Miller et al. for chick cartilage collagen.[95] Basically, purified preparations of soluble collagens, reconstituted fibrils and tendons, and extracts of sternal of tibial cartilages are cleaved with CNBr in a manner similar to that described by Bornstein and Piez.[53] The proteins are suspended in 70% formic acid (10 mg/ml), the vessel is flushed with nitrogen, and 500-fold molar excess (relative to the methionine content) of CNBr is quickly added. The reaction is allowed to proceed for 4 hours at 30°. Any insoluble material is removed by centrifugation at 30,000 g, and the supernatant is lyophilized. The dried peptides are redissolved in 50 ml of water and relyophilized to ensure complete removal of CNBr and formic acid. After CNBr cleavage, the peptides are fractionated on a series of carboxymethyl (CM) cellulose columns followed by columns of phosphocellulose. The authors of this survey have not had any personal experience with these particular procedures and the reader is referred to the references listed for further details.

F. Amino Acid Analyses

The original method of Piez and Morris[96] has been modified by several procedures which reduce the time required for complete amino acid analyses. The method of Miller and Piez[91] is the most commonly used. More recently, Mashburn and Hoffman have devised an automated technique for the analyses of polysaccharide and protein in cartilage.[97] Minor modifications have been introduced by Osborne et al.[98] for collagen, and by Gerber and Kemp for elastin.[99] All these methods are reproducible and readily set up by someone familiar with the automatic amino acid analyzer (Beckman-Spinco, Model 120 C, or its equivalent). The detailed procedures are adequately described by the authors of the references provided. A wide variety of new approaches to methods for the preparation of protein hydrolyzates are available.[100] A fluorometric procedure for amino acid assay also has been developed by S. Udenfriend et al.[101]

[93] K. A. Piez, E. A. Eigner, and M. S. Lewis, Biochemistry **2**, 58 (1963).
[94] A. H. Kang, K. A. Piez, and J. Gross, Biochemistry **8**, 1506 (1969).
[95] E. J. Miller, J. M. Lane, and K. A. Piez, Biochemistry **8**, 30 (1969).
[96] K. A. Piez and L. Morris, Anal. Biochem. **1**, 187 (1960).
[97] T. A. Mashburn, Jr. and P. Hoffman, Anal. Biochem. **36**, 213 (1970).
[98] R. M. Osborne, R. W. Longton, and B. L. Lamberts, Anal. Biochem. **44**, 317 (1971).
[99] G. E. Gerber and G. D. Kemp, J. Chromatogr. **71**, 361 (1972).
[100] H. D. Spitz, Anal. Biochem. **56**, 66 (1973).
[101] S. Udenfriend, S. Stein, P. Bohlen, and W. Dairman, in "Chemistry and Biology of Peptides," p. 655. Ann Arbor Sci. Publ. Ann Arbor, Michigan, 1972.

G. Preparation of Human Tissue Obtained by Biopsy for Amino Acid Analysis

1. Skin

The subcutaneous fat is removed by mechanical scraping, the skin is minced finely with scissors and weighed. It is extracted 24 hours in the cold with gentle shaking in 0.45 N NaCl in 20 mM Na$_2$HPO$_4$ buffer, pH 7.0 (10 ml per gram of tissue). The supernatant is removed, the fragments rinsed with the same solvent, and the rinse added to the extract. Neutral-salt-soluble collagen (micromoles of hydroxyproline per gram wet weight) can be measured on this if desired. The fragments are next extracted 24 hours in 0.5 M Na citrate, pH 3.8. The supernatant is discarded. The insoluble residue is washed with water, and freeze-dried. Appropriate aliquots are hydrolyzed in sealed tubes preferably for 20 hours in 6 N HCl at 108°.

2. Cancellous Bone

The bone is fragmented, washed several times with Krebs-Ringer bicarbonate buffer, pH 7.4, vigorously shaken to rid the material of marrow, blotted, and weighed. The fragments are extracted overnight in the cold with 0.45 M NaCl in 20 mM Na$_2$HPO$_4$ buffer, pH 7.0, and the supernatant is discarded. The fragments are rinsed several times with water and then extracted in 10% EDTA, pH 7.4, for 6 days with shaking in the cold. The EDTA is changed daily. The now demineralized matrix is washed several times with distilled water and freeze-dried. The best procedure is either grinding the bone manually with a mortar and pestle which are thoroughly chilled with dry-ice, or in a Wigl-Bug Shaker (VWR Scientific) also chilled with dry-ice. Aliquots of the powder are hydrolyzed as described for the skin.

H. Collagen Cross-Linking

1. Introduction

A primary step in collagen cross-linking is the formation of peptide-bound α-aminoadipic-δ-semialdehyde (allysine) by deamination and oxidation of the ε-carbon atom in lysine. In collagen, these lysine (or hydroxylysine) derived aldehydes can then react to form cross-links by either an aldol condensation with a second allysyl residue or by Schiff base formation with the ε-amino group of other lysyl or hydroxylysyl residues. As indicated in the section of lysyl oxidase (Section V,2), the formation of allysine in collagen is catalyzed by the enzyme lysyl oxidase.[102] Lathyrogens such as β-aminopropionitrile (BAPN) block lysyl oxidase, thereby preventing the formation of collagen and elastin cross-links (Section III,B). Animals fed lathyrogens develop connective tissue disorders, such as dissecting aneurysms, scolioses, and exostoses because of the decreased collagen tensile strength which ensues.

Aldehyde have been known to occur in collagen since their detection by Paz et al., in 1965,[103] who adapted the specific colorimetric reaction of Sawicki et al.[104] Aldehydic links have been determined also with the thiosemicarbaside-binding technique developed by Rojkind and Gutierrez.[105] Since both reactions are useful in preliminary studies of the effect of hormones on collagen cross-linking,[106] details of each are provided below.

More sophisticated methods for the measurement of specific cross-link derivatives are available. These include characterization of the aldehydic precursors and their Schiff cross-links following either cyanogen bromide cleavage of purified soluble collagens or reduction of prepared insoluble collagen with $NaBD_4$ and $NaBT_4$.[107-110] Preparative and analytical chromatography, amino acid analyses, isolation of the radioactive components and conversion of these to volatile derivatives with their analysis by mass spectrometry[109-112] are all in use. Peptides resulting from cyanogen bromide cleavage are studied under the experimental conditions of in vitro aging, incubation with extracts containing lysyl oxidase, and incubation with a variety of copper chelators or enzyme inhibitors.[48,113] These and more recently described cross-links of collagen and elastin are reviewed in detail by Gallop et al.[2]

Measurement of cross-links in insoluble collagen has been useful in

[102] R. C. Siegel and J. R. Martin, *J. Biochem.* **245**, 1653 (1970).
[103] M. A. Paz, O. O. Blumenfeld, M. Rojkind, E. Henson, C. Furfine, and P. M. Gallop, *Arch. Biochem. Biophys.* **109**, 548 (1965).
[104] E. Sawicki, T. R. Hauser, T. W. Stanley, and W. Elbert, *Anal. Chem.* **33**, 93 (1961).
[105] M. Rojkind and A. M. Gutiérrez, *Arch. Biochem. Biophys.* **131**, 116 (1969).
[106] D. H. Henneman, *Proc. Int. Congr. Endocrinol. 4th, 1972, Excerpta Med. Found. Int. Congr. Ser.* p. 1109.
[107] M. L. Tanzer, *J. Biol. Chem.* **243**, 4045 (1968).
[108] G. L. Mechanic and M. L. Tanzer, *Biochem. Biophys. Res. Commun.* **41**, 1597 (1970).
[109] G. L. Mechanic, P. M. Gallop, and M. L. Tanzer, *Biochem. Biophys. Res. Commun.* **45**, 644 (1971).
[110] A. J. Bailey, C. M. Peach, and L. J. Fowler, *Biochem. J.* **117**, 819 (1970).
[111] M. A. Paz, E. Henson, R. Rombauer et al., *Biochemistry* **9**, 2123 (1970).
[112] M. A. Paz, P. M. Gallop, O. O. Blumenfeld et al., *Biochem. Biophys. Res. Commun.* **43**, 289 (1971).
[113] K. Deshmukh and M. E. Nimni, *Biochemistry* **10**, 1640 (1971).

the diagnosis of clinical disease in man. For example, quantitative changes in cross-links and their precursors in bone collagen have been reported in vitamin D deficiency and a lack of reducible intermolecular cross-links in dermal collagen is reported in Ehlers-Danlos syndrome.[19,114] Future investigations may well demonstrate other examples.

A most useful technique, both rapid and inexpensive, has been devised by N. R. Davis for the determination of lysine-derived collagen cross-links.[115] This type of measurement enables investigators to establish that glycosylated hydroxylysine residues participate in the biosynthesis of bone collagen cross-links.[116]

From a functional standpoint, collagen which lacks the appropriate cross-links has a decrease in tensile strength. This can be determined on skin, blood vessels, and even the uterus, by the use of an Instron machine. The technique is described below (Section V,H,4).

2. Measurement of Total Aldehydic Residues[103,104]

CHEMICAL PRINCIPLE. Aldehydes are converted to highly colored tetra-azopentamethine cyanine dyes by a specific and sensitive reaction with N-methyl benzothiazolone hydrazone hydrochloride (MBTH).

REAGENTS

1% MBTH: 100 mg of N-methyl benzothiazolone hydrazone (Aldrich Chem. Co.) is dissolved in 10 ml of distilled water, filtered, and stored in the cold for not more than 10 days.

$FeCl_3$, 0.2%: 200 mg of $FeCl_3$ are dissolved in 100 ml of distilled water on the day of use.

Acetone, reagent grade

Glycine-HCl buffer: Approximately 980–985 ml of 0.1 M glycine (7.51 g/liter water) are adjusted to pH 4.0 with 15–20 ml of 0.1 N HCl.

Acetaldehyde standard: prepare a stock standard of 10 mM (10 μmoles of acetaldehyde per milliliter). Dilute this 1:100 with water on day of use and run 0.1–1.0-ml aliquots through the procedure (0.01–0.1 μmole ml).

Purified collagen (Section V,A).

PREPARATION OF COLLAGEN GEL. Purified collagen (5–10 mg of lyophilized powder) which has been reconstituted (see Section V,A) is dialyzed against 0.1 M glycine·HCl buffer, pH 4.0, for 24–48 hours in the cold. The solution is next gelatinized by heating at 60° for 30 minutes.

---

[114] G. L. Mechanic, *Biochem. Biophys. Res. Commun.* **47**, 267 (1972).
[115] N. R. Davis, *Biochem. Biophys. Res. Commun.* **52**, 877, (1973).
[116] D. R. Eyre and M. J. Glimcher, *Biochem. Biophys. Res. Commun.* **52**, 663 (1973).

PROCEDURE. To 1.0 ml of the dialyzed, gelatinized material (equivalent to 5–10 mg of collagen) add 0.2 ml of 1% MBTH. Heat the mixture at 100° for 3 minutes. Cool the tube to room temperature and add 2.5 ml of 0.2% FeC_3. Allow mixture to stand at room temperature for 5 minutes. Add 6.5 ml of acetone, mix, and read absorbance at 670 nm in a Coleman, Jr. spectrophotometer. Run a reagent-glycine-buffer blank and a series of standards 20 nmoles to 1.0 µmole/ml of acetaldehyde) with each series of determinations. The standard curve is linear up to 0.1 µmole of acetaldehyde per milliliter.

It is essential that the pH value of the collagen gel and MBTH be between pH 3 and 4. Acetate and citrate ions (in concentrations over 50 mM) and phosphate ions (in concentrations over 10 mM) interfere with the color reaction. Hence, most purified preparations of collagen require initial dialysis against glycine·HCl buffer, pH 4.0.

Neutral-salt collagen has about 0.9 µmole acetaldehyde equivalent per 100 mg of collagen, acid-soluble collagen about 1–2 µmole/100 mg, and cysteamine-soluble collagen about 2–3 µmole/100 mg. Since the weight of the collagen studied might vary from solution to solution when dialyzed against glycine·HCl buffer, the content of aldehyde can be expressed as micromoles of acetaldehyde per 100 µmoles of hydroxyproline. For this, an aliquot of the same glycine-dialyzed, gelatinized collagen sample is hydrolyzed in a sealed tube in 6 N HCl for 16 hours at 115°, and the content of hydroxyproline is determined (see Section VII).

3. Incubation of Native Collagen with Thiosemicarbazide (TSC)[105]

CHEMICAL PRINCIPLE. The aldehydes present in collagen react with TSC to form both saturated and α-β unsaturated aldehyde thiosemicarbazones. The kinetics of the reaction of binding TSC to the protein is followed by observing the appearance of the saturated aldehyde thiosemicarbazone (λ_{max} 265) and α-β unsaturated thiosemicarbazone (λ_{max} 295) absorbance peaks in a Beckman DB spectrophotometer.

REAGENTS AND EQUIPMENT
   Thiosemicarbazide (TSC), 10 mM: 91.13 mg TSC/100 ml, Eastman Organic Chemicals.
   Citric acid–sodium citrate buffer, 0.1 M, pH 4.3
   Beckman DB spectrophotometer
   Lyophilized purified collagen

PROCEDURE. Dissolve the lyophilized, purified collagen (about 5 mg/ml) preparation of 0.1 M citric acid–sodium citrate buffer, pH 4.3, with stirring overnight in the cold. Remove undissolved particles by centrifugation (100,000 g) for 1 hour in a Spinco Model L ultracentrifuge.

To 1.0 ml of the clear, reconstituted collagen solution add 0.2 ml of 10 mM TSC. Mix and immediately determine the absorption spectra between 250 and 350 nm (zero-hour reading). Prepare a blank by adding 0.2 ml of 10 mM TSC to 1.0 ml of 0.1 M citric acid–sodium citrate buffer, pH 4.3. Incubate the tubes (collagen samples and citric acid–sodium citrate buffer–TSC blank) at 20° and read the spectra after 1, 3, 5, 10, and 24 hours. Saturated thiosemicarbazones have a peak absorbance at 265 nm and the α-β unsaturated thiosemicarbazones at 295 nm.

In order to subtract the UV absorbancy of the collagen itself a second set of tubes for each sample of collagen is prepared: 1.0 ml of the same original collagen solution is diluted to 1.2 ml by the addition of 0.2 ml of distilled water, incubated at 20°, and the spectra (250–350 nm) determined at 1, 3, 5, 10, and 24 hours. All readings of these tubes (i.e., those incubated with 0.2 ml of H_2O) are read against a blank of 1.0 ml citric acid–sodium citrate buffer to which 0.2 ml of water has been added. The absorbancy reading of the collagen per se is subtracted from that of the collagen plus TSC.

Model aldehyde and ketone thiosemicarbazones are prepared as follows: a 2 M excess of a carbonyl compound (acetaldehyde, acetone, crontonaldehyde) is added to a solution of TSC in water. After warming in a water bath at 50° for 15 minutes, the solutions are cooled and kept at 4° for 2 hours. The precipitate which forms is filtered, washed, and air-dried. The compounds are finally recrystallized from methanol or ethyl acetate petroleum ether. Absorbance spectra at 265 and 295 nm are determined on known molar concentrations of these compounds. The model compounds have a molar extinction coefficient of 14×10^3 for the saturated thiocarbazones. About 0.76 μmole of the saturated aldehyde thiosemicarbazone is formed per 100 mg of collagen.

4. Preparation of Skin for Stress-Strain Testing

PRINCIPLE

A standardized strip of skin (or blood vessel) is stretched under constant conditions until it breaks in the center. The force in grams required to do this represents the tensile strength of the tissue.

SPECIAL EQUIPMENT

A model TTBM Instron instrument is the instrument of choice. The Instron, type "C-cell" with a 0.5–25.0 K range is appropriate for guinea pig and rat skin. Other types of "cells" with higher ranges are needed for human skin.

A cutting device which cuts from a sample of skin several strips of standardized size (e.g., approximately 2–3 cm along the vertical axis and 3/4 cm along the horizontal axis). Several single-edged razor blades set in a plastic holder are appropriate.

PROCEDURE. A sample of full-thickness, shaven abdominal skin (8 × 2 cm) is laid on a cutting board being sure that the 8-cm length is along the cephalocaudol axis and that the midline aponeurosis is not included. The skin is stripped free of subcutaneous tissue and fat, wiped on both sides with gauze soaked in acetone, and pinned to the cutting board. Care should be taken to keep the skin slightly moist with a few drops of saline (0.15 M NaCl). It must not be stretched when pinned to the board. The cutting device is placed on the skin and hammered evenly through the skin.

The strips (2–3 per sample) are placed in boiling distilled water for exactly 2 minutes and immediately quenched in cold water. The sample is then air-dried and stored at room temperature for 1–2 days before being tested, or it is immediately tested on the strip-strain apparatus (Instrom). The latter procedure is preferred.

After the tensile strength has been measured on each strip (several per skin sample), the skin is minced, defatted (acetone for two 1-hour periods, followed by ether for 4 hours with shaking in the cold), and dried *invacuo* (24 hours at 110°). It is next hydrolyzed in sealed tubes in 6 N HCl at 115° for 16 hours. Hydroxyproline is measured on the hydrolyzate (Section VII).

The results are expressed as the force in grams per centimeter of skin per micromole of hydroxyproline in that centimeter of skin that is required to break the strip in the midline. In rat skin the normal control value is 90.26 ± 4.3 g cm per micromole of hydroxyproline.[117]

VI. Measurement of DNA Content

Rates of collagen biosynthesis, turnover, or degradation should be expressed in terms of the number of cells participating in the metabolic process if comparisons of rates between tissues are to be made.[5]

Many methods for measuring DNA have been described. Of these the method of W. S. Schneider *et al.*[118] has proved reliable, sensitive, and convenient in the authors' hands. The procedure is based on the finding that nucleic acids are preferentially soluble in hot trichloroacetic acid while other tissue compounds are not. It is applicable to homogenates of soft tissues (uterus, granuloma, prostate, ovary, etc.) and to chips of cancellous bone. As little as 50 mg of the latter can be studied. All tissues, in particular bone, are first rinsed free (with physiological saline) of marrow or blood cells. Soft tissues are first homogenized in 0.15 M NaCl

---

[117] D. H. Henneman, *Clin. Orthop.* **83**, 245 (1972).

in 20 mM phosphate, pH 7.4 (about 2 g of tissue per 10 ml). One milliliter of the resultant homogenate (about 20% solution) is mixed with 2.5 ml of cold 10% trichloroacetic acid (TCA) and centrifuged (1000 g for 1 hour). The sediment is washed and centrifuged twice with 10% TCA. Bone is first extracted with 0.1 N NaOH (2.0 ml/200 mg) with gentle shaking overnight, at room temperature. The weak alkaline extract of bone is centrifuged (35,000 g for 1 hour at 4°), and an equal volume of 20% TCA is added to the supernatant. The resultant precipitate is centrifuged (100 g for 1 hour), and washed and centrifuged twice more with cold 10% TCA (500 g for 20 minutes). The tissue sediment (either bone or soft tissue TCA extract is carried through the procedure as outlined by W. C. Schneider.[118]

VII. Measurement of Proline and Hydroxyproline and Their ^{14}C-Labeled Derivatives

A. Proline[119]

CHEMICAL PRINCIPLE. A characteristic red color develops when proline interacts with ninhydrin in acidic solutions. Interfering basic amino acids (lysine, hydroxylysine, and ornithine) are removed by shaking protein acid hydrolyzates with Dowex 50W-X8 No. 1930 (200–400 mesh) resin.

MATERIALS

Permutit: Dowex ion-exchange resin (AG 50W-X8, 200–400 mesh, Bio-Rad).

Ninhydrin reagent: 5 g of ninhydrin, 120 ml of glacial acetic acid, and 80 ml of 6 M phosphoric acid are combined, heated at 70° to dissolve, and cooled to room temperature before using. Can be stored at 4° in a dark bottle for several weeks.

Benzene (ACS-Fisher)

Glacial acetic acid

Phosphoric acid (6 M) (85% purity; 1.7 specific gravity; 407 ml/liter for 6 M solution.

PROCEDURE. Aliquots of acid hydrolyzates of tissue are diluted to contain 5–20 μg of proline. In general, about a 500-fold dilution per 100 mg of dried, defatted tissue is needed. The solutions (within the pH range of 1–7) are shaken with about one-tenth their weight of Dowex for 5 minutes. In general, about one Pasteur-pipette tip-full is sufficient, unless large amounts of lysine or ornithine are present (as in the incubation

[118] W. C. Schneider, this series, Vol. 3 [99].
[119] W. Troll and J. Lindsley, *J. Biol. Chem.* **215**, 655 (1955).

medium—see Section IV,B), at which time more Dowex is required. The mixture is allowed to stand for 5 minutes to let the Dowex settle. At least 6 ml is treated with Dowex since 5 ml of the supernatant is carried through the remainder of the procedure.

Five milliliters of the supernatant is transferred to a screw-capped tube (200 × 25 mm Kimax) with white rubber-lined cap; 5 ml of glacial acetic acid and 5 ml of ninhydrin reagent are added, and the mixture is heated in a vigorously boiling water bath for 1 hour. The tubes are cooled to room temperature in an ice bath; 5 ml of benzene is added, and the tubes are mixed vigorously on a Vortex for 1–2 minutes.

An aliquot of the benzene layer is transferred to a Coleman tube, and the absorbance read on a Coleman Jr. spectrophotometer at 515 nm. A reagent blank with 5 ml of water is carried through the procedure. A standard curve of 5–20 µg/ml is run with each series of determinations. The method is reliable within 1–2%.

B. Hydroxyproline: Stegemann Procedure[120]

CHEMICAL PRINCIPLE. Hydroxyproline is oxidized with chloramine-T in a weak-acid buffer (pH 6.0) to pyrrole-2-carboxylic acid. The pyrrole is next formed in the presence of perchloric acid which destroys the chloramine-T and prepares the material for the formation of a chromophore with p-dimethylaminobenzaldehyde. The method is sensitive to 10 nmoles of hydroxyproline. It is particularly useful with relatively pure preparations of collagen extracts and insoluble residues despite the possibility of errors introduced during the oxidation step.

MATERIALS

NaOH and HCl for neutralization, 0.01–2.0 N

Charcoal-resin mixture for clarification of urine hydrolyzates and any others with high salt or humin content (see below, method of Prockop-Udenfriend technique for preparation).

Chloramine-T solution (50 mM): 2.82 g of chloramine-T is dissolved in 40 ml of cold water; 60 ml of ethylene glycol monomethyl ether and 100 ml of acetate buffer (see below) are added, mixed, and stored in a brown bottle. Solution keeps about 1 month stored in the cold.

Acetate–citrate buffer, pH 6.0: Dissolve 50 g of citric acid in 500 ml of water. Add 12 ml of glacial acetic acid, 120 g of sodium acetate·3 H$_2$O, and 30 g of NaOH. Bring to 1 liter with water, check the pH, and add a few drops of toluene as a preservative. Keep several months when stored in the cold.

[120] H. Stegemann, *Hoppe-Seyler's Z. Physiol. Chem.* **311**, 41 (1958).

p-Dab solution (Ehrlich's reagent): Dissolve 20 g of p-dimethylaminobenzaldehyde in 200 ml of ethylene glycol monomethyl ether. Store in a dark bottle. Keeps approximately 1 month when stored in the cold.

Perchloric acid, 4 M: Dilute 325.43 ml of a 70–72% solution or 379.65 ml of a 60–62% solution to 1 liter with water.

PROCEDURE. The dried, defatted tissue or dried extract of soluble collagen is hydrolyzed with 6 N HCl in a sealed tube for 16 hours at 115°. About 1 ml of acid per total extract per gram of soft tissue and about 2 ml of acid per 25–50 mg of dried, defatted insoluble residue are needed. The hydrolyzate of soluble collagen extract is clarified with charcoal-resin if the extract has not been dialyzed against distilled water prior to hydrolysis. Insoluble residues may need clarification if the humin content is high. Urine hydrolyzates always require clarification. About 1 g of resin per 10 ml of diluted hydrolyzate is added and mixed, and the solution is centrifuged (15,000 g for 30 minutes). An aliquot of the supernatant (1–5 ml) is neutralized to pH 7.0. If the hydrolyzate contains only small amounts of hydroxyproline, it should be evaporated to dryness and redissolved in water twice. Neutralization becomes more accurate or is unnecessary as a consequence of this step. Final neutralization should be accomplished with a minimum of either 0.01 N NaOH or 0.01 N HCl; 0.2% phenolphthalein in absolute alcohol is used as indicator. To 2 ml of this neutralized diluted sample, add 1 ml of chloramine-T, mix, and hold at room temperature for 20 minutes. Add 2 ml of 4 M perchloric acid, mix, and hold at room temperature for 5–7 minutes. Add 1 ml of p-Dab solution, mix, and place tube in a 60° water bath for exactly 15 minutes. Cool the tubes in ice water, transfer to Coleman tubes, and read absorbancy at 550 nm in a Coleman Jr. spectrophotometer.

Run a reagent blank with water and dilute standards containing 5, 10, and 20 μg per 2 ml of solution. Stock standard hydroxyproline is prepared by dissolving 25 mg of hydroxy-1-proline per 100 ml of 10^{-3} N HCl. Dilute the stock solution 1:10 for analysis.

The dilutions required for any given hydrolyzate depend upon the tissue, the type of extract, and the species of animal. Skin has more soluble collagen than bone per gram wet weight of tissue, and the rat has more soluble collagen per unit weight than the dog or monkey. Man has the least of all primates. Normal rat skin (depending upon the age of the rat) has aproximately 5 μmoles of hydroxyproline per gram of neutral salt-soluble collagen and about 5–10 μmoles per gram of acid-soluble collagen. The amount of collagen soluble in neutral salts and acid in bone is 1–3 μmoles/g in the rat and usually less than 1 μmole/g in man. The amount of collagen (i.e., hydroxyproline) soluble in any given solvent varies with the administration of hormones. 17β-Estradiol, for example, produces a marked reduction of acid-soluble collagen.

C. Hydroxyproline: Prockop and Udenfriend Procedure[121]

1. General Procedure

CHEMICAL PRINCIPLE. Hydroxyproline is oxidized to pyrrole-2-carboxylic acid in the presence of a measured excess of alanine (so that other amino acids or similar substances do not influence the yield of pyrrole). Interfering substances are removed after oxidation by extraction with toluene before the decarboxylation to pyrrole by heat. The pyrrole produces a red color with Ehrlich's reagent. Samples may contain 0.01–0.10 μmole of hydroxyproline. The sensitivity of this analysis allows measurement of 0.5 μg of hydroxyproline in solutions containing over 50 mg of other amino acids. If greater sensitivity is desired, the pyrrole may be extracted into a small volume of dilute acid and the chromaphore developed in an aqueous medium.

MATERIALS

Potassium borate buffer (pH 8.7): 61.84 g of boric acid and 225 g of potassium chloride are mixed into about 800 ml of distilled H_2O, the pH is adjusted with about 40–50 ml of KOH (60%), and the final volume is diluted to 1 liter with distilled water.

Alanine solution: 10 g of alanine is dissolved in about 90 ml of distilled water, the pH is adjusted to 8.7 with KOH (5%), and the final volume diluted to 100 ml with distilled water.

Chloramine-T solution: 0.2 M in 2-methoxyethanol made fresh daily (2.817 g/50 ml)

Sodium thiosulfate: 3.6 M (893.45 g/liter) can be stored for several weeks at room temperature when layered with toluene.

Ehrlich's reagent (25%): 27.4 ml concentrated sulfuric acid is added slowly to 200 ml of absolute alcohol and cooled. 120 g of Ehrlich's reagent (p-dimethylaminobenzaldehyde) is added to 200 ml of absolute alcohol in a separate beaker. The acid-ethanol solution is added with stirring to the Ehrlich's–alcohol solution. The mixture can be stored in the cold for several months. Crystals that form are redissolved by placing the reagent for a few minutes in warm water (±40–50°).

Phenolphthalein: 0.2 g/100 ml of absolute ethanol

Charcoal–resin mixture: 20 g of analytical grade ion-exchange resin (Ag 1-X8, 200–400 mesh) is mixed with 10 g of Norit A (Fisher).

[121] D. J. Prockop and S. Udenfriend, *Anal. Biochem.* **1**, 228 (1960).

The mixture is washed several times with 6 N HCl in a coarse sintered-glass funnel and dried to a fine powder with frequent rinsing and suction with a 1:1 ethanol:ether mixture.

Stock hydroxyproline solution: 1 mg/ml, i.e., 1000 mg L-hydroxyproline/liter MW = 131.1 standards. Dilute stock to 10 μg/ml and run 0.5–2.0 ml aliquots diluted to 5.0 ml with water (5–20 μg), in addition to blank.

PROCEDURE. The tissue sample is hydrolyzed in a sealed tube in 6 N HCl at 115° for 16 hours. Dried, defatted fragments or dried, evaporated extracts are hydrolyzed by the addition of 1–5 ml of 6 N HCl; solutions are hydrolyzed with an equal volume of concentrated HCl. To clarify the hydrolyzate, it is purified with the charcoal–resin mixture up to a volume of 1 ml, as needed. This is mixed well and centrifuged (15,000 g for 30 minutes). An aliquot of supernatant (free of resin) is evaporated to dryness and redissolved in water to a final volume of 5.0 ml. The volume of the aliquot of supernatant depends upon the amount of collagen in the hydrolyzate (see comments above on dilutions under the Stegemann procedure for hydroxyproline, see Section VII,B).

Add an excess of KCl (2 spoonfuls), and 1 drop of 0.2% phenolphthalein and adjust pH to 8.7 by adding 0.5% KOH dropwise. Combine two parts of potassium borate buffer (pH 8.7) and one part of 10% alanine solution (pH 8.7) in volumes sufficient to add 1.5 ml of the mixture per sample to be analyzed (this mixture is made fresh for each daily use). After addition of the alanine-borate buffer the solution is mixed well. Add 1.0 ml of 0.2 M chloramine-T and let stand at room temperature for exactly 20 minutes. Stop the oxidation with the addition of 3 ml of 3.6 M sodium thiosulfate. If necessary, resaturate with solid KCl.

Add 10 ml of toluene, cap the tubes and shake horizontally for 5 minutes. Centrifuge briefly to separate phases and discard the toluene (upper) layer. Repeat toluene extraction and again discard the toluene layer. Place the capped tubes in boiling H_2O for 30 minutes. Cool. Add exactly 7 ml toluene. Cap the tubes and shake for 5 minutes. Centrifuge briefly if necessary. Transfer 5 ml of the toluene layer into Coleman tubes, add 2.0 ml of 25% Ehrlich's reagent quickly. Shake. Leave samples at room temperature for 30 minutes and read absorbance at 560 nm on a Coleman Jr. spectrophotometer.

2. *Procedure for Hydroxyproline Determination in Urine Samples*

The reagents are the same as those described above for the Prockop and Udenfriend hydroxyproline technique.

Hydrolyze 4.0 ml of urine with 4.0 ml concentrated HCl in sealed tubes for 16 hours at 115°. Clarify the hydrolyzate with the charcoal-resin mixture up to a volume of 1 ml, as needed. *Mix well* and centrifuge for 30 minutes at 12,000 g. Pipette duplicate 1.0 ml samples from the supernatant fluid into screw-top tubes (Teflon-lined caps, 110 × 20 mm). Evaporate to dryness. Resuspend in 2.0 ml of H_2O. Add enough HCl to supersaturate samples. Combine 3 parts alanine solution (10%) with 7 parts borate buffer. Add 1.0 ml of the mixture and mix well. Add 0.6 ml of freshly made chloramine-T solution (0.2 M). Mix and allow to stand 20 minutes exactly. Stop the oxidation with 2.0 ml of 3.6 M sodium thiosulfate. Mix immediately.

Add 4.0 ml of toluene. Shake horizontally for 3 minutes. Centrifuge briefly. Remove and discard toluene layer. Add 6.0 ml of toluene. Cap tubes and place in boiling water bath for 30 minutes. Centrifuge briefly if necessary. Pipette 4.0 ml aliquot of toluene layer (avoiding H_2O or KCl contamination) into Coleman tubes. Add 2.0 ml of Ehrlich's reagent. Shake and let stand at room temperature for 15 minutes. Measure absorbance at 560 mm in a Coleman Jr. spectrophotometer.

Duplicate samples from the same hydrolyzate agree within 1–2%. The excretion of hydroxyproline should be measured on a 24-hour collection of urine, preferably for at least three consecutive days. The patient should be on a low gelatin diet during this period. The urine should be collected and kept cold with added preservative (either toluene or about 1% of the daily urinary volume of concentrated HCl). If nondialyzable peptides are to be measured, a 200-ml aliquot of the 24-hour total volume should be frozen until analysis. Nondialyzable peptides containing hydroxyproline are measured on a 200-ml aliquot of urine which has been evaporated to dryness and hydrolyzed following exhaustive dialysis against distilled water in the cold.

D. [^{14}C]Hydroxyproline: Juva and Prockop Procedure without Modification[122]

Materials

0.2 M Chloramine-T (Eastman Organic Chemicals) prepared in distilled water just before use

Sodium pyrophosphate (Fisher, ACS) 0.2 M, adjusted to pH 8.0 with HCl

Sodium thiosulfate (Fisher, ACS), 3.6 M, dissolved in distilled water and stored under toluene at room temperature (1 month)

1.0 M Tris buffer, pH 8.0 (adjusted with HCl).

---

[122] K. Juva and D. J. Prockop, *Anal. Biochem.* **15**, 77 (1966).

Silicic acid (Mallinckrodt No. 2847, 100 mesh): 1.5 g suspended in 5 ml of toluene

p-Dimethylaminobenzaldehyde, or Ehrlich's reagent (p-Dab) (Fisher, ACS). See method of Prockop and Udenfriend.

Phosphor solutions: (1) toluene scintillation solution: 4 g of POP, 50 mg of POPOP (both of Packard) to 1 liter with toluene or (2) 15 g of 2.5 diphenyloxazole (Pilot Chemicals) and 50 mg of p-bis[2-(5-phenyloxazolyl)]benzene (Pilot Chemicals) dissolved in 1 liter of toluene.

Pyrrole (Fisher) for standard solution: several aliquots of 0.1 μmole/ml are prepared by dilution with toluene and stored at $-20°$ for several months.

Carrier solutions: prepare concentrated stock of L-proline and L-hydroxyproline (General Biochemicals or Sigma) such that 0.1 ml contains 1.0 mg L-proline and 2.0 mg L-hydroxyproline.

PROCEDURE. Samples for assay are hydrolyzed in sealed tubes at 115° with 6 N HCl for 16 hours. If significant amounts of humin or salts are present, clarify with charcoal-resin mixture (see hydroxyproline method of Prockop and Udenfriend). Evaporate the hydrolyzates to dryness and redissolve in distilled water (2–4.0 ml). If extracts of human bone or skin are being analyzed (200–500 mg wet weight), dilute to 1.0 ml and take aliquot containing the maximum number of counts (less than 80,000 dpm of ^{14}C or 150,000 dpm of ^{3}H or more concentrated collagen solutions). Place aliquot in screw-capped test tube (200 × 25 mm o.d.) and add 1.0 mg of L-proline and 2.0 mg of L-hydroxyproline (carrier solution from concentrated stock). Adjust pH to 8.0 with NaOH solution and dilute to 8.0 ml with distilled water.

Add 6 ml of 0.2 M sodium pyrophosphate buffer, pH 8.0, and oxidize sample by adding 1.0 ml of 0.2 M chloramine-T solution. Let stand 20 minutes at room temperature. Terminate reaction by adding 6.0 ml of 3.6 M sodium thiosulfate solution. Adjust pH of solution to faint pink with sodium hydroxide (0.1–1.0 N NaOH) after adding a drop of phenolphthalein solution. Add 4 ml of 1.0 M Tris buffer, pH 8.0 (at 24°) and saturate with an excess of NaCl. Add 10 ml of toluene, seal with Teflon-lined screw cap and shake vigorously (Burrell wrist shaker in horizontal position) for 10 minutes. Centrifuge briefly (1000 g for 5 minutes) to separate layers, remove toluene phase with suction, and discard. Recap tube and place in boiling water bath for 25 minutes. Cool to room temperature and, if necessary, add additional NaCl to maintain saturation with salt.

Add exactly 12 ml of toluene (smaller exact amounts can be added if desired) and again shake vigorously for at least 5 minutes. Centrifuge briefly to separate layers. Pipette a known amount (e.g., 10 ml) of upper toluene phase on to a short silicic acid column. Column is prepared by pouring a slurry of 1.5 g of silicic acid in 5 ml of toluene into a 10×300 mm glass column with a coarse sintered-glass disk; a small amount of sand or glass wool or both is added to prevent clogging. After placing 10 ml of the toluene extract on the column, immediately start collecting the eluate in a 25-ml volumetric flask. Elute further with a 5-ml and then a 10-ml volume of fresh toluene and bring to 25 ml in the volumetric flask.

Place 20 ml of the 25-ml pooled eluate in a counting vial, and add 1 ml of phosphor solution (2) (see Materials). Count in a liquid scintillation counter with suitable internal standards or externally standardized quench curves for the particular phosphor solution added in order to calculate counting efficiency. Smaller aliquots of the eluate (e.g., 15 ml of the 25-ml volume) can be taken and 3.0 ml of the scintillation solution containing POP-POPOP [see Materials, phosphor solution (1)] added for counting.

For the colorimetric assay of pyrrole, a 0.1-ml aliquot of the toluene extract is diluted to 5.0 ml with toluene, and 2.0 ml of the Ehrlich's reagent is added and rapidly mixed. The color is developed at room temperature in the dark for 30 minutes. Absorbance is measured at 560 mm against a reagent blank. The amount of pyrrole is calculated by reference to suitable control standards of 0.02 and 0.04 μmole run at the same time. Recovery controls containing 1 mg of L-proline and 2 mg of L-hydroxyproline and known amounts of L-[^{14}C]proline (80,000 dpm) or [3,4-^{3}H]proline (150,000 dpm) are run through the assay and column procedure with each series of samples. If 10 ml of the 12-ml toluene extract of both the standard and the unknown are placed on the column, this dilution factor is the same for both.

CALCULATIONS

$$\text{Observed cpm} \times 100/E \times 25/20 \times 12/10 \times 5/4 \times 100/R = \text{dpm}$$

of [^{14}C]hydroxyproline in aliquot taken from evaporated, diluted (H$_2$O) hydrolyzate where E = efficiency of counting systems; $25/20$ = amount of column eluate counted; $12/10$ = amount of toluene extract placed on column; $5/4$ = correction for the loss of carboxyl carbon in the conversion of uniformly labeled [^{14}C]hydroxyproline to [^{14}C]pyrrole; $100/R$ = percent recovery of hydroxyproline as pyrrole in the assay.

The silicic acid column simplifies the assay procedure for [^{14}C]hydroxyproline, or [^{3}H]hydroxyproline, and improves reproducibility within 1%. Of primary importance is the elimination of error introduced by the presence of radioactive compounds other than [^{3}H]- or [^{14}C]pyrrole in the final toluene extract. It must be emphasized that the aliquot assayed should contain less than 0.5 µmole of hydroxyproline and not more than 80,000 dpm of ^{14}C (150,000 dpm of ^{3}H).

Recently, Bondjers and Björkerud[123] have devised a most sensitive method for the measurement of hydroxyproline in connective tissues on the nanogram level. Experience with the procedure is as yet limited. Nonetheless, it gives promise of usefulness.

E. Paper Chromatographic Separation of [^{14}C]Proline and [^{14}C]Hydroxyproline (Method of McFarren[124])

CHEMICAL PRINCIPLE. Proline and hydroxyproline, present in an acid hydrolyzate which has been evaporated to dryness and redissolved in water, can be separated by a single descending paper chromatogram in a system of m-cresol:boric acid buffer (1:4) at pH of 8.4. If both amino acids are labeled with ^{14}C, the strips can be cut and counted in a liquid scintillation counter without elution into the phosphor solution.

MATERIALS

Precut chromatographic paper (Whatman No. 1. 5¾ × 18 inches cut into 6 strips; style ERS from A. H. Thomas, Philadelphia).

Boric acid buffer (0.2 M, pH 8.4): a 4-liter solution is prepared by dissolving 12.380 g of boric acid in 1000 ml of 0.2 M KCl (14.920 g); 2000 ml of distilled water is added and the pH is adjusted to 8.4 with 0.2 N NaOH (±160 ml). Distilled water is added to a final volume of 4 liters.

Chromatographic system (single descending period of 16–19 hours): 1 part of m-cresol, 98% pure (e.g., 100 ml) is equilibrated for at least 6 hours in a separatory funnel with 4 parts of boric acid buffer (0.2 M, pH 8.4) (e.g., 400 ml). The lower, m-cresol phase is placed in the trough and the upper, boric acid phase is placed in the bottom of the chromatocab or jar.

PROCEDURE. The pieces of paper (5¾ × 18 inches) are dipped in boric acid buffer, pH 8.4, and air dried. They are marked by drawing a pencil line 9–10 cm from the upper edge (wide margin end) which crosses each strip about 1 inch below the cuts which divide the paper into six strips. The unknowns and standards are spotted on this line. The size of the spot should not exceed 1 cm; smaller spots are preferred.

Hydrolyzates of collagen extracts or tissue fragments which are labeled with [^{14}C]proline and [^{14}C]hydroxyproline are evaporated to dryness and redissolved in distilled water. Aliquots of this (25–100 µl

[123] G. Bondjers and S. Björkerud, *Anal. Biochem.* **52**, 496 (1973).
[124] E. F. McFarren, *Anal. Chem.* **23**, 168 (1951).

containing ±1000 cpm, of which at least 400 cpm are hydroxyproline) are placed on the line or origin. A single 16–19 hour descending chromatogram is run.

The papers are removed from the jar and hung in a well-ventilated hood to air-dry. The portion of the paper that was immersed in the m-cresol is blotted with paper toweling to avoid dripping. Protective latex gloves should be used when working with m-cresol. Drying takes about 3–4 days, at which time the counting efficiency is about 50%.

After drying, the papers are cut into strips which extend 2 inches below and 2 inches above the R_f of proline and the R_f of hydroxyproline. Initially, if desired, the reproducibility of such R_f values can be verified by running [^{14}C]hydroxyproline as well as [^{14}C]proline standards. Each strip (about 3 inches long) is rolled and placed in a counting vial with 15 ml of scintillation counting solution (4 mg of POP and 50 mg of POPOP in 1 liter of toluene) and counted in a liquid scintillation counter.

Standards are spotted and run at the same time (e.g., 10,000 dpm of [^{14}C]proline) in order to ascertain not only the recovery and counting efficiency, but also the degree to which high proline counts might overlap and contaminate the hydroxyproline counts. The R_f of proline under the described conditions is 0.63–0.65 and of hydroxyproline 0.23–0.21. Overlap of proline with hydroxyproline is less than 1%. Insoluble collagen in acid hydrolyzates of well-rinsed, dried, and defatted tissue fragments generally contains enough hydroxyproline relative to proline for accurate counting of hydroxyproline to be accomplished. If extracts of soluble collagen are purified, or if they have been dialyzed vs. distilled water prior to hydrolysis, the specific activity of hydroxyproline can be determined within 1–2%. The amino acid residues (and hence the radioactivity) do not elute from the paper into the toluene scintillation solution. If the specific activity of the material is not high, or if large amounts of proline counts compared to hydroxyproline counts are present, the method of Juva and Prockop (already described) is more sensitive.

VIII. Measurement of Protein with the Folin Phenol Reagent

A. Original Lowry Method[125]

CHEMICAL PRINCIPLE. Proteins in solution react with copper at an alkaline pH. When the Folin phenol reagent is added it is reduced and a color develops. The reagent is reactive at pH 10 for a short time only,

---

[125] O. H. Lowry, N. J. Rosebrough, A. L. Farr, and R. Randall, *J. Biol. Chem.* **193**, 265 (1951).

hence its addition and mixing must be as nearly instantaneous as possible (see procedure).

MATERIALS

Reagent A: 2% Na_2CO_3 in 0.1 N NaOH

Reagent B: 0.5% $CuSO_4 \cdot 5H_2O$ in 1% sodium or potassium tartrate

Reagent C, alkaline copper solution: 50 ml of reagent A is mixed with 1 ml of reagent B. The mixture is discarded after 1 day.

Dilute Folin phenol reagent: Dilute 2 N Folin-Ciocalteu phenol (Fisher) to a 1 N solution with distilled water.

Stock standard solution of versatol serum (Fisher): Dilute stock 1:100 and run 0.02–0.2 ml as standards; 0.2 ml of standards, reagent blanks, and unknown protein solutions are run through procedure.

PROCEDURE. Proteins that are in solution or are readily soluble in dilute alkali can be measured. A 0.2-ml aliquot of the protein solution (containing 5–100 µg of protein) is pipetted into a 10-ml test tube; 1 ml of reagent C is added, mixed well, and allowed to stand for 10 minutes at room temperature. Then 0.10 ml of dilute Folin reagent is added very rapidly and mixed within a second or two. The solution is allowed to stand 30 minutes or longer, after which it is transferred to cuvettes and the absorbance read on a Zeiss spectrophotometer or its equivalent at 750 mm.

B. Hartree Modification[126]

CHEMICAL PRINCIPLE. There are two disadvantages to the Lowry protein method. First, the color yields of different proteins vary considerably; and second, the relationship between color yield and protein concentration is not linear. The modification developed by Hartree eliminates the second disadvantage. This is accomplished by increasing the concentration of the alkaline copper tartrate reagent and by raising the reaction temperature to 50°.

MATERIALS

Solution A: 2 g of potassium sodium tartrate and 100 g of Na_2CO_3 are dissolved in 500 ml of 1 N NaOH and diluted to 1 liter with water. Can be stored at room temperature in polyethylene bottle for 6 months.

Solution B: 2 g of potassium sodium tartrate and 1 g of $CuSO_4 \cdot 5 H_2O$ are dissolved in 90 ml water and 10 ml of 1 N NaOH is added. This can be stored at room temperature in a polyethylene bottle for 6 months.

[126] E. F. Hartree, *Anal. Biochem.* **48**, 422 (1972).

Solution C: 1 volume of Folin-Ciocalteu reagent is diluted with 15 volumes of water. The solution (prepared daily) should be between 0.15 N and 0.18 N when an aliquot is titrated to pH 10 with 1 N NaOH. If the acidity exceeds 0.18 N it can be adjusted with 1 N NaOH.

PROCEDURE. Protein solutions containing 5–100 μg of proteins are diluted to 1.0 ml with water in a 10-ml test tube (13 mm diameter); 0.9 ml of Reagent A is added. A water blank (1.0 ml) and standards are set up in the same fashion. All tubes are placed in a water bath at 50° for 10 minutes and cooled to room temperature; 0.1 ml of reagent B is added. The tubes are left at room temperature for at least 10 minutes, then 3.0 ml of reagent C is added forcibly and rapidly to ensure mixing within 1 second. The tubes are again heated at 50° for 10 minutes and cooled to room temperature. Absorbance is read at 650 nm in 1 cm cuvettes in a Zeiss spectrophotometer.

This procedure when applied to the range of 15–110 μg protein results in a direct proportionality between the weight of protein and absorbancy at 650 mm. Reproducibility is within 2%.

The cuvettes may become etched from the alkaline solutions. They should be rinsed with 6 N HCl before the final rinsing of the day.

Acknowledgment

Supported in part by a grant from The John A. Hartford Foundation, Inc., by U.S. Public Health Service Grants AM 15158 from the National Institute of Arthritis, Metabolism and Digestive Diseases, and RR 5591 from the General Research Support Branch, Division of Research and Resources.

Author Index

Numbers in parentheses are reference numbers and indicate that an author's work is referred to, although his name is not cited in the text.

A

Aakraag, A., 292
Aaronson, S. A., 281
Abelson, H. T., 278, 279, 281
Acs, G., 241, 283
Adams, B. J., 23
Adams, R. A., 198
Adesnik, M., 279, 283
Ahmed, K., 177, 198
Akiyoshi, H., 187
Albersheim, P., 23
Alder, A. J., 232
Aldridge, W. G., 23, 29(40)
Alfert, M., 136
Algranati, I. D., 179
Allfrey, V. G., 93, 97, 98(1), 106, 132, 145, 174, 177, 178, 179(7), 181(7), 185(7), 186, 190(7), 198
Anderson, E. C., 46, 49(7)
Anderson, J. M., 292
Anderson, K. M., 15
Anderson, S. L., 206, 207(21)
Anderson, T. F., 68, 70(16)
Anfinsen, C. B., 266, 268(1), 269(1)
Ansevin, A. T., 107, 109, 146
Ansley, H. R., 136
Appleman, M. M., 289
Arias, I. M., 252, 260, 264
Armentrout, S. A., 274
Arnold, E. A., 153
Aro, H., 344
Aronson, R. B., 355
Asboe-Hansen, G., 358
Ash, J. F., 285
Askenasi, R. S., 357, 358(86)
Attardi, G. J., 78, 82(17), 278
Attramadal, A., 292
Aurbach, G. D., 323
Aviv, H., 272
Axelrod, J., 285
Axén, R., 268

B

Bachmann, L., 6, 10(6), 18(4), 19(4)
Bader, J. P., 283
Baggiolini, M., 125
Bagi, G., 100
Baglioni, C., 289
Bailey, A. J., 361
Baker, R. F., 279
Balhorn, R., 141, 142
Baliga, B. S., 287
Balinsky, J. B., 262
Ballentine, R., 325
Barker, K. L., 292
Barker, S. B., 328
Barondes, S. H., 265
Barrett, A. J., 352, 353(79)
Barrett, I. D., 133
Barrnett, R. J., 21
Bartley, J., 109, 149
Barton, A. D., 161
Bauer, E. A., 349, 352(77)
Baxter, J. D., 292
Beard, N. S., Jr., 274
Beer, M., 24
Beerers, H., 111
Beiser, S. M., 303, 305(3, 4)
Bekhor, I., 160, 183
Belt, W. D., 24
Bender, M. A., 56, 78
Bendich, A., 78
Bennett, L. L., Jr., 274, 287
Bensch, K. G., 285
Benya, P., 341, 342(58)
Berenbaum, M. C., 277
Berger, K., 341, 342(58)
Berlin, C. M., 246, 249, 252, 262(14, 27), 265(27), 266(14)
Berlowitz, L., 136
Bernard, J., 281
Bernardi, G., 94, 160

Bernhard, W., 13, 19, 25, 36(29), 37, 38(76), 39, 68
Bertina, J. R., 277
Bess, L. G., 120
Bessman, S. P., 111
Beychok, S., 232
Bhuyan, B. K., 289
Bilek, D., 132
Billing, R. J., 93, 99
Bird, A., 12
Birge, S. J., Jr., 316
Birks, J. B., 294, 296
Birnstiel, M. L., 12
Bishop, J. O., 63
Björkerud, S., 374
Black, M. M., 136
Blanquet, P., 21
Platt, L. M., 281
Bloch, D. P., 41, 136
Block, K. J., 349, 351(75), 352(75)
Blount, J. F., 115
Blout, E. R., 237
Blumenfeld, O. O., 309, 361(2), 362(103)
Blumenkrantz, N., 353, 354(84), 355(84), 356(84), 358(84)
Bock, K. W., 259
Böhlen, P., 116, 359
Boime, I., 272
Boiron, M., 281
Boivin, A., 62
Bollum, F. J., 42, 275
Bolund, L., 136
Bondjers, G., 374
Bonner, J., 93, 97, 98(4), 99, 100, 101, 106, 107, 108(16), 114(16), 119(16), 120(16), 127(16), 131(16), 132(16), 135(16), 138, 144, 145, 146, 148(1), 149(16), 150(16), 153(24), 160, 174, 183, 227, 228, 229(16)
Bootsma, D., 49
Borenfreund, E., 78
Borisky, G. G., 285
Bornstein, P., 285, 336, 338, 359(53)
Borun, T. W., 138
Botchan, M., 80, 82(19)
Bourgeois, S., 185
Bouteille, M., 6, 8, 10(3), 12(16), 13(3), 15(3), 19, 27(3), 36(3), 41(3)

Bradley, M. O., 285
Bradshaw, W. S., 279
Brahms, J., 232, 235
Brahms, S., 235
Brandt, J., 287
Brandt, W. F., 103
Bransome, E. D., Jr., 298, 299, 300, 301, 302
Brantmark, B. L., 109
Bray, G. A., 183, 188(21)
Bresnick, E., 283
Brewer, E. N., 41
Briggs, L., 22
Bright, H. J., 344
Brinkley, B. R., 65, 88, 285
Britten, R. J., 95
Brockman, R. W., 287
Broquist, H. P., 277
Brown, I. N., 277
Brown, R. F., 41, 289
Brumm, A. F., 77
Brunner, R., 274(23), 275, 289
Brutlag, D., 146
Buchanan, D. L., 253, 254(30), 257
Budd, A. L., 296
Budd, G. C., 72, 75(19)
Bücher, T., 274
Bunn, H. F., 257
Burka, E. R., 289
Burke, D. C., 283
Burkholder, G. D., 78
Burton, K., 77, 173
Busch, H., 82, 84, 85(30), 106
Bustin, M., 120, 137, 192, 194, 196
Butcher, R. W., 289
Butler, J. A. V., 103, 106, 109(26), 116, 117(26, 68), 124, 127(26), 227

C

Caldwell, D., 215
Cameron, A., 146, 162, 173, 174(6)
Cameron, I. L., 54, 59, 60(9), 61(9)
Campbell, G. L., 279
Campbell, J. C., 273
Cantor, C. R., 222
Cantor, K. P., 64, 78

Carnes, J. D., 72
Caro, L. G., 6, 10(7)
Carpenter, J. C., 287
Carroll, A., 177, 190(11)
Carroll, D., 228
Carter, H., 302
Cartwright, L. N., 283
Cassim, J. Y., 221
Chalkley, R., 93, 94(3), 98, 103, 106, 108(16), 109, 114(16), 119(16), 120(16), 127(16), 131(16), 132(16, 88), 135(16, 88), 138, 139, 141, 142, 144, 148(1), 149, 153(19), 154(19)
Chambon, P., 283
Chan, V. L., 283
Chan-Curtis, V., 24
Chang, C.-K., 278, 281
Charanich, S., 281
Chignell, C. F., 222
Chignell, D. A., 222
Chorazy, M., 78
Chrambach, A., 157
Chrisholm, E. M., 264
Chu, L. L. H., 292
Church, R. B., 145
Chytil, F., 176, 192, 193(11), 194, 195(10, 11), 196, 197(10)
Clark, W. B., 279
Clayton, R. M., 273
Cleaver, J. E., 59, 60(11), 277, 286, 289
Coffey, R. J., 350, 351(78)
Coggeshall, R., 25
Cogliati, R., 32
Cognetti, G., 138
Cohen, A. L., 68, 281
Cohen, B. I., 110
Cohen, L. H., 115, 141, 142(7)
Cohen, P. P., 262, 281, 327
Cohn, M., 292
Cole, A., 64, 78, 80
Cole, R. D., 103, 120, 132, 137
Coleman, A., 281
Coleman, J., 281
Coleman, M. K., 158
Coleman, W. H., 289
Collins, A., 262
Colombo, B., 289
Conconi, F. M., 289
Connolly, T. N., 87

Conover, J. H., 63, 72(1)
Corry, P. M., 64, 80
Creasey, W., 285
Creighton, M. O., 123
Croissant, O., 12
Crook, L. E., 281
Cruft, H. R., 120
Cuatrecasas, P., 266, 267, 268(1), 269(1, 9), 270(6, 9), 271, 272(9), 273(9)
Culotti, J., 45
Cunningham, L. W., 358
Cutroneo, K. R., 283
Czajkowski, R. C., 115

D

Dahmus, M. E., 98, 99, 101, 106, 108(16), 114(16), 119(16), 120(16), 127(16), 131(16), 132(16), 135(16), 138, 144, 148(1)
Dairman, W., 115, 116, 359
Darnell, J. E., 279, 283
Darzynkiewicz, Z., 136
Dashman, T., 261, 266(46)
Dastugue, B., 177, 198
Dauguet, C., 12
Davidson, D., 61
Davidson, N., 228
Davies, B. J., 127
Davies, P. T., 299
Davis, A. T., 285
Davis, B., 165, 192, 273
Davis, N. R., 362
Davis, R. C., 232
Davison, P. F., 106
Dawson, R. M. C., 274(22), 275
DeBarnard, B., 257
DeBernardo, S. L., 115
Decker, K., 289
Dedmon, R. E., 303
deDuve, C., 257, 265(38)
Dehm, P., 313, 322
Deinhardt, F., 303
De Lange, R. J., 103
DeLeon, A., 264
Delius, H., 153
Delovitch, T. L., 273
DeLuca, H. F., 327
De Nooij, E. H., 116

Derenzini, M., 28
Desai, L. S., 198
Deshmukh, A., 336, 337(48), 340(48), 361(48)
Deshmukh, K., 336, 337(48, 49, 50), 340(48, 49, 50), 361(48)
De Sombre, E. R., 274, 292
Deterding, J. H., 299
Dethlefsen, L. A., 279
Dev, V. G., 303, 305(3)
Deysson, G., 274, 277, 285
Dickermann, H. W., 93
Diegelmann, R. F., 285, 322, 340(22), 341(56), 343(56)
DiMarco, A., 281
Dingle, J. T., 334, 352, 353(79)
Dingman, C. W., 148
Dixon, G. H., 103, 104(10)
Djordjevic, B., 281
Dobbs, H. E., 125
Doenecke, D., 94
Doida, Y., 49
Donnelly, G. M., 83
Dorfman, A., 315
Doty, P., 153, 154(30)
Dounce, A., 106, 109
Doyle, D., 241, 245, 252, 260(24)
Drew, R. M., 48
Druyan, R., 257
Drysdale, J. W., 157
Duerksen, J. D., 93
Dunker, A. K., 133
Dupuy-Coin, A. M., 6, 13(3), 15(3), 19, 27(3), 36(3), 41(3)

E

Eagle, H., 303
Edelman, I., 291, 292
Edwards, L. J., 106, 124(24)
Ehrlich, H. P., 285, 336
Eichberg, J., 112
Eigner, E. A., 359
Eisen, A. Z., 349, 350, 351(78), 352(77)
Eisler, W. J., 296, 299
Ekenberg, E., 93
Elbert, W., 361, 362(104)
Elger, W., 291, 292
Elgin, S. C. R., 107, 146, 149(16), 150(16), 153(24), 159, 174

Elliot, D. C., 274(22), 275
Elliot, W. H., 274(22), 275
Elrick, R. H., 299
Elsas, L. J., 289
Engelberg, J., 49, 56
Epstein, C. J., 130
Epstein, E. H., Jr., 336
Erbe, R. W., 345
Erlanger, B. F., 303, 305(3, 4)
Erlichman, J., 202, 203
Ernbäck, S., 268
Esponda, P., 21
Estensen, R. D., 285, 286
Etherington, D. J., 353
Evans, K., 132
Eyre, D. R., 362
Eyring, H., 215

F

Fakan, S., 13
Fakanova, J., 32
Falvey, A. K., 270
Fambrough, D., 98, 101, 106, 108(16), 114(16), 119(16), 120(16), 127(16), 131(16), 132(16), 135(16), 138, 148
Fan, H., 278, 283, 290
Fang, S., 292
Farr, A. L., 77, 110, 111(43), 112(43), 113(43), 173, 375
Fasman, G., 192, 230, 232, 235, 236(30), 237
Faulstick, H., 283
Fedak, S. A., 316
Feigelson, P., 261, 266(46)
Felicetti, L., 289
Fesler, K. W., 262
Figen, J. F., 265
Finlayson, G. R., 157
Firket, H., 49
Fischer, G. M., 323
Fitton-Jackson, S., 314, 315, 317
Fiume, L., 283
Flanagan, B., 326, 327(33), 330(35)
Foley, G. E., 198
Ford, J. D., 358
Forrester, S., 106, 107(25), 145
Forro, F., 59
Fotherby, K., 274, 292
Fouts, J. R., 112

AUTHOR INDEX

Fowler, L. J., 361
Fox, R., 358
Franceshini, P., 78
Franz, H. E., 289
Franz, M., 289
Franzblau, C., 355
Freedland, R. A., 263
Freeman, M. V. R., 303
Frenster, J., 33, 80, 93, 97, 98(1)
Friedkin, M., 285
Friedman, D. L., 41, 44
Friedman, P. A., 274, 281
Friedman, R. M., 283
Fritz, P. J., 246
Froehner, S. C., 146
Fudenberg, H. H., 193
Fujimura, F., 106, 108(16), 114(16), 119(16), 120(16), 127(16), 131(16), 132(16), 135(16), 138, 144, 148(1)
Fujimura, G., 98
Fuller, W., 232
Furfine, C. 361, 362(103)
Furlan, M., 109

G

Gabelman, N., 63, 72(2)
Gaillard, C., 160
Gaillard, P. J., 331
Galavazi, G., 49
Gallop, P. M., 309, 361(2), 362(103)
Gallwitz, D., 277
Gan, J. C., 254
Ganschow, R. E., 245, 265
Garcia, A. M., 62
Garlick, P. J., 252
Garner, G. E., 68
Garrard, W. T., 99, 100
Garrels, J. I., 149
Garro, A. G., 305
Gautier, A., 23, 24(43), 27, 28(43), 32
Gautschi, J. R., 286, 289
Geiger, P. J., 111
Gelfant, S., 59
Gellert, M. F., 112
Gerber, G. E., 359
Gerhardt, B., 111
Gershey, E. L., 177, 198
Geschwindt, I. I., 136
Geuskens, M., 13

Giacomoni, D., 78
Giese, H., 29
Gilmour, R. S., 145, 160
Gimenez-Martin, G., 21
Giro, M. G., 160
Glass, R. D., 260
Glimcher, M. J., 362
Goldberg, I. H., 274, 281, 289
Goldhaber, P., 331
Golditch, M., 341, 342(58)
Goldsmith, H., 232
Good, N. E., 87
Goodfriend, T. L., 192
Goodridge, A. G., 246, 250(17)
Gottesfeld, J. M., 99, 100
Gottlieb, D., 274
Gould, H. J., 168
Granboulan, N., 7
Granboulan, P., 6, 7
Granner, D., 142
Grant, M. E., 309, 314
Gratzer, W. B., 133
Gray, W. R., 159
Graziano, S. L., 153
Greenberg, D. M., 275
Greenberg, J. R., 278
Greenfield, N., 235, 236(30)
Greengard, O., 241
Gregory, J. D., 112
Griffiths, G., 168
Grob, D., 265
Grollman, A. P., 278, 281, 287, 289
Gronow, M., 168
Grosjean, M., 221
Gross, J., 322, 338, 349, 351(75), 352(75), 359
Gross, P. R., 138
Grossman, A., 262, 265(48)
Grossman, L., 272
Grower, M. F., 298
Guerrier, C., 28
Gulyas, S., 48
Gutiérrez, A. M., 361, 363(105)

H

Haase, G., 8
Hahn, J. D., 291, 292
Hale, A. J., 62
Hall, B. K., 310

Halme, J., 309, 343(4), 344, 345(4)
Halter, P., 264
Hamilton, L. D., 232
Hamilton, M. G., 78, 283
Hamlin, J., 273
Handa, D. T., 255
Handschumacher, R. E., **277**
Hanes, S., 54, 56(25)
Hansson, V., 292
Hardesty, B., 287, 289
Harper, E., 349, 350(73), 351(73, 75), 352(73, 75)
Harris, A., 254. 292
Harris, T. E., 55
Hartley, J. W., 281
Hartley, R. W., Jr., 252, 265(28)
Hartman, B. K., 159
Hartree, E. F., 178, 376
Hartwell, L. H., 45
Harvey, A. M., 265
Harvey, C. L., 289
Hauser, T. R., 361, 362(104)
Hay, A. J., 283
Hayatsu, H., 110
Haywood, L. D., 238
Hearst, J. E., 64, 78, 80, 82(19)
Heath, D. A., 323
Heidelberger, M., 192
Heidema, J., 177, 190(11)
Heinricks, W. L., 292
Hendel, R. C., 281
Henneman, D. H., 310, 321, 323, 326(5), 361, 365(5)
Henson, E., 361, 362(103)
Herd, J. K., 112
Hermann, W. L., 292
Hershko, A., 286, 287
Hey, A. E., 106, 124(24)
Hiatt, H. H., 278
Hill, R. J., 153, 154(30)
Hindennach, I., 153
Hirsch, A. H., 202
Hirsch, M., 290
Hirschhorn, K., 63, 72(1, 2)
Hjertén, S., 267
Hnilica, L. S., 103, 105(7), 106, **107**(7), 108(7), 109, 120, 124(24), 146, **172**
Hodge, L. D., 283
Hoffman, P., 359
Hohman, P., 132

Hohmann, P., 142
Holme, G., 273
Holmes, A. W., 303
Holmes, D. S., 99
Holmquist, G., 63
Holt, S. J., 36
Holtzer, H., 279, 285, 286, 289
Hooper, C. W., 232
Hooper, J., 103
Hord, G., 107
Horwitz, A. L., 315
Horwitz, M. S., 281
Horwitz, S. B., 278, 281
Housman, D., 287, 289
Housten, L. L., 158
Howard, A., 44, 58
Hsu, T. C., 64, 76, 78, 88, 278, 285
Huang, R. C., 98, 101, 106, 108(16), 114(16), 119(16), 120(16), 127(16), 131(16), 132(16), 135(16), 138, 144, 146, 148(1), 153, **227**
Huberman, J. A., 78, 82(17), 98, 106, 108(16), 114(16), 119(16), 120(16), 127(16), 131(16), 132(16), 135(16), 138, 144, 148(1)
Hughes, W. L., 59
Hultin, T., 169
Humphrey, R., 64, 65(10), 278
Hunter, A. L., **138**
Hurlbert, R. B., 77
Hutchison, D. J., 78
Hutton, J. J., Jr., 343, 345(60), 346, 348(68)
Huxley, H. E., 23, 28(42)

I

Inoue, S., 80
Iorio, R., 62
Isaacs, M. A., 93
Isenberg, I., 237
Ishida, H., 177, 198
Ivarie, R. D., 278
Iveter, K. J., 292
Iyar, V. N., 281
Izawa, S., 87

J

Jackson, H., 292
Jacob, J., 5, 12

AUTHOR INDEX

Jacobsen, M., 287, 289
Jacques, P., 353
Jacquillat, C., 281
Jain, S. C., 281
James, D. W. F., 106
James, F., 274, 292
Jandl, J. H., 257
Janowski, M., 96, 97, 98(3)
Jeanteur, P., 12
Jeffay, H., 254
Jeffrey, J. J., 349, 350, 351(78), 352(77)
Jensen, E. V., 274, 292
Jensen, R., 93, 94(3), 98, 106, 108(16), 109, 114(16), 119(16), 120(16), 127(16), 131(16), 132(16), 135(16), 138, 144, 148(1), 149
Jericijo, M., 109
Jirgenson, B., 215
Johns, E. W., 103, 105, 106, 107(25), 108, 109(13, 26, 31), 116, 117(26, 68, 69), 124, 127(26), 133, 139, 145, 227
Johnson, A. W., 103, 107
Johnson, C. B., 94
Johnson, G., 132
Johnson, L. W., 246
Johnson, R. T., 63
Johnson, W. C., Jr., 215, 216(5), 235(5), 237
Jones, A. R., 292
Jones, B. F., 265
Jones, K. M., 274(22), 275
Jones, K. W., 63
Jost, J. P., 246
Jukes, T. H., 277
Jung, G., 8
Juva, K., 371

K

Kahwanago, L., 292
Kai, K., 110
Kakefuda, T., 83
Kalberer, F., 125, 127(81)
Kamiyama, M., 177, 198
Kane, R. E., 80, 83(22), 87(22)
Kang, A. H., 322, 338, 359
Kaplan, J., 192
Kaplowitz, P. B., 198
Kasten, F. H., 277
Katunuma, N., 244

Kauzman, W., 216
Kawasaki, T., 160
Kedes, L. H., 138
Kedinger, C., 283
Kefalides, N. A., 314, 336, 357, 358(86)
Keilin, D., 178
Keller, S., 341
Kelly, J. W., 136
Kemp, G. D., 359
Kemp, J. D., 279
Kemp, R. G., 289
Kenney, F. T., 241, 246, 251, 262(23), 266(23)
Kersten, H., 274
Khairallah, E. Z., 246
Kidwell, W. R., 41, 44
Kihlman, B. A., 274, 277, 281, 285
Kiho, Y., 267
Killander, D., 136
Killias, U., 23
Kim, K.-H., 281
Kim, Y. S., 252, 262(26), 263(26), 264(26)
Kimsey, J., 292
King, J., 157
King, T. P., 158
Kinkade, J. M., 120
Kish, V. M., 177, 198, 203, 206(7)
Kisieleski, W. E., 161
Kivirikko, K. I., 344, 345
Klein, L., 323
Klein, P. D., 296, 299
Klein, W. J., Jr., 303
Kleinsmith, L. J., 145, 177, 190(11), 198, 203, 206(7)
Klenow, H., 279
Kletzien, R. F., 285
Klevecz, R., 77
Knight, S. G., 289
Knox, W. E., 244
Kobayashi, K., 244
Kobayashi, Y., 301
Koch, A. L., 253, 254(31)
Koch, F. C., 325
Koller, T., 24, 68
Kominani, E., 244
Kominami, S., 244
Kopin, I. J., 285
Kopriwa, B. M., 7, 10(11)
Korner, A., 287, 289
Korson, R., 29

Korzybski, T., 274, 281, 289
Kossel, A., 102
Kostraba, N. C., 177, 198
Kotzin, B. L., 279
Kowszyk-Gindifer, Z., 274, 281, 289
Kram, R., 279, 286(51), 287(51), 289
Kramer, M., 291, 292
Krane, S. M., 345, 358
Krause, M. O., 48
Krauze, R. J., 146, 162, 173, 174(6)
Krebs, E. G., 202
Kruh, J., 177, 198
Kuff, E. L., 95
Kumar, A., 281
Kumar, K. V., 41
Kung, G. M., 160, 183
Kunitz, M., 113
Kuntzman, R., 257
Kuriyama, Y., 246
Kurylowicz, W., 274, 281, 289
Kuter, D. J., 154

L

La Cour, L. F., 21
Ladoulis, C. T., 24
Laemmli, U. F., 154, 157, 165, 173
Lagunas, R., 111
Laird, C. D., 305
Lamberts, B. L., 359
Landridge, R., 232
Lane, J. M., 322, 359
Langan, T. A., 145, 177, 198
Langreth, S. G., 41
Larsson, A., 277
Laurence, D. J. R., 115
Laval, M., 6, 13(3), 15(3), 27(3), 36(3), 41(3)
Layman, D. L., 347
Layne, E., 111, 113(44)
Leblond, C. P., 59, 61
Leder, P., 272
Leduc, E. H., 19, 36(29)
Lee, K. L., 262
Legrand, M., 221
Lehrer, H., 192
Leighton, F., 257, 265(38)
Leimgruber, W., 115, 116
Leskowitz, S., 242
Lettré, H., 10

Levin, W., 257
Levine, L., 137, 192
Levinson, B., 289
Levinson, B. B., 279, 283
Levy, S., 174
Lewis, M. S., 359
Li, J. B., 244
Li, M. J., 237
Liao, S., 292
Lieberman, I., 41, 289
Lilienthal, J. L., Jr., 265
Lin, S., 285
Lin, S.-Y., 287
Lindell, T. J., 283
Lindh, N. O., 109
Lindsay, D., 138
Lindsley, J., 366
Lipmann, F., 274, 289
Liss, M, 191, 194
Litt, G. J., 302
Little, E. P., 254
Liu, H. C., 215
Liu, T.-Y., 174
Lodish, H. F., 287, 289, 290(62)
Loftfield, R. B., 254
Longton, R. W., 359
Love, R., 136
Lowry, D. R., 281
Lowry, O. H., 77, 110, 111(43), 112(43), 113(43), 173, 375
Lucas-Lenard, J., 274
Luck, J. M., 103, 113(11), 119(11)
Ludueña, M. A., 285
Luguvoy, J. K., 328
Lynch, W. E., 41

M

McCaman, M. W., 115
McCarthy, B. J., 93, 94, 96, 97, 98(2, 3), 145, 305
McConaughy, B. L., 94, 97, 98(2), 305
McConnell, D. J., 99
McFarren, E. F., 374
MacGillivray, A. J., 146, 160, 162, 164, 165, 168(10), 173, 174(6)
McGuise, P. M., 283
Machek, G., 274(23), 275, 289
Machicao, F., 122, 123(76)

AUTHOR INDEX 387

Mackevicius, F., 161
McKnight, G. S., 250
McMeekin, T. L., 325
MacRay, E. K., 19
Maekawa, T., 53
Maestre, M. F., 233, 234, 235(29)
Magun, B. E., 136
Mahieu, P., 49
Maio, J. J., 64, 80, 82(20)
Maizel, J. V., 133, 134(95), 154
Makrasch, L. C., 112
Malawista, S. E., 285
Malkin, M., 289
Mamont, P., 287, 289
Manchester, K. L., 112
Mandl, I., 341
Maral, R., 281
Margolis, F., 261, 266(46)
Marinozyi, V., 19, 23, 24(43), 28(43), 37
Marks, P. A., 289
Marlow, D. P., 68
Marmur, J., 185
Marshall, I., 252
Martin, G. R., 339, 347
Martin, J. E., 292
Martin, J. R., 361
Martini, O. H. W., 168
Marushige, K., 97, 98(4), 101, 106, 108(16), 114(16), 119(16), 120(16), 127(16), 131(16), 132(16), 135(16), 138, 144, 145, 146, 148(1)
Marver, H. S., 262
Marvin, D. A., 232
Masburn, T. A., Jr., 359
Mathews, M. B., 287, 289
Mathison, G. E., 112
Maudsley, D. V., 301
Mavrides, C., 262, 265(48)
Mechanic, G. L., 322, 361, 362(19)
Meetz, G. D., 19
Meihlac, M., 283
Mendelsohn, J., 78, 82, 83(15), 88
Mendensohn, M. L., 54, 55(24), 56(24), 59, 61(12), 62
Mentré, P., 24, 29(48, 49)
Messier, B., 59
Messinger, K., 287
Miller, D. A., 303, 305(3)
Miller, D. W., 159
Miller, E. J., 336, 339, 358, 359(91)

Miller, J. V., Jr., 268, 269(9), 270(9), 271, 272(9), 273(9)
Miller, O. J., 303, 305(3, 4)
Miller, R., 289
Mirsky, A. E., 62, 93, 97, 98(1), 114, 177, 178, 198
Mitchell, W. M., 176
Mitchinson, J. M., 277, 285
Mitchison, J. M., 45
Mitsugi, K., 289
Miyagi, M., 187
Miyazawa, Y., 94
Mizel, S. B., 285, 286
Mizushima, S., 153
Modak, S. P., 13
Moldave, K., 272
Mommaerts, W. F. H. M., 232
Monneron, A., 4, 25, 37, 38(76)
Montgomery, J. A., 274
Moore, D. E., 78, 82, 83(15)
Moore, D. S., 232, 234(28), 235(28)
Moore, S. J., 174
Morales, M. F., 112
Morris, L., 359
Morris, P. W., 283
Mosteller, R. D., 287, 289
Moulé, Y., 4
Moyne, G., 29
Mueller, G. C., 41, 42, 43, 44, 277, 283
Mukherjee, B. R., 78
Mundkur, B., 23
Munro, H. N., 287
Muramatsu, M., 84
Murphy, W., 278
Murray, K., 227, 228
Mussini, E., 343, 345(60)

N

Nachtwey, D. S., 54, 59, 60(9), 61(9)
Nagai, Y., 338
Nakane, P. K., 197
Nakata, Y., 283
Nameroff, M., 279
Nandi, S., 292
Narayanan, A. S., 347
Nasser, D. S., 94, 96, 97, 98(3)
Natelson, S., 328
Nathans, D., 289

Neary, M. P., 296
Neelin, E. M, 124
Neelin, J. M., 124
Neimann, I., 333
Neithlich, H. W., 265
Nemer, M., 138
Nero, Z., 315
Neumann, F., 291, 292
Newman, J. F. E., 278
Nichols, G., Jr., 316, 326, 327(33), 330(35)
Nimni, M. E., 336, 337(48, 49, 50), 340(48, 49, 50), 361(48)
Nishiura, J. T., 94
Noland, B. J., 115
Nomura, M., 153
Nooden, L. D., 174
Nordman, C. E., 281

O

Ochoa, S., 179
Ohga, Y., 200, 206(15), 207(15)
Ohlenbusch, H., 98, 106, 108(16), 114(16), 119(16), 120(16), 127(16), 131(16), 132(16), 135(16), 138, 144, 148(1), 228
Oka, T., 249, 251(20)
Okada, S., 49
Okazaki, K., 279
Oliver, D., 139, 142
Olivera, B. M., 98, 106, 108(16), 114(16), 119(16), 120(16), 127(16), 131(16), 132(16), 135(16), 138, 144, 148(1), 228
O'Malley, B. W., 62, 172, 174(1)
Omura, T., 246
Ono, T., 273
Oravec, M., 281
Ornstein, L., 127
Orth, G., 12
Osborn, M., 203, 248
Osborne, R. M., 359
O'Sullivan, J., 112
Overgaard-Hansen, K., 279
Ozaki, M., 153

P

Painter, R. B., 48, 59
Palacios, R., 273
Palacow, I., 95
Palade, G. E., 246, 259
Pallotta, D., 136
Palmer, J. S., 323
Palmiter, R., 249, 251(20), 273
Pankalainen, M., 344
Panyim, S., 103, 109, 131, 132(88), 135(88), 139, 149, 153(19), 154(19)
Parker, R. P., 299
Parry, N., 22
Pastan, I., 283
Patel, G. L., 107
Patroni, M., 232
Paul, J., 145, 146, 160, 162, 173, 174(6)
Paul, R., 281
Pavich, M., 95
Paweletz, N., 10
Pawlowski, P., 136
Paz, M. A., 361, 362(103)
Peach, C. M., 361
Peacocke, A. B., 228
Pease, D. C., 36, 67, 73(13)
Peck, W. A., 287, 316
Pelc, S. R., 44, 58
Peng, C. T., 296
Penman, S., 278, 279, 281, 283, 290(31)
Perdue, J. F., 285
Perl, W., 289
Perlman, S., 278, 281, 283, 290
Perry, R. P., 278
Pery, Z. M., 60
Pestka, S., 274, 289
Peterkofsky, B., 281, 285, 322, 340(22), 341(56), 343(56)
Peterman, M. L., 78
Peters, D., 29
Peters, M. A., 112
Petersen, D. F., 46, 49(7)
Peterson, E. A., 95
Pfeifer, A., 60
Phillips, D. M. P., 103, 106, 108, 109(26, 31), 117(26), 124, 127(26), 139
Piez, K. A., 336, 338, 339, 358, 359(53, 91)
Pincus, J. B., 328
Pinnell, S. R., 345, 347, 358
Pitot, H. C., 246
Plagemann, P. G., 279, 285, 286
Platt, T., 159
Platz, R. D., 177, 198, 206(7)
Poccia, D. C., 153, 154(30)

Pogo, A. O., 178
Pollister, A. W., 114
Poole, B., 255, 257, 265
Porath, J., 268
Porter, G. A., 292
Pothier, L., 198
Potter, V. R., 60, 77
Poulos, T. L., 175
Powers, M. J., 339
Prensky, W., 63
Prescott, D. M., 47, 48(8), 51(9), 53(9), 56, 78
Price, P. A., 174, 175
Price, P. M., 63, 72(1, 2)
Price, V. E., 252, 265(28)
Prockop, D. J., 309, 313, 314, 322, 344, 345, 353, 354(84), 355(84), 356(84), 358(84), 369, 371
Pronczuk, A. W., 287
Pruitt, K. M., 246
Puck, T. T., 46, 47, 48(8), 49(7), 53(8), 54(8), 55(8)
Puvion, E., 21

Q

Quastler, H., 53, 54(23), 58, 59(2), 60(2)
Quie, P. G., 285

R

Rabinowitz, M., 257
Raisz, L. G., 333
Rall, S. C., 137
Ramp, W. K., 322, 362(19)
Randall, R., 77. 110, 111(43), 112(43), 113(43), 173, 375
Rao, P. N., 49, 63
Rasch, E., 23
Rasch, G., 52
Rasmussen, P. S., 103, 113(11), 119(11)
Raspé, G., 274
Ravel, J. M., 289
Rechcigl, M., 241, 252, 262, 265(28)
Recher, L., 22
Reddi, A. H., 274, 292
Reeck, G. R., 95, 98
Rees, K. R., 281

Regoeczi, E., 294
Reichard, P., 277
Reichman, M., 278, 283
Reid, B., 45
Reimann, E. M., 202
Reiner, J. M., 253, 254(29)
Revel, J.-P., 278
Revel, M., 278
Rexroth, A. K., 257, 258(43)
Reynolds, E. S., 25
Reynolds, J. J., 334
Rhoads, R. E., 250, 346, 348(69)
Rich, A., 267
Riches, P. G., 164
Richter, K. H., 146, 165
Richwood, D., 173, 174(6)
Rickwood, D., 146, 162, 164, 165, 168(10)
Riddle, M., 177, 198
Rieke, W. O., 141
Rifkind, R. A., 289
Riggs, A. D., 185
Righetti, P., 157, 254
Rigler, R., 136
Riley, R. F., 158
Ringertz, N. R., 136
Ris, H., 62, 87
Robbins, E., 48, 138
Roberts, N. E., 346, 348(69)
Roberts, W. K., 278, 289
Robins, E., 115
Robison, G. A., 289
Rogers, P., 317
Rojkind, M., 361, 362(103), 363(105)
Rombauer, R., 361
Ronzio, R. A., 279
Roodyn, D. B., 84
Rosbash, M., 278, 283
Rosebrough, N. J., 77, 110, 111(43), 112(43), 113(43), 173, 375
Rosen, O. M., 202, 203
Rosenberg, L. E., 289
Rosenberg, M., 285
Rosenbloom, J., 358
Rosenfeld, L., 215
Rothstein, H., 59
Rothstein, M., 275
Rousseau, G., 292
Rowe, W. P., 281
Roy-Burman, P., 274
Roychoudhury, R., 41

Rubin, C. S., 203
Ruddon, R. W., 206, 207(21)
Rudock, D., 264
Rueckert, R. R., 42, 133
Ruiz-Carrillo, A., 132
Rumke, P., 191
Rumke, R., 194
Rusch, H. P., 41
Rutschmann, J., 125, 127(81)
Rutter, W. J., 279

S

Sajdera, S. W., 112
Sakore, T. D., 281
Salditt, M., 279, 283
Salpeter, E. E., 6, 18(4), 19(4)
Salpeter, M. M., 6, 10(6), 18(4), 19(4)
Salzman, N. P., 78, 82, 83(15), 88
Samejima, K., 115
Samuels, H. H., 291, 292
Sandberg, A. L., 191, 194
Sanders, T. G., 279
Sanger, J. W., 285
Satake, K., 103, 113(11), 119(11)
Sato, H., 285
Sawicki, E., 361, 362(104)
Saxena, V. P., 236
Schachman, H. K., 112
Schaffhausen, B., 232
Scherrer, K., 273
Schildkraut, C. L., 64, 80, 82(20)
Schimke, R. T., 241, 245, 246, 249, 250, 251(20), 252(13), 257(13), 260(24), 262(14, 27), 265(27), 266(13, 14), 273
Schluderberg, A., 281
Schmid, R., 265
Schmid, W., 72, 283
Schmitz, H., 77
Schneider, W. C., 365(118), 366
Schneir, M., 341, 342(58)
Schofield, J. G., 285
Schreck, R. R., 303, 305(4)
Schreyer, M., 32
Schroeder, T. E., 285
Schwartz, S., 265
Scolnick, E. M., 281
Scott, W. A., 278, 290
Seeds, W. E., 232

Segal, H. L., 252, 262(26), 263(26), 264(26)
Segrest, J. P., 358
Sehon, A. H., 273
Seifter, S., 309, 349, 350(73), 351(73), 352(73), 361(2)
Sekeris, C. E., 146, 165, 283
Sellinger, O. Z., 262
Servis, R. E., 110
Sevall, J. S., 99, 174
Sevčik, V., 274
Shambaugh, G. E., 262
Shapiro, A. L., 133, 134(95), 154
Shapiro, R., 110
Sharpe, S. E., III, 299, 300, 301, 302
Shaw, L. M. J., 146
Shaw, M. W., 277, 281
Shaw, P. D., 274
Shelton, K., 174, 179
Shephard, G. R., 115
Sherbaum, O., 52
Sheridan, W. F., 21
Sherman, F. G., 53, 54(23), 58, 59(2), 60(2)
Sherod, D., 132
Sheth, K., 115
Shields, D., 283
Shields, R., 287, 290
Shih, T. Y., 228, 229(16), 230, 237
Shooter, K. V., 106
Shyamala, G., 292
Sibatani, A., 187
Siebert, G., 106
Siegel, R. C., 347, 361
Siekevitz, P., 246, 259
Sies, H., 274
Siev, M., 279
Sih, C. J., 289
Silverman, D. J., 112
Simard, R., 12, 274, 281
Siminovitch, L., 283
Simons, K., 344, 345
Simpson, R. T., 95, 98, 174
Simson, P., 106, 109(26), 117(26), 124, 127(26)
Sinex, F. M., 355
Singer, R. H., 278, 290(31)
Singh, R. M. M., 87
Sisken, J. E., 55, 56(30), 83
Sjöqvist, A., 169

Skehel, J. J., 283
Slayter, E. M., 211, 212(1)
Sluyser, M., 191, 194
Smart, J. E., 146, 228
Smetana, K., 106, 136
Smith, B. C., 93
Smith, E. L., 103
Smith, J., 95
Smith, J. W., 21
Smith, K. D., 145
Smith, M. A., 241
Smithies, O., 124, 127
Sobell, H. M., 281
Sober, H. A., 95, 98, 174
Socher, S. H., 62
Somers, C. E., 78, 278
Sommer, K. R., 103, 139
Sonnenbichler, J., 122, 123(76)
Spelsberg, T. C., 109, 146, 172, 174(1), 175(12), 176, 192, 194, 195(10), 196, 197(10)
Spencer, E. M., 158
Spencer, M., 232
Spiker, S., 139
Spiro, M. J., 309, 353
Spiro, R. G., 309, 353
Spitz, H. D., 359
Spohrand, G., 273
Spooner, B. S., 285
Sporn, M. B., 148
Springer, A., 285
Spudich, J. A., 285
Staehelin, T., 270
Stange, J. L., 257, 258(43)
Stanley, T. W., 361, 362(104)
Stanners, C. P., 52, 54(20)
Stecher, P. G., 274(21), 275
Steel, G. G., 54, 56(25)
Steffen, J., 47, 48(8), 53(8), 54(8), 55(8)
Stegemann, H., 367
Steggles, A. W., 172, 174(1)
Stein, S., 116, 359
Stein, W. H., 174
Steinbeck, H., 291, 292
Steiner, J. W., 60
Stellwagen, R. H., 103, 120, 279, 281, 283
Sterling, W. R., 252, 265(28)
Stewart-Blair, M. L., 289
Stocher, E., 60
Stockdale, F., 279

Stoffler, G., 153
Stollar, B. D., 137, 191, 192(4), 194, 196
Stone, J., 115
Strasser, F. F., 277
Stuart, R. J., 21
Stubblefield, E., 62, 64, 65(8, 10), 72, 76, 77, 78(4), 80(4), 82, 83(4), 86, 87(4), 88, 285
Studdert, D. S., 232
Studier, W. F., 99
Subirana, J. A., 103
Suhadolnik, R. J., 274(24), 275, 277, 283, 289
Summers, W. P., 283
Summerson, W. H., 328
Sung, M., 127
Sutherland, E. W., 289
Suzuki, H., 185
Swaneck, G. E., 292
Swart, C., 283
Sweeney, E. W., 246, 249, 262(14), 266(14)
Sweet, R., 289
Swick, R. W., 255, 257, 258(43)
Swift, H., 23, 62
Szepesi, B., 263
Szybalski, W., 281

T

Taechman, L. B., 60
Takahaski, M, 54, 55(24), 56(24), 59, 61(12)
Takai, S.-I., 289
Takeda, M., 200, 206(15), 207(15)
Takeuchi, T., 322
Talbot, D. N., 159
Tan, C. H., 187
Tanzer, M. L., 361
Tappel, A. L., 346, 348(68)
Tarantola, V. A., 252, 265(28)
Tartof, K. D., 278
Tashjian, A. H., Jr., 331
Tata, J. R., 161, 283
Tates, A. D., 10
Tavitian, A., 283
Taylor, E. L., 285
Taylor, E. W., 285
Teich, N., 281
Teng, C., 41

Teng, C. S., 177, 179(7), 181(7), 185(7), 186, 190(7), 198
Teng, C. T., 177, 179(7), 181(7), 185(7), 186, 190(7), 198
Tengi, J. P., 115
Ten Haaf, F. E. L., 296
Terasima, T., 51
Thiéry, J. P., 23, 29
Thoa, N., 285
Thomas, C. A., Jr., 94
Thomas, W., 279, 283
Thompson, E. B., 268, 269(9), 270(9), 271, 272(9), 273(9)
Till, J. E., 52, 54(20)
Tinoco, I., Jr., 215, 216(5), 222, 235(5)
Tishler, P. V., 130
Tobey, R. A., 46, 49(7)
Todaro, G. J., 281
Todd, K., 12
Tolmach, L. J., 51
Tomkins, G. M., 279, 281, 283, 286(51), 287(51), 289, 290, 291, 292
Toole, B. P., 322
Tournier, P., 39
Toverud, S. U., 322, 362(19)
Trasher, J. D., 59, 60(5)
Traut, R. R., 153
Treffers, H. P., 192
Trelstad, R. L., 322
Trevithick, F. R., 123
Troll, W., 366
Truman, D. E. S., 273
Tschudy, D. P., 262
Tsuchiya, J., 53
Tsvetikov, A. N., 103, 113(11), 119(11)
Tuan, D. Y. H., 101, 106
Tuan, P., 228
Tunis-Schneider, M. J. B., 233, 234, 235(29)
Turkington, R. W., 177, 198
Turner, J. C., 302
Turner, L. V., 112
Turner, M., 277

U

Udenfriend, S., 115, 116, 159, 343, 345(60), 346, 348(68, 69), 359, 369
Uenoyama, K., 273
Umana, R., 109
Umeda, T., 41, 289
Umezawa, H., 274(20), 275
Unhjem, O., 292
Uretsky, S. C., 283
Ussing, H. H., 257

V

Vaes, G., 349, 352(76), 353
Vallejo, C. G., 111
van den Broek, H. W. J., 174
Van Slyke, D. D., 355
Van't Hof, J., 59
Van Tubergen, R. P., 6, 10(7)
Van Vunakis, H., 192
Velick, S. F., 246
Venable, J. H., 25
Vendrely, C., 46, 62
Vendrely, R., 62
Verbit, L., 220, 221(7)
Vessell, E. S., 246
Vesterberg, O., 170
Viñuela, E., 133, 134(95), 154, 179
Voelkel, E. F., 331
von Berswordt-Wallrabe, R., 291, 292
von Hippel, P. H., 112
Von Holt, C., 103

W

Wagner, R., 285
Wagner, T. E., 232, 234(28), 235(28)
Wakabayashi, K., 107
Waksman, B., 242
Walch, R. J., 136
Walsh, D. A., 202
Wang, C. H., 298
Wang, S., 107
Wang, T. Y., 84, 106, 145, 177, 198
Wangh, L., 132
Warburton, D., 303, 305(3, 4)
Ward, D. N., 103
Ward, M., 137, 191, 192(4), 194
Ward, V. L., 287
Warner, J. R., 281
Wasserman, E., 137

AUTHOR INDEX

Wasserman, F., 161
Wataya, Y., 110
Watson, M. L., 23, 24, 29(40)
Waxman, S., 63, 72(2)
Webber, C., 294
Weber, K., 154, 159, 203, 248
Webster, P. L., 61
Weigele, M., 115, 116
Weinberg, F., 283
Weinberg, R., 279, 283
Weiner, A. W., 159
Weintraub, H., 279, 286, 289
Weisberger, A. S., 274
Welch, A. D., 277
Wells, B., 21
Wessells, N. K., 285
Westenbrink, H. G. K., 116
Weston, P. D., 352, 353(79)
Wetlaufer, D. B., 236
White, E. L., 246
Whitescarner, J., 22
Whitmore, G. F., 48, 283
Widholm, J., 98, 106, 108(16), 114(16), 119(16), 120(16), 127(16), 131(16), 132(16), 135(16), 138, 144, 148(1)
Widnell, C. C., 161
Wieland, O., 283
Wieland, T., 283
Wigle, D. T., 103, 104(10)
Wilhelm, J. A., 103, 107, 146
Wilkens, M. H. F., 232
Wilkes, E., 83
Wilkes, M, 99
Wilkins, W. H. F., 232
Williams, J. A., 285
Williams, J. G., 133
Williams, M. A., 18
Williams-Ashman, H. G., 274, 292
Wilson, E. M., 174, 175(12)
Wilson, L., 285, 286
Wilson, M. R., 232
Wilson, R. F., 100

Wilt, F. H., 93, 96(6)
Wimber, D. E., 59
Winget, G. D., 87
Winter, W., 87
Wisse, E., 10
Wittman, H. G., 153
Wolf, G., 254
Wolff, J., 285
Woods, J. F., 317
Wooten, G. F., 285
Wray, W., 64, 65(8, 10), 77, 78(4), 80(4), 82, 83(4, 8), 86, 87(4), 88
Wrenn, J. T., 285
Wu, R. S., 281

Y

Yamada, K. M., 285
Yamamoto, N., 187
Yamamura, H., 200, 206(15), 207(15)
Yang, J. J., 221
Yang, J. T., 222, 227(12)
Yanowitz, I. S., 289
Yasmineh, W. G., 93
Yotsuyanagi, Y., 28
Young, B. R., 286, 289
Young, K. E., 153
Yphantis, D. A., 159
Yunis, J. J., 93

Z

Zabin, I., 273
Zenses, M., 76
Zetterberg, A., 136
Zipser, D., 267
Zubay, G., 23, 28(42)
Zweidler, A., 141, 142(7)
Zylber, E., 283

Subject Index

A

Acid concentration, in histone extraction, 107
Acid extraction of chick bone collagen, 339
Acid phosphatase, associated with cell fractions, 86
Acid production, of bone, 328
Acridine orange, DNA fixation, 33
Acriflavine, in electron microscope cytochemistry of nucleic acids, 24, 32
Acrylamide gel electrophoresis, of chromatin protein kinases, 202
Actinomycin D
  ^{3}H-labeled, binding, 12
  RNA inhibitor, 278, 280
Adenosine, in DNA synthesis, 42
Adenylate kinase, associated with cell fractions, 86
Adsorbent, for affinity chromatography, 267
Affinity chromatography
  adsorbents for, 267
  column preparation in, 270
  enrichment of polysomes synthesizing specific protein, 266, 271
Alcohol dehydrogenase, associated with cell fractions, 86
Aldehydic residue, total, measurement, 362
Alkaline copper reagent, in quantitative histone determination, 110
Alkaline phosphatase, associated with cell fractions, 86
α-Amanitin, RNA inhibitor, 282
Amberlite chromatographic determination of histones, 119
Amethopterin
  DNA inhibitor, 276
  in DNA synthesis, 42
Amino acid
  analysis, 359
    of nonhistone chromosomal proteins, 158
  preparation of human biopsy tissue, 360
  in histone fraction from calf thymus, 105
Aminopterin, DNA inhibitor, 276
(8-Anilino-1-naphthalenesulfonic acid (ANSA), in histone assay, 115
Anisomycin, protein inhibitor, 287, 288
Antiandrogen, 292
Antibody, see also specific antibody
  preparation, in protein determination, 243
Antichromosomal protein, 191
Antiestrogen, 292
Antigen, for immunization of protein, 242
Antigen–antibody complex, disrupted, electrophoresis, in SDS-acrylamide gels, 248
Antigenicity, measurement by quantitative microcomplement fixation, 195
Antiglucocorticoid, 292
Antihormone, 291, 292
Antimineralocorticoid, 292
Antinucleoside (antinucleotide)
  in study of nucleic acid in cells, 302
  reaction with human metaphase chromosomes, 304
Aorta, preparation for collagen studies, 311
Aurin tricarboxylic acid, protein inhibitor, 287, 288
Autoradiography, electron microscope (EM ARG)
  coating in, 6
    dipping method of, 7
    loop method of, 6
  of chromosomes, 71, 72
  cytological procedures, 5
  exposure, 9
  grain density, results, 18
  interpretation of data, 5
  localization of proteins and nucleoproteins, 5
  quantitative analysis of results, 15

sectioning, 6
vessels, washing, 3
5-Azacytidine, RNA inhibitor, 278, 282

B

Basement membrane, carbohydrate composition, 353
Beer–Lambert law, 217
Bicarbonate buffer solution, 327
Biopsy tissue, human, preparation for amino acid analysis, 360
Bone
 cancellous, preparation for amino acid analysis, 360
 cell isolation, mechanical disruption of, 317
 collagen
  extraction, chick, 339
  metabolic studies, 326
  methods of organ culture, 331, 333, 334
  specific, 328
 defatting, drying, and demineralization, 312
 enzymatic digestion of, 316
 isolation of cells, 316, 317
 oxygen consumption, 328
 preparation for collagen studies, 311
 procollagenase, mouse, 352
Bovine serum albumin, phosphorylated, coupling of histones, 192
5-Bromo-2'-deoxyuridine, DNA inhibitor, 278, 280
BSA, see Bovine serum albumin
Buffer, see also specific types
 in chromosome isolation, 77, 87
 interference, in quantitative histone determination, 112

C

Calcitonin, effect on collagens, and administration, 319
Calcium deficiency, 321
Camptothecin, inhibitor of DNA and RNA, 278, 280
Carbohydrate, hydroxylysine-linked, 353

Cartilage
 isolation of cells, 314
 method of isolation, 315
 preparation for collagen studies, 311
Catalase, associated with cell fractions, 86
CD, see Circular dichroism
Cell
 in circular dichroism, 229
 isolated, suspensions, 312
Cell culture
 exponential, execution point, 46
 synchronized, execution point, 48
Cell cycle
 phases, 58
 in vivo analysis
  colcemid in, 61
  colchicine in, 61
  cytochemistry, quantitative, 62
  by direct observation, 58
  isotopic labeling, 59
  metabolic inhibitors and, 61
  percent labeled mitosis curve, 60
Cell density analysis, in exponential cultures, 46
Cell division
 action of chemical agents on cycle, 44
 cell nucleus and, 1–89
 doubling time, 55
 execution point, 45–46, 48
 inhibitors, 279, 284
 in absence of, 56
 termination point, 45
 measurement in absence of steady-state cell cycle conditions, 49
 under steady-state cell cycle conditions, 52
Cell movement, inhibitors, 279, 284
Cell synchronization, in chromosome isolation, 77
Cellogel electrophoresis, in quantitative histone assay, 122
Chromatin
 constituent components, 145
 separation, in analysis of chromosomal nonhistone proteins, 162
 differential centrifugation, 94
 fractionation, 93
 by agarose gel filtration, 96
 on hydroxyapatite, 160

magnesium chloride and, 98
  in template active and inactive regions, evidence for, 99
  procedure, 98
  fractions, properties, 100
  in isolation and characterization of nonhistone chromosomal proteins, 144
  preparation
    in analysis of chromosomal nonhistone proteins, 162
    Bonner method of, 144
  structure and function, 104
  template active and inactive regions
    chemical composition, 100
    isolation, 97
  thermal elution from hydroxyapatite, 94
Chromatin protein kinase, 198
  assay of activity, 201
  fractionation, 199
    substrate specificities and effects of cyclic AMP, 205
    typical results, 203
  heterogeneity, 206
  lability, 207
Chromatin proteins, see Proteins
Chromatography, see also specific types
  DNA-cellulose, 181
  in quantitative histone assay, 117
Chromosome
  autoradiography of, 71
  critical point drying, 68
  DNase digestion, 87
  in interphase cells, 63
  isolated
    preparation for electron microscopy, 67
    stability, 87
  isolation, 64
    buffers, 77
    universal medium, 76
  metaphase
    composition, 82
    fractionation, 81
    isolation, 65
      methods and properties, 78, 80
      parallel procedures with mitotic apparatus and nuclei, 75, 79

preparation, 305
reaction with antinucleoside antibodies, 304
whole, electron microscope autoradiography procedure, 73
α-Chymotrypsin, protein digestion, concentration and incubation period, 38
Circular dichroism (CD), 214
  analysis of nucleoprotein complexes, 209
  difference, 225
  ellipticity in analysis, 216
  experimental parameters, 227
  instrumentation, 221
    calibration, 227
    modifications, 224
    slit width of instrument, 230
  optical parameters, 229
  presentation of data, 231
  principles of measurement, 222
  spectra, relationship with structure, 231
Circularly polarized light and circular dichroism analysis, 212
Clomiphene citrate, antiestrogen, 292
Colcemid
  inhibitor of chromosome movement, 279, 284
  in vivo cell cycle analysis, 61
Colchicine
  inhibitor of chromosome movement, 279, 284
  in vivo cell cycle analysis, 61
Collagen
  acid-soluble
    biosynthesis in adult bone, 329
    purification, 338
  biosynthesis
    in bone, 328
      formation rate, 330
      resorption rate, 330
    rate of, 309
  chick bone, extraction, 339
  composition, 335
  cross-linking, 360
  cysteamine-soluble, purification, 337
  hormone effect on, 307
    on degradation, 349
    experimental procedures, 319
    metabolic studies, 324

labeled
  [^{14}C]glycine, 350
  in vivo, 322
  lyophilized, purified, reconstitution, 340
  native, incubation with thiosemicarbazide, 363
  neutral-salt-soluble
    extraction, 336
    purification
      ^{14}C-labeled, 337
      in supernatant, 337
  purification, 335
    by digestion with purified collagenase, 340
  radioactive, assay in presence of other proteins, 342
  subunits, chromatographic separation, 358
Collagenase
  digestion, in measurement of [^{14}C]hydroxylysine, 354
  hormone effect on, 349
  purification, 341
    assay of radioactive collagen in presence of other proteins, 342
    digestion of collagen, 340
    partial, 350
  separation from inhibitory serum antiproteases, 352
  tadpole, zymogen, 351
Column chromatography, of phosphorylated nonhistone proteins, 200
Complement fixation, quantitative, 195
Copper deficiency, 321
Cordycepin, RNA inhibitor, 279, 282
Cortisone, antiglucocorticoid, 292
Creatine kinase, associated with cell fractions, 86
Critical point drying of chromosomes, 68
  apparatus, 70
Cyclic adenosine monophosphate, effect on protein kinases, 205, 206
Cycloheximide, protein inhibitor, 287, 288
Cyproterone, antihormone, 291, 292
Cysteamine extraction, 338
Cysteamine-soluble collagen, *see* Collagen(s)
Cytochalasin B, inhibitory properties, 284, 286

Cytosine arabinoside, DNA inhibition, 276
Cytosol, requirement, in DNA replication, 43

D

Dansylation of proteins, 159
Daunomycin (DNA, RNA inhibitor), 281
Defatting of tissue, for collagen studies, 312
Demineralization of tissue, for collagen studies, 312
Denaturation, method, 306
2'-Deoxyadenosine, DNA inhibitor, 276
Disulfide-containing reagents, in histone determination, 111
Deoxyribonuclease (DNase)
  digestion of chromosomes and nuclei, 87
  isolation of acidic chromatin proteins, 174
  protein digestion, concentration and incubation period, 38
Deoxyribonucleic acid (DNA)
  association with protein, in preparation of nucleoprotein complexes, 227
  binding of phosphorylated acidic chromatin proteins, 181
    comparison of procedures, 190
    measurement, 187
  bone content, 328
  collagen content, measurement, 365
  electron microscope (EM) cytochemistry, 28
  geometric parameters, 233
  localization
    labeling conditions, 12
    techniques, 28
  methylation, 186
  removal, in SE preparation of nonhistone chromosomal protein, 152
  replication-chromosome-condensation cycle, 45
  replication from HeLa cells, nuclear system for, 41
  structure, relationship with CD spectra, 232

synthesis
  assay, 42
  inhibitors, 275, 280
Deoxyribonucleotide, requirement, in DNA replication, 43
Developers, photographic, 9
DNA, *see* Deoxyribonucleic acid
DNA-cellulose chromatography, 181
DNA polymerase, associated with cell fractions, 85, 86
DNase, *see* Deoxyribonuclease
Don-C cloned strain, 76
Drying of tissue, for collagen studies, 312
Dye-histone interactions, in direct quantitative determination, 115

E

ECTHAM-cellulose, chromatography on, 95
Electron microscope autoradiography, *see* Autoradiography, electron microscope
Electron microscopy
  of chromosomes, 67
    autoradiography, 72
  cytochemistry
    methods staining both nucleic acids, 23
    of nucleic acids, 22
    of nucleoproteins, 19
    localization, 19
Electrophoresis
  disc gel, analysis of nonhistone chromosomal proteins, 154
  of disrupted antigen–antibody complex in SDS-acrylamide gels, 248
  in quantitative histone assay, 122
  in polyacrylamide gels, 127
  SDS polyacrylamide gel, of chromatin nonhistone proteins, 165
Electrophoretic elution, in analysis of nonhistone chromosomal proteins, 158
Emetine, protein inhibitor, 287, 288
Emulsion, coating, in electron microscope autoradiography, 6
  interference color, 8

Enzyme, *see also* specific substances
  associated with cell fractions, 86
  decay of activity following effect of inducing agents, 261
    following inhibition of protein synthesis, 262
  degradation in animal tissue, analysis, 241
  digestion
    interpretation of results, 39
    procedure, 36
  half-life, and hormone-stimulated rate of synthesis, 262
    in short-term incorporation of isotope, 251
  hormone-stimulated rate of synthesis, 262
  hydrolytic, 353
  immunotitration, 243
  level, steady state kinetics, 252
  lysosomal, hormone effect on, 349
  peptidylproline and peptidyllysine hydroxylase activity, 343
  rate of turnover, estimation, 260
  return to steady state following irreversible inhibition, 264
  synthesis in animal tissue
    altered rate, 251
    analysis, 241
17α-Estradiol, antiestrogen, 292
Estrogenic steroid, 320
Ethamoxytriphetol, antiestrogen, 292
Ethylenediaminetetraacetic acid (EDTA) staining method, 25
Euchromatin, disperse, distribution, 33

F

Fluorescamine, 115
Fluorescein conjugate, 303
Fluorescence, nuclear, of mouse L cells, 303
Fluorimetry, in direct quantitative histone assay, 114
5-Fluoro-2'-deoxyuridine, DNA inhibitor, 276
1-Fluorodinitrobenzene (FDNB), in histone assay, 115

Fluoxymesterone, antiglucocorticoid, 291, 292
Folin–Ciocalteau reagent, 111, 112
Folin phenol reagent, measurement of protein, 375
Formaldehyde fixation, 4
Formamide denaturation, 306
Fusidic acid, protein inhibitor, 287, 288

G

G_1 phase, in measurement of termination point, 50, 52
G_2 phase, in measurement of termination point, 51, 54
Galactosylhydroxylysine collagen, 353
Gamma counter, 294
Gel filtration, in quantitative histone assay, 120
Glucose-6-phosphate dehydrogenase, associated with cell fractions, 86
Glucosyltransferase, 353
Glutaraldehyde fixation, 4
α-Glycerolphosphate dehydrogenase, associated with cell fractions, 86
[^{14}C]Glycine-labeled collagen, preparation, 350
Gold latensification, 10
Growth hormone, effects of, 320
Guanidine extraction of chick bone collagen, 340

H

HAPTA technique, 27
Hartree modification, of Lowry measurement of protein, 376
HeLa cell, DNA replication, nuclear system, 41
Heme precursor, single administration, in measurement of protein degradation, 257
Hexokinase, associated with cell fractions, 86
Histone
 AKP, 104
 analysis
  cytochemical measurements, 136
  methods, 102
 calf thymus fractions, amino acid composition, 105
 in chromatin fractions, 101
 composition, 102
 coupling with human serum albumin, 191
  with nucleic acids, 192
  with phosphorylated bovine serum albumin, 192
 dialysis, 108
 electrophoresis, Bonner method of, 127
 fractionation
  chemical, 116
  method of, 131
  quantitative, 109, 116
 heterogeneity, 102
 immunochemical characteristics, 191
 immunochemical reactions, 137
 isolation, 105
 KSA, 104
 nomenclature, 103–104
 phosphorylation
  assay for given histone fraction, 139
  detailed analysis, within given histone fraction, 142
 proteolytic degradation, 109
 quantitative determination, 110
  chromatography, 117
 radioactivity, determination, 125, 129
 selective dissociation from chromatin in circular dichroism, 228
 selective phosphorylation, assessment methods, 138
 separation, method of, 133
 staining, 19
Histone–dye interaction, in direct quantitative determination, 115
Hormone, see also Antihormone, specific substances
 effect on collagens, 307
  experimental procedures, 319
  metabolic studies, 324
  preparation, dosage, routes of administration, 319
Human serum albumin (HSA), coupling of histones, 191
Hydroxyapatite
 in chromatin fraction, 160
 in chromatin separation, 162
 thermal elution from, 94

SUBJECT INDEX

17-Hydroxycorticosteroid, effect on collagens, preparation, dosage, route of administration, 319
Hydroxylysine
 ^{14}C-labeled, measurement of, 353
 total aglycosylated plus nonglycosylated, 357
 glycosylated, separation, 358
Hydroxylysine-linked carbohydrates, 353
17α-Hydroxyprogesterone, antiglucocorticoid, 291, 292
Hydroxyproline
 ^{14}C-labeled
 Juva and Prockop procedure, 371
 measurement
 paper chromatography, 374
 procedures, 367, 369, 371
 in urine samples, 370
Hydroxyurea, DNA inhibitor, 276

I

Immunization procedure for proteins, 242
 chromosomal, 193
Immunodiffusion, 246
Immunoelectrophoresis, 246
Immunofluorescence, indirect, 307
Immunogen, chromosomal protein as, 191
 testing of nuclear origin, 196
Immunoprecipitate, specificity, 249
Immunoprecipitation, for isolation of labeled protein, 246
 controls, 250
 quantitative, 245
Immunotitration of enzyme activity, 243
Incubation
 in bone collagen metabolic studies, 326
 enzymatic, 37
 in soft tissue collagen metabolism studies, 324
Inhibitor
 metabolic, *in vivo* cell cycle analysis, 61
 in study of hormone mechanisms in cell culture, 273
Insulin, effect on collagens, preparation, dosage, route of administration, 320

Iron, colloidal, 21
Isocitric dehydrogenase, associated with cell fractions, 86
Isoelectric focusing gel, 157
Isotope
 continuous, method of protein degradation determination, 257
 double, determination of heterogeneity of protein degradation, 258
 labeling, *in vivo* cell cycle analysis, 59
 reutilization, in measurement of protein degradation, 254
 short-term incorporation into enzyme, 251
 single, method of protein degradation determination, 253

L

Labeling conditions
 for DNA localization, 12
 for protein localization, 15
 for RNA localization, 13
Labeling index analysis, in exponential cell cultures, 48
Lactic dehydrogenase, associated with cell fractions, 86
Lathyrogenic compounds, 321
Lead salt, in ribonucleoprotein staining, 24
Lens, isolation of cells, 314
Light scattering, parameter in circular dichroism, 230
Liquid scintillation counting, 295
Lowry method
 original, of measurement of protein with Folin phenol reagent, 375
 of quantitative determination of proteins, 110
Lysine,^{14}C-labeled, separation from ^{14}C-labeled hydroxylysine, 357
Lysine–tryptophan ratios, determination, in histone analysis, 137
Lysosomal enzymes, *see* Enzymes, lysosomal
Lysyl oxidase, 321
 activity of enzymes, 347

M

Magnesium chloride, in chromatin fractionation, 98
Magnesium deficiency, 321
2-Mercapto-1-(β-4-pyridethyl)benzimidazole, RNA inhibitor, 282
N-Methyl benzothiazolone hydrazone hydrochloride, 362
17α-Methylnortestosterone, antihormone, 291, 292
17α-Methyltestosterone, antiglucocorticoid, 291, 292
Microtubular disruptive drugs, 322
Mitomycin C (DNA, RNA inhibitor), 281
Mitosis, in measurement of termination point, 49, 52
Mitotic apparatus, 83
  isolation, 84
  problems, 88
  stability of, 87
  parallel isolation procedures with metaphase chromosomes and nuclei, 75
Mitotic cell accumulation function, rate of increase, 56
Mitotic index analysis, in exponential cell cultures, 47
Mouse bone and skin, procollagenase, 352
Mouse L cell
  microscopy, 304
  nuclear fluorescence, 303

N

Nafoxidine, antiestrogen, 292
Nitrile, 321
Nonhistone chromosomal protein, 144
  analysis, 165
    chemical techniques, 158, 161
    disc gel electrophoresis, 154
  fractionation
    ion exchange chromatography, 160
    QAE-Sephadex, 164
  immunochemical characteristics, 192
  isolation and characterization, 160
    general strategies, 145
  phosphorylated, column chromatography, 200
  phosphorylation $in\ vitro$, 164
  SDS isolation, 165
  SE preparation, 150
    controls and reproducibility, 153
Nuclease, digestion of proteins, 35
Nucleic acid
  antinucleoside and antinucleotide antibodies, 302
  coupling with histone complexes, 192
  electron microscope cytochemistry, 22
  methods staining both nucleic acids, 23
Nucleic acid-associated proteins, staining, 21
Nucleoprotein
  autoradiographical and cytochemical localization, 3–41
  circular dichroism spectra, analysis, 237
  complexes, circular dichroism analysis, 209
  preparation for, 227
  light-scattering, of solutions, 225
Nucleus
  DNase digestion, 87
  isolation, 84
    procedures with metaphase chromosomes and mitotic apparatus, 75
    stability, 87
  preparation, in analysis of chromosomal nonhistone proteins, 161
Nutritional deficiency, 321

O

Optical rotation, 215
Organ culture, in bone collagen metabolism studies, 331
Osmium ammine, 32
Osmium tetroxide fixation, 5
Ovary, preparation for collagen studies, 311

P

Pactamycin, protein inhibitor, 287, 288
Papain, protein digestion, concentration and incubation period, 38

Paper chromatography, of proline and hydroxyproline, 374
Parathyroid hormone, preparation, dosage, route of administration, 319
Parlodion solution, 7
D-Penicillamine, 321
Pepsin, digestion of proteins, 35
Peptidyllysine, activity of enzymes, 343
Peptidylproline, activity of enzymes, 343, 346
pH
  in chromosome isolation, 65
  in histone fractionation, 132
Phenidon developer, 10
Phenol extraction, analysis of phosphorylated acidic chromatin proteins, 179
p-Phenylenediamine, 10
Phosphocellulose columns, preparation, 199
Phosphoglucomutase, associated with cell fractions, 86
6-Phosphogluconate dehydrogenase, associated with cell fractions, 86
Phosphorus deficiency, 321
Phosphotungstic acid, 21, 27
Photooxidation, 307
  denaturation, 306
  microscopy in, 307
Piperazine-N,N'-bis(2-ethane sulfonic acid)monosodium monohydrate (PIPES), isolation buffer, 77, 87
Plane-polarized light, and circular dichroism analysis, 209
Polyacrylamide gel, in phosphorylation assay for given histone fraction, 141
Polyacrylamide gel electrophoresis of histones, 127
Polysome
  preparation for affinity chromatography, 270
  synthesizing specific protein, enrichment with affinity chromatography, 266, 271
Polyuridylic acid, protein inhibitors, 287, 288
Precipitin reaction, quantitative, 245, 248
Preelectrophoresis of polymerized gels, 131

Procollagenase, mouse bone and skin, 352
Progesterone, antihormone, 291, 292
Proline
  analogs, 322
  measurement, 366
  separation from hydroxyproline, method, 374
[^{14}C]L-Proline, 322
  paper chromatography, 374
Pronase
  digestion of proteins, 35
  concentration and incubation period, 37
Prostate, preparation for collagen studies, 311
Protease inhibitor, in histone extraction, 109
Protein, *see also* specific substances
  chromatin, 101
    acidic phosphorylated
      binding to DNA, 181
      comparison of procedures, 190
      isolation and characterization, 171, 177, 181
      measurement, 187
      reagents, for, 171
    membrane filter analysis, 185
    sucrose gradient analysis, 185
  chromosomal, immunochemical characteristics, 191
  dansylation of, 159
  degradation, measurement, 253
  electron microscope, autoradiographic localization, labeling conditions, 15
  enzymatic digestion of, 33
    concentration and incubation periods, 37
  enzyme, in synthesis and degradation, 241
  immunologically reactive, determination, 242
  measurement with Folin phenol reagent, 375
  rate of turnover, effect of isotope reutilization, 255
  structure, relationship with CD spectra, 235
  substrate, radioactive, preparation, 342
  synthesis, inhibitors, 241, 286, 288

Pulse labeling, *in vivo* cell cycle analysis, 59
Purging, dry light, in circular dichroism, 231
Puromycin, protein inhibitor, 289
Pyridoxamine phosphate gels, preparation, 269

Q

QAE-Sephadex fractionation, of nonhistone proteins, 164
Quenching, in liquid scintillation counting, 295
  methods of correction, 297

R

Radioisotope experiment, double label
  design, 293
  instrument settings, 299
  mathematics, 296
  sample preparation, 302
Ribonuclease (RNase), protein digestion, concentration and incubation period, 38
Ribonucleic acid (RNA)
  localization, labeling conditions, 13
  synthesis, inhibitors, 275, 280, 282
Ribonucleoprotein (RNP), staining, in electron microscope cytochemistry of nucleic acids, 24
Ribosomes, purified, affinity chromatography, 272
RNA, *see* Ribonucleic acid
RNA polymerase associated with cell fractions, 85, 86
Ruthenium red, in electron microscope cytochemistry of nucleic acids, 32

S

S phase, in measurement of termination point, 51, 53
Saline citrate solution, in isolation of histones, 106
Salt extraction analysis of phosphorylated acidic chromatin proteins, 177

Sampling, in electron microscope autoradiography and cytochemistry, 3
Scanning, of histone gels, 129
Scintillation counting, 294
  liquid, 295
SE method, of nonhistone chromosomal protein preparation, 150
  controls and reproducibility, 153
Skin
  mouse, procollagenase, 352
  preparation for amino acid analysis, 360
    for collagen studies, 311
    for stress-strain testing, 364
Sodium dodecyl sulfate (SDS)-acrylamide gel electrophoresis, 133, 146
  of acidic chromatin protein, 173
  disrupted antigen–antibody complex, 248
  of nonhistone chromosomal proteins, 146
    controls and reproducibility, 148
    reagents, 172
    two-dimensional, 168
Sodium dodecyl sulfate-phosphate gels, 156
Sodium dodecyl sulfate-tris-glycine gels, 156
Sodium fluoride, protein inhibitor, 289
Sparsomycin, protein inhibitor, 287, 289
Spectroscopy, circular dichroism, relationship with structure, 231
Spirolactone, antihormone, 292
Spironolactone, antihormone, 291, 292
Staining, antibody, fluorescent, 304
Starch gel electrophoresis, in quantitative histone assay, 124
Steady state kinetics of enzyme levels, 252
Steroid, effect on collagens, preparation, dosage, route of administration, 319
Stress-strain testing of skin, preparation, 364
Sucrose, in histone determination, 111
Sucrose gradient analysis, of phosphorylated acidic chromatin proteins, 185
Sulfate deficiency, 321
Sulfhydryl-containing reagent, in histone determination, 111

T

Tadpole collagenase, zymogen, 351
Tendon, isolation of cells, 313
Testosterone, antiglucocorticoid, 292
Thallium ethylate, 24, 29
Thiosemicarbazide, incubation of native collagen, 363
Thymidine
  DNA inhibitor, 276
  ^{3}H-labeled
    localization of DNA, 13
    in vivo cell cycle analysis, 59
Thyroid hormone, effect on collagens, preparation, dosage, route of administration, 320
Tissue
  fragment, bone, in collagen metabolic studies, 326
  preparation, in study of hormone effect on collagen, 310
  soft, collagen metabolism studies, 324
    defatting, drying and demineralization, 312
Tissue culture
  cell line and, in chromosome isolation, 76
  in collagen degradation studies, 349
Toyocamycin, RNA inhibitor, 283
Trichloroacetic acid, in quantitative histone determination, 113
Triton gel system, phosphorylation of histone fractions, 142
Trypsin
  digestion of proteins, 35
    concentration and incubation period, 38

Turbidimetric procedures, in quantitative histone determination, 113
Tyrosine
  content, of histones, in direct quantitative determination, 114
  fluorescence, in direct quantitative histone determination, 114

U

Uranyl salts, in DNA localization, 28
Urea, in histone fractionation, 132
Uridine diphosphate (UDP)-glucose, 353
Uterus, preparation, for collagen studies, 311

V

Vinblastine, inhibitor of chromosome movement, 279, 284
Vitamin C deficiency, 322
Vitamin D deficiency, 321

W

Wavelength scanning speed, relation with time constant, in circular dichroism, 230
Wilson purified hormone, preparation, dosage, route of administration, 319

Z

Zymogen of tadpole collagenase, 351